Unmasking
the
Natural Universe

RICHARD FREEMAN

ISBN: 978-0-6455774-0-2

I am proud to have Richard as my brother.
This book is a testimony to Richard's intellectual and personal attributes.

Richard has had a lifelong passion for both cosmology and fishing. Indeed he is proud to say he has lived a working life at sea as a professional fisherman, totally by choice.

His love of the sea gave him an exposure to the wonders of the universe. Frequently beyond the sight of land he endured a hard working life of long days and nights while alone at sea where his mind and thoughts became absorbed and drove him to seek rational answers to the mysteries and contemporary theories, possibly attached to our own genesis.

His being has passionately and thoughtfully sought out answers and explanations through decades of deep research and an uncanny level of common sense.

This work should be reviewed objectively - the reader may well be enlightened and possibly surprised by the conclusions.

Chris Freeman AM.
B.Com. FICD. FFIN. FDIA.
Past Adjunct Professor University of QLD.

Dedicated to our Mother, Joan Freeman.

Mum enjoyed her time as a Journalist writing columns for country papers as well as the occasional book. Mum would tap away at her type writer day and night. My heart aches when I recall how Mum spent so much of her time rewriting manuscripts because she had discovered simple typing errors. Mum would have much enjoyed the word processors of today.

Mum always said to me "Richard always do your very best whether you succeed or not. What matters most is that you can always look back and be proud that you have done your very best".

Richard Freeman.

CONTENTS

ACKNOWLEDGMENTS

I express my gratitude and thanks to retired English teacher and friend Vivienne Braund.

I wish to thank the many dedicated Scientists and Physicists, including those below, whose discoveries have 'unknowingly' helped in compiling the evidence used throughout this book.

I include a special mention to Prof. Dr. Alfred Leitenstorfer and his team from the University of Konstanz in Germany who have *"manipulated pure nothingness and observed the fallout".*

A special mention to Physicist and cosmologist Paul Davies who has that special approach of looking pass the boundaries of theoretical physics. I also wish to include Prof. Derek Leinweber and team at the University of Adelaide who have revealed *"it is extraordinary, empty Space turns out to be the thing that gives us all, most of our mass".*

Stephen Hawking's life will forever be an inspiration to all. Creating a universe which *'total energy remains zero'* as expressed in his book 'A Brief History of Time' plays an important role in the modeling in this book as does his theory *'Hawking's Radiation'*. I have many of his books.

I express my gratitude to CERN related sites, which are also a reliable source of, based on fact, cutting edge information. I adore the honesty and frankness of these sites.

I express my gratitude to NASA for their 'no claim to copyright' images and for providing a most reliable source of factual information and text. I simply adore everything about NASA sites.

NASA images and text are legally in public domain. I also extend my gratitude to the **European Southern Observatory** as a source of images. All other images have been created by the author.

Last but not least, my loving thanks to all of my family.   Book updated 01/08/2025.

[000] **Superscript characters refer to reference URL. This book is written in layman's terms but provides over 150 web addresses to the relevant scientific evidence of every subject which allows one to read the directly associated science, experimental evidence or physics by simply clicking on a link. The eBook links are alive and in the paper back the links begin at page 404.**

I began collecting data for this model over three decades ago and began putting pen to paper well over two decades ago. My goal was to base this Model on **existing science and knowledge**. Cover: author's image of the Milky Way Galaxy; 10/04/2021, Pentax, f3.5, 20 sec, ISO 200.

The Natural Universe provides links to <u>overwhelming evidence</u> and provides <u>unprecedented</u> <u>correlation</u> of <u>all</u> subjects while solving numerous <u>mysteries not explainable</u> from a Big Bang.

(CNN) Renowned Theoretical Physicist Michio Kaku has said; *"we may have to rewrite all the text books about the beginning of the universe. It takes **<u>many billions of years</u>** to create a Galaxy like the Milky Way but the James Webb Space Telescope has identified six Galaxies that exist **<u>half</u> <u>a billion years</u>** after the Big Bang that are up to **<u>ten times bigger</u>** than the Milky Way Galaxy".* He says; *"**<u>something is wrong</u>**, we may have to revise our theory of the creation of the universe".*

<u>Only the Natural Universe</u> can account for the startling images revealed by the James Webb Space Telescope: The Natural Model is **<u>the only model</u>** which correctly predicted the **<u>elongated</u> shapes** of adolescent Galaxies. The Natural Model is **<u>the only model</u>** which <u>correctly predicted</u> large matured Galaxies existing alongside adolescent Galaxies in the early Universe. **<u>Only the</u> <u>Natural Universe</u>** can form the very first Galaxies **<u>before the time</u>** of the proposed Big Bang.

<u>The only model which provides realistic, detailed and superbly correlated answers:</u>

Unlike the Big Bang, the Natural Universe tells **<u>exactly how Gravity gradually arose</u>** from empty Space and began building a <u>Naturally Forming Universe</u>. (Chapter 5, page 78 to 83.)

Unlike the Big Bang, the Natural Universe **<u>obeys the laws of physics</u>** and <u>at the very creation of</u> <u>Matter</u> correctly accounts for its <u>mandatory</u> creation of Anti-matter.

Unlike the Big Bang, the Natural Universe explains the <u>unexpected</u> James Webb Space Telescope discoveries of Galaxies that **<u>shouldn't be visible</u>** due to the <u>Big Bang's creation of cosmic fog</u>.

Unlike the Big Bang, the Natural Universe **<u>actively creates</u>** its own dimensions of Space and time.

Unlike the Big Bang, the Natural Universe **<u>rationally explains how</u>** a Quantum particle resides in all places at the same time and **<u>rationally explains</u>** Quantum particle entanglement.

Unlike the Big Bang, the Natural Universe allows the use of Einstein's Special Relativity to completely and **<u>rationally unravel</u>** wave and particle duality.

Unlike the Big Bang, the Natural Universe **<u>tells exactly what Dark matter is</u>** and how it was created from a <u>known source</u> capable of providing Dark matter with its <u>invisible</u> properties.

Unlike the Big Bang, the Natural Universe explains exactly why all Galaxies **<u>have to be</u> accelerating** due to Gravity alone and has <u>no use</u> for the hypothetical Dark energy.

Bottom line: The Natural Universe tells how the <u>whole</u> universe <u>really</u> works.

1 A BRIEF OF THE NATURAL MODEL.

NASA's James Webb Telescope was launched on Dec. 25, 2021. Would this mighty Telescope reveal the weird and wild shapes of numerous colliding adolescent Galaxies as theorized from a Big Bang <u>or</u> would first Galaxies be forming by conforming to Matter <u>being created</u> and ejected from the poles of Super Massive Black Holes like my Natural Model? Would there be any massive Galaxies too matured to have possibly formed from a Big Bang?

The telescope's deepest images were revealed on 11 July 2022. [121] Page heading reads; **'Webb telescope spots old, massive Galaxies that shouldn't exist'**. The Galaxies are so massive they should not be possible under current cosmological theory. ***"It's bananas, these Galaxies should not have had time to form"*** said Erica Nelson assistant professor of astrophysics at CU Boulder. The images of Galaxies too massive and matured to have plausibly come from a Big Bang obviously conform to a Natural Universe which can form the first Galaxies <u>before the time</u> of the proposed Big Bang.

The images also failed to reveal the weird and wild shapes of numerous colliding adolescent Galaxies theorized as a way to quickly grow Galaxies from a Big Bang. [112] Clearly conforming to the Natural Model, the James Webb Space Telescope has recently discovered a very distant Galaxy from just 330 million years after the said Big Bang which *"appears elongated, almost like a <u>peanut</u>"*. Others have described *"early galaxies were <u>frequently</u> shaped like surfboards and pool noodles"*. The descriptions clearly match the way a Natural Universe ejects its Matter <u>created</u> at poles of SMBHs. That is, if newly created Matter is ejected in <u>opposite directions</u> from Quasar like jets at the poles of a Super Massive Black Hole it is logical to first create an elongated Galaxy which <u>makes no sense from a Big Bang</u>.

First forms an elongated Galaxy

[116] Link describes "a molecular gas fountain in the ancient (early) universe"; the Astronomers have discovered a Quasar, J2054-0005, powered by a Super Massive Black Hole which they say is *"spewing out molecular gas, the raw material needed to form new stars".* The Quasar was observed by the Atacama Large Millimeter/submillimeter Array (ALMA) in Northern Chile. For <u>a Natural Model</u> this provides outstanding evidence of how a Galaxy <u>creates its own Matter</u> at the poles of its central Super Massive Black Hole. The Astronomers are actually <u>directly detecting</u> the Matter being ejected.

[103] **85% of all mass and 90% of our Milky Way Galaxy's mass is in the form of invisible and not directly detectable Dark matter. Consequently, it is surely common sense to begin a universe with Dark matter, rather than a Big Bang which <u>only accounts</u> for the remaining 15% of all Matter. This is why my Natural Universe begins by first creating Dark matter from a <u>known source</u> which is capable of providing it with its invisible properties.**

Modeling is heavily based on evidence from well over 500 reference links to research and observations; over 150 of which are provided in this book. The Natural Model is a **physical reality model** which unmasks how our universe began and physically operates. The Natural Model does not introduce new inexplicable forces, unexplained energies, incomprehensible multiple dimensions or write new physics. The model gathers the <u>predictions</u> of successful theories, astronomical data, the results of scientific experiments and <u>true observations</u> and then the model <u>begins with a clean slate</u> to build a universe designed to unravel the universe's greatest mysteries.

^{022; 103 video.} **Why do all stars and Galaxies <u>reside</u> in regions <u>richest</u> with Dark matter? Answer; Dark matter will provide the crucial rudimentary mechanism for creating and powering the protons which create the hydrogen atoms to make stars. Being based on a Quantum vacuum fluctuation will provide Dark matter with its invisible, <u>virtual properties</u> and is why all attempts <u>have failed</u> to detect a Dark matter 'particle'.**

Dark matter acting as a <u>virtual proton</u> will play an overwhelming role in physically unraveling each of the following <u>unsolved mysteries</u>: <u>Tells how Gravity</u> slowly arose from seemingly <u>empty Space</u> and <u>began building</u> the Natural Universe. **<u>Tells how</u>**, <u>without</u> Dark energy, Gravity is accelerating Galaxies. **<u>Tells how</u>** all aspects of the CMB radiation are explained without a Big Bang. **<u>Tells physically how</u>** Einstein's Gravity works and exactly how Quantum Gravity works. **<u>Tells how</u>** all Black Holes, including Super Massive Black Holes, are created without the current, incorrect notion of infinity which causes an erroneous singularity. **<u>Tells what stays</u>** to provide mass and Gravity within all Black Holes. **<u>Tells why</u>** Gravity is mysteriously weak. **<u>Tells why</u>** the vacuum of Space-time is not a true vacuum but a false vacuum. **<u>Tells how</u>** all mass is created. **<u>Reconciles Gravity with all mass</u>** and at Quantum scales. **<u>Tells what</u>** provides the passing of time. **<u>Tells exactly how</u>** we are all provided with our own unique Dimensional Reality. **<u>Tells exactly how</u>** empty Space became dimensional Space-time. **<u>Tells how</u>** all Matter is created without the Matter and Anti-matter asymmetry problem. **<u>Tells how</u>** all types of Galaxies are first formed. **<u>Tells exactly how</u>** a Quantum particle can be in all places at the same time. **<u>Tells why</u>** simply observing or measuring physically changes a fuzzy wave into a particle.

Don't expect this model to be like any you have seen before. To solve the universe's most baffling mysteries a Natural Model is required to provide a **totally and completely new understanding of the universe** and different to the proposal that everything was instantly given from total nonexistence at a Big Bang. When things are given this way there is little reason to explain how they were created. Many mysteries have remained unsolvable by way of a Big Bang. [001; 008] A Big Bang explains only 4.6 % of the contents of the universe. A basic flaw of a Big Bang is the laws of physics and the validated results of many experiments absolutely confirm this same 4.6% relating to all visible Matter should have been completely annihilated by Anti-matter within the first second of the Big Bang. [066] Dark matter and Dark energy account for the remaining 95.4% of the contents of the universe, however, 'dark' means Physicists have no idea what it is or where it came from.

My Natural Model is designed to explain the entire contents of the universe and is based on a desire to fully explain the source of Gravity. The Natural Universe begins with the smallest configurations from empty Space. [014] The energy of Quantum vacuum fluctuations has been explained by the uncertainty principle first put forward in 1927 by German Physicist Werner Heisenberg. It says at every tiny point of empty Space there are temporary changes in energy which occasional change into mass, in the form of virtual particle and anti-particle pairs which mostly unite and mutually annihilate. But, if forced apart they can avoid annihilation and become real particles.

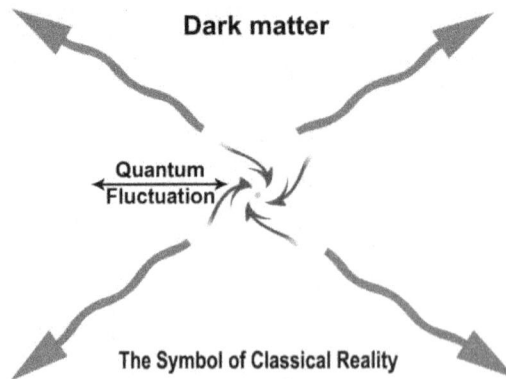

Dark matter

Quantum
Fluctuation

The Symbol of Classical Reality

The symbol of classical reality. Note: Every explanation will be correlated directly to the simplicity of this symbol. The modeling will not deviate from this symbol no matter how <u>bold a prediction</u> appears. This provides <u>unprecedented</u> correlation throughout the Model which builds a universe where every topic reinforces every other topic. **There are no unrelated theories, be it the mass of a proton, or the acceleration of whole Galaxies, <u>all are directly correlated.</u>** Quantum vacuum fluctuations <u>spontaneously arise</u> from every tiny point of empty Space which <u>provides a known source</u> for the COST (Dark matter) to be sourced in the massive quantities required to create a universe. The following six lines of text unlock answers to how <u>everything</u> in a Universe is created and operates from the <u>tiniest of scales</u>.

Based on a Quantum vacuum fluctuation the COST (Dark matter) fluctuates from an annihilating phase of a pair of virtual particles to a tiny phase of virtual mass provided by a state of <u>empty Space</u>. The negative energy (density) of virtual mass attracts the dimensions of Space-time and its powerful collapsing and reappearing state creates tiny Gravitational waves which provide new <u>entangled</u> dimensions to empty Space.

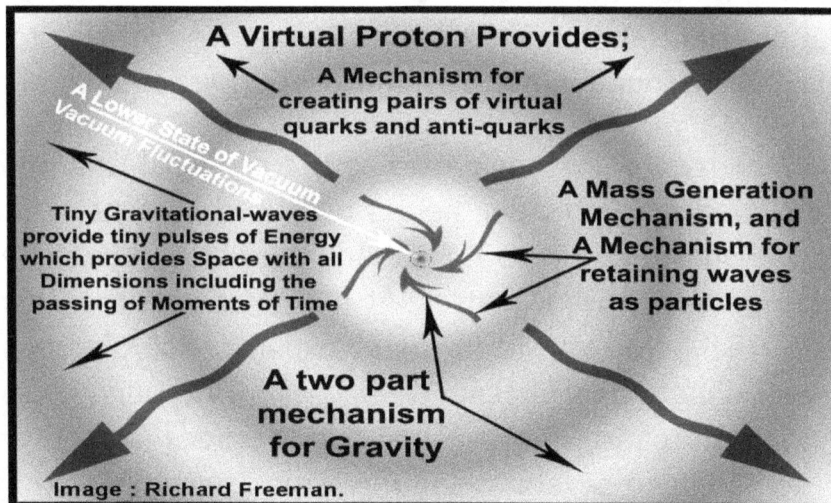

A Virtual Proton Provides;

A Mechanism for creating pairs of virtual quarks and anti-quarks

A Lower State of Vacuum Vacuum Fluctuations

Tiny Gravitational-waves provide tiny pulses of Energy which provides Space with all Dimensions including the passing of Moments of Time

A Mass Generation Mechanism, and A Mechanism for retaining waves as particles

A two part mechanism for Gravity

Image : Richard Freeman.

Note: <u>Negative energy density</u> will be abbreviated to <u>Negative energy</u>. **Dark matter** in this model is a tiny **C**onfiguration **O**f **S**pace and **T**ime called the COST. The COST is not a particle. Sophisticated detectors designed to

5

directly detect a Dark matter 'particle' have truly detected nothing. The Large Hadron Collider has also failed to find 'particle' based evidence of Dark matter. Exposed only by its Gravity, Scientists have labeled invisible Dark matter a ghost particle. No known actual 'particle' has such properties. Fittingly, the COST in its Dark matter state is also invisible, is exposed only by its Gravity and may best be called a virtual proton.

Acting as a virtual proton the COST is an amplified, permanently cyclic state of a Quantum vacuum fluctuation which, similar to a regular proton, fluctuates from a phase of coming and going pairs of virtual sea quarks to a tiny negative phase of 'virtual mass'. [011] Like found in a regular proton, virtual mass is a tiny phase of vacuum deeper than dimensional Space. Virtual mass will create a proton's regular mass. Virtual mass allows the COST to not violate the laws of energy conservation since the lower state phase provides negative energy which cancels the positive phase so as the sum-total mass-energy remains zero. This will allow a universe of Matter to be created in a way its sum-total mass-energy remains zero. By acquiring three additional Matter quarks the COST will be transformed from its Dark matter state of a neutral, virtual proton into a regular positive proton and **is why all Galaxies are found within the densest regions of Dark matter.**

The COST will be responsible for providing the essential Quantum vacuum fluctuations powering the beating hearts of all protons and neutrons. The energy of both phases of the COST is variable which allows the COST to match the properties of Dark matter. At higher speeds the energy phase increases and the phase of virtual mass expands its volume below the ground-state of dimensional Space. [015; 016 false vacuum] Although I first modeled this deep vacuum phase more than two decades ago I have been astonished to discover several scientific experiments have now clearly discovered evidence of this tiny phase of deeper vacuum, however, Scientists have no idea as to why it should exist and have labeled it as an 'astonishing phenomenon' and 'one of the biggest mysteries in physics'.

The modeling will show why the <u>unique properties</u> of Gravitational waves allow tiny Gravitational waves to be the prime contender for <u>providing dimensions to Space</u>. The 'fluctuations' of virtual mass mimics a very minuscule, rapidly collapsing and reappearing Black Hole which is ideal for creating extraordinary tiny, positive, Gravitational waves which will <u>provide entangled dimensions to Space</u>. Like expanding ringlets from raindrops on a pond the extraordinary tiny, positive, Gravitational waves carry energy from the energy event which created them. This entangles the created dimensions to Space with the energy event from which they are created. <u>Entanglement</u> will allow time dilation and length contraction to be unique to the speed and Gravitational position of a COST. A far outer void of empty Space attracts the tiny Gravitational waves at the speed of light. The positive energy of the tiny Gravitational waves <u>raises</u> the ground state of empty Space to the ground state of dimensional Space-time.

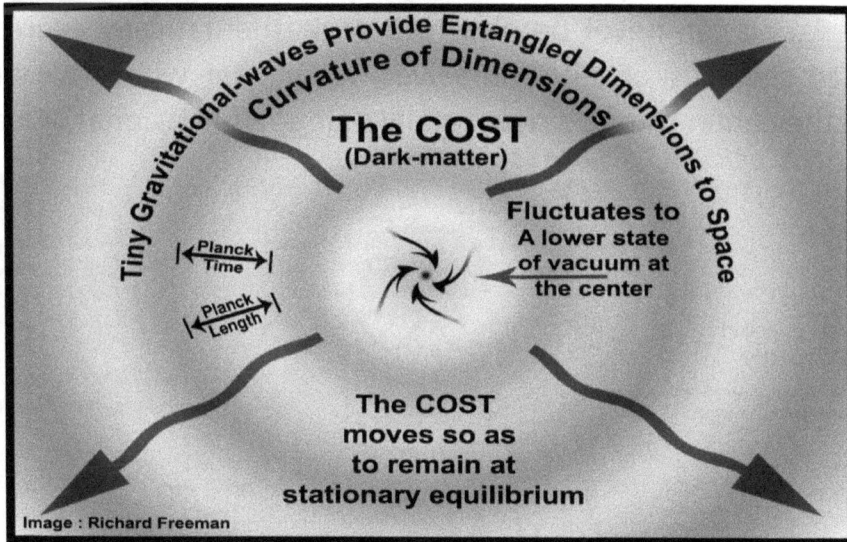

Image: The modeling will show how the COST, by providing **<u>only the primary dimension of length,</u>** will provide us all with a complete entangled Dimensional Reality from where we are <u>locked</u> into the traditional predictions of Albert Einstein's Relativity. The distance across a tiny Gravitational wave provides Planck length, the smallest allowed distance within a Dimensional Reality and the passing of the tiny Gravitational wave provides Planck time, the shortest moment of passing of time within a Dimensional Reality. This is why, like Einstein explained, *'dimensional Space and the passing of time cannot be separated'*, since, one is the wavelength of the wave and the other is the passing of the wave, hence, Space-time.

Numerous COST units are transformed from their virtual proton state into nucleons to form <u>all atoms</u>. Because Gravitational waves expand on a flat plane it will require <u>many thousands of atoms to come together</u> to build a complete entangled, Dimensional Reality free of weird Quantum behavior. [018] Scientists say it is puzzling as they have no reasoning why this threshold of Quantum behavior should exist at the size of an object which has a few thousand atoms. Entanglement will permanently <u>lock everyone of us at the center</u> of our very own unique, Dimensional Reality built by the very atoms in our bodies. This is why, even though our universe <u>will have a center and an outer edge</u>, it can never be observed or mathematically expressed that way. We all correctly observe ourselves, no matter where we are, to be located at the center of one's own <u>totally unique</u> dimensional universe.

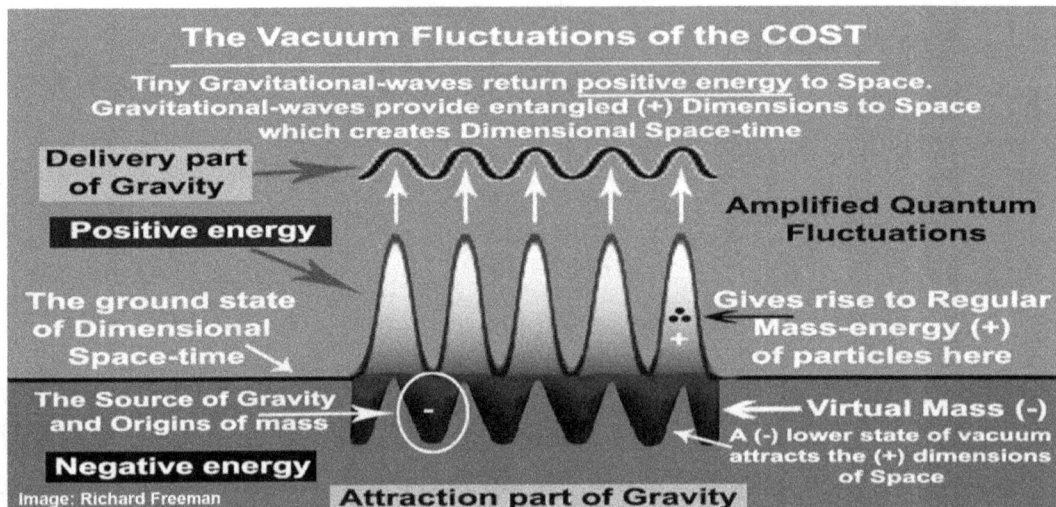

Where our mass comes from; the particles within a proton account for only 1% of a proton's mass. When the COST is transformed from a virtual proton to a regular proton an <u>inherited</u> deep vacuum phase of <u>virtual mass</u> will appear between quarks providing a strong interaction and give rise to the remaining 99% of regular mass and provide a source of Gravity. <u>One teaspoon</u> of neutron star material weighs around four billion tons which is like more than 26,500 ocean liners as large as the luxurious Queen Mary 2 all in one teaspoon. [002] It has been discovered that the pressure inside of a tiny proton peaks <u>ten times higher</u> than within a neutron star. To provide such tremendous pressure at the core of a proton one obviously requires a source which is a step further than a neutron star. The very next step from a neutron star is a Black Hole, or as in this case, a **minuscule** fluctuating Black Hole. **Virtual mass mimics a tiny, rapidly collapsing and reappearing Black Hole** and being of <u>a lower state of vacuum</u> than dimensional Space is negative energy which <u>attracts the positive dimensions of Space-time</u> which <u>rush in at light speed</u>. This causes the quarks of protons and neutrons to initially, at <u>near light speed,</u> spiral tightly towards the center of a proton where the resistance pressure peaks higher than within a neutron star.

[029] When the quarks of a proton are closest the phase of virtual mass diminishes as does the strong force being created which momentarily allows the quarks to move farthermost apart before a new phase of virtual mass is immediately reproduced in full by the next fluctuation. [003] The <u>near light speed</u> excitement and extreme agitation of a proton's particles creates pandemonium. The <u>total energy</u> given from one complete cycle of the near light speed agitation of particles represents a proton's <u>regular mass</u> (mass-energy) which is different to virtual mass. The process naturally provides the proton with spin and oscillations at a rate that's related to the regular mass which is being created by the rapidly collapsing and reappearing phase of virtual mass. The combined resource of protons and neutrons, all phasing to virtual mass, maintain the structural integrity of nucleus of atoms heavier than a hydrogen atom; mutual electromagnetic repulsion of protons is not a problem for the <u>deep vacuum</u> of virtual mass. Because the negative energy density state of virtual mass attracts the positive

dimensions of Space-time it is also the source of the <u>strong</u> attraction part of Gravity, thus, wherever one finds regular mass one will also find Gravity.

[023] When all particles are crushed out of existence inside of a Stellar Black Hole virtual mass is what remains behind to provide the Black Hole's mass and Gravity. Super Massive Black Holes are made directly from the fluctuations of virtual mass and provide the powerhouses to transform virtual protons into regular protons. Because the escape speed inside of a Black Hole exceeds the speed of light, the tiny, expanding Gravitational waves which provide entangled dimensions to Space cannot now escape from a COST. This provides an area inside of a Black Hole where, without <u>entangled</u> dimensions, the area inside of a Black Hole is effectively dimensionless and without a passing of time. [084] Inside of a Black Hole there is now an area where every place is in the same dimensionally nowhere place which mirrors a singularity without the <u>incorrect notion of infinity</u> which causes an erroneous, mathematical singularity. Since the inside is effectively dimensionless this area can only be measured from the outside.

How Gravity works and why it is mysteriously very weak: Protons and neutrons make up the nucleus of atoms. Powered by a Quantum vacuum fluctuation and residing at the hearts of Dark matter, protons and neutrons, a tiny, <u>lower state</u> phase of virtual mass mimics a minuscule, rapidly collapsing and reappearing Black Hole which naturally, like all Black Holes, <u>attracts the positive dimensions of Space</u> and in doing so provides the **strong attraction part** of Gravity. The rapidly collapsing and reappearing phase of virtual mass creates very tiny Gravitational waves which provide entangled dimensions to Space and provide the **variable delivery part of Gravity** which mirrors Einstein's General Relativity. The tiny Gravitational waves travel eternally at the speed of light which allows Gravity to propagate throughout all Space at the speed of light. The curvature of dimensions becomes warped, by being stretched, towards the attraction phase of virtual mass of other COST units. [020] *See link:* ***Planck time and Planck***

---

length are linked to the ultimate limit of quantization and directly to Gravity. **Planck length is where smooth Gravity becomes quantized.** (At scales smaller than Planck nothing is measurable and is where **Quantum Gravity** will take over). The wavelength of a tiny Gravitational wave provides a Planck length and the passing of the wave provides Planck time, thus, Space and time cannot be separated. A stretched, tiny Gravitational wave maintains Planck value which will allow the speed of light to be self observed as a constant. Gravity is delivered by contracting the distance across a 'stretched' wave to maintain a Planck length which causes Gravitational length contraction and causes objects to 'fall' by continually repositioning their stationary equilibrium position at the center of their entangled dimensions of Space. A Planck value is preserved because it is derived from a **single passing pulse of energy** carried by the passing wave. The obtained 'speed' times an object's 'mass' provides a falling object with momentum while it remains at stationary equilibrium, restrained at the center of numerous tiny units of expanding energy. **The variable delivery part of Gravity has a restraining part which operates in the opposite direction of a falling object causing Gravity to be very weakly delivered and finally solves exactly why Gravity is far weaker than other forces.**

A stretched out wave takes longer to pass which is why applying Gravity slows the passing of time. Slowing the passing of time allows Gravity to stretch and so warp the curvature of the dimensions of Space-time without the actual act of stretching the wave exceeding the speed of light which is not allowed. Because a stretched out wave maintains Planck time one's own rate of time, although slower, appears unchanging and normal. **Thus, passing time allows the continuous light speed renewal of your entangled dimensions to Space which allows your time dilation and length contraction factors to be refreshed, at the amazing speed of light, to precisely match your speed and Gravitational position in the universe**.

Virtual mass is a tiny phase of pure empty Space within Dark matter, protons and neutrons which provides negative energy for the source of the attraction part of Inner Gravity. A far and seemingly endless, outer void of empty Space provides an enormous resource of negative energy for the source of the attraction part of Outer Gravity. Both Inner and Outer Gravity are delivered by the very same tiny Gravitational waves which actively provide entangled dimensions to Space. Outer Gravity accelerates Galaxies outwards without inexplicable, hypothetical Dark energy expanding the Space between Galaxies. Galaxies are simply free falling in the direction the universe is expanding. Dark energy is a form of energy inserted by Scientists in an attempt to explain why Galaxies have unexpectedly been observed to be accelerating. Inner and Outer Gravity are created as one of the same.

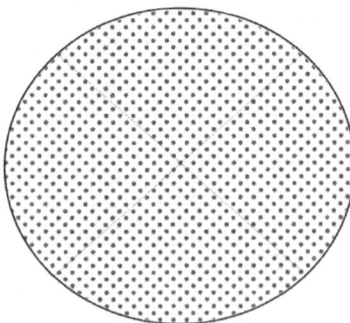

Imagine each dot is a Galaxy. If we were to move a dot from near the center outwards it becomes nearer to the dot ahead, now the dot ahead needs to move outwards at a greater speed to provide true expansion, all of which continually snowballs as we move closer to the outer edge. Because this model uses Gravity to accelerate Galaxies, our position in the universe, the distance between Galaxies and the mass of other Galaxies will affect the speed. Called the Hubble Tension, the different speeds of the expansion of the universe have now been confirmed by the James Webb Space telescope. Scientists now say *"the universe appears to be expanding at bafflingly different speeds depending on where we look"*. With a Natural Model, Galaxies have to be accelerating away which is observed to be true.

The Natural Model will use actual experimental evidence to explain how Gravity, powered by nothing more than empty Space, slowly arose from squeezing numerous Quantum vacuum fluctuations and began building a Naturally Forming Universe. In doing so the model flows fluently from the creation of the first COST units (Dark matter) to the creation of Galaxies. The positive energy of all created Matter is canceled by the negative energy of virtual mass which is the source of Gravitational Attraction and is responsible for creating regular mass. This allows a universe to begin from zero mass-energy empty Space and continue to maintain a sum-total zero mass-energy universe which is required to preserve the conservation of energy law. **Acting as a virtual proton, the COST provides the vital invisible Gravity and mass generation engines for regular protons.**

Brief: How Super Massive Black Holes (SMBHs) are made before Galaxies: Gravity brings virtual protons (Dark matter) together creating super large, clouds of virtual protons. Now armed with stronger Gravity virtual protons migrate to a rotating center which becomes exceptionally dense and where strong Gravity prevents the tiny Gravitational waves escaping from virtual protons. Without the escaping tiny Gravitational waves the passing of time stops and virtual protons are finally free to come tightly together which causes a near light speed Gravitational collapse. The Gravitational collapse forms a SMBH at the center of a disk shaped <u>Dark Galaxy.</u> Virtual protons from the thick, dense disk will naturally continue to migrate to the center of the Galaxy where virtual protons will be <u>used</u> within Matter creation cores at the poles of the central SMBH causing much of the <u>thick dark disk</u> to be <u>replaced</u> with a disk shaped, bright Galaxy of stars. An outer halo of Dark Matter <u>remains today as undeniable evidence</u> of how virtual protons (Dark matter) were sourced, so removed, from a central area of a large cloud of virtual protons to create a bright Galaxy of stars and its central SMBH.

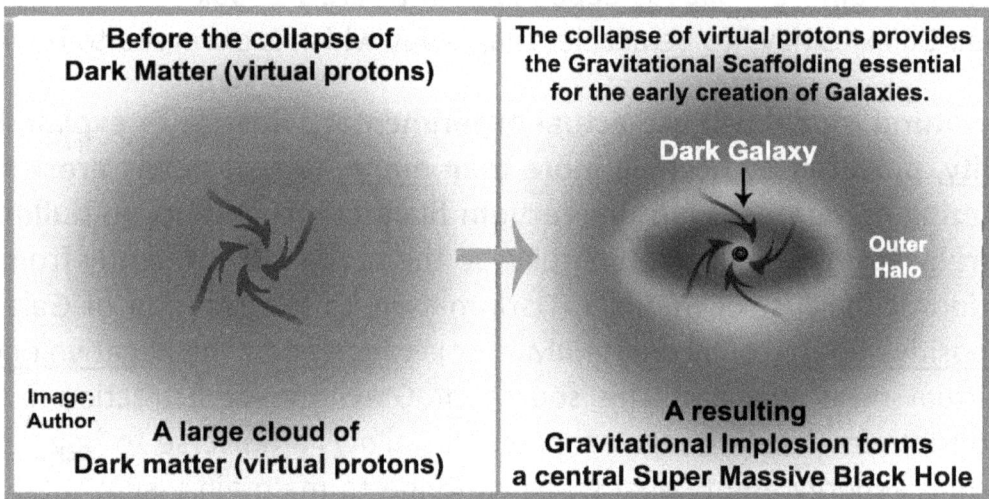

Before the collapse of Dark Matter (virtual protons)	The collapse of virtual protons provides the Gravitational Scaffolding essential for the early creation of Galaxies.
	Dark Galaxy
	Outer Halo
Image: Author — A large cloud of Dark matter (virtual protons)	A resulting Gravitational Implosion forms a central Super Massive Black Hole

[059, 091, 104] Why is the James Webb Space Telescope alarmingly finding matured, large Galaxies where they *"expected only to find tiny, young, baby Galaxies"*. Answer: Without a Big Bang, Matter production likely began well <u>before</u> the said Big Bang. The CMB radiation, the large number of old metal poor stars and the observation that Quasar activity peaked in the early universe indicates Matter production very likely <u>peaked</u> at a similar time to the proposed Big Bang. However, unlike a Big Bang, regular Matter within a Natural Universe is actually born in dense pulses <u>inside</u> of already created dark Gravitational scaffolds which were <u>imperative</u> to allow the observed rapid beginning of the formation of regular disk shaped Galaxies. [004] Without this <u>essential</u> Gravitational scaffolding Galaxies like our Milky Way could not have existed as their stars would fly away due to lack of Gravity.

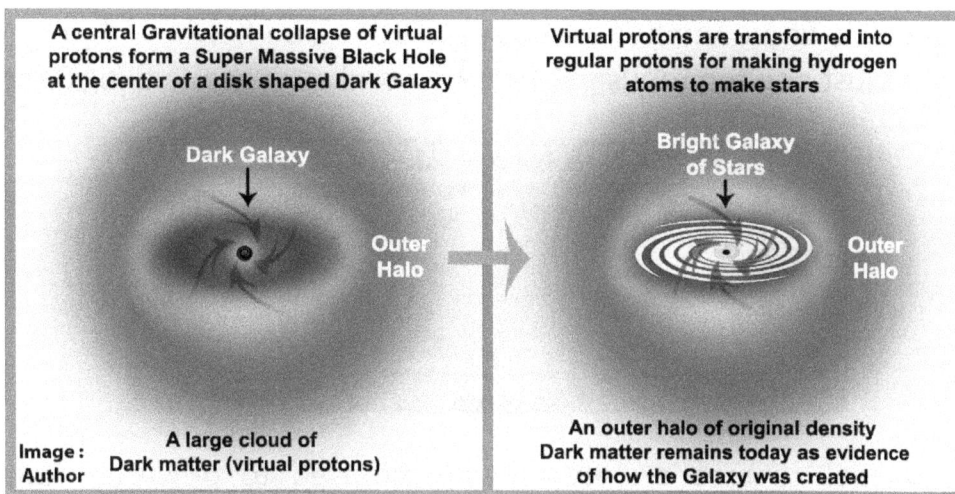

A central Gravitational collapse of virtual protons form a Super Massive Black Hole at the center of a disk shaped Dark Galaxy	Virtual protons are transformed into regular protons for making hydrogen atoms to make stars
Dark Galaxy / Outer Halo	Bright Galaxy of Stars / Outer Halo
Image: Author — A large cloud of Dark matter (virtual protons)	An outer halo of original density Dark matter remains today as evidence of how the Galaxy was created

All Matter was not likely derived from a colossal burst of an unknown-to-man form of energy appearing from a never to be revealed source at a super-natural like event of a Big Bang. Because of the tight bond between Mass and Gravity it is fitting that the massive Gravitational resource of Super Massive Black Holes is used to create all regular Matter. The model will show why the source of both mass and Gravity is pure empty Space.

The transformation of virtual proton to regular protons necessitates the production of mirror anti-particles. A virtual proton produces pairs of virtual particles in the form of quarks and anti-quarks, however, without three additional Matter quarks a virtual proton in its Dark matter state is not a regular proton. As virtual protons become exposed to the accretion disks and steered to the poles of Super Massive Black Holes their speed and energy increases to at least 3 pairs of virtual quarks and anti-quarks which persist to annihilate each other. The Black-body spectrum of the Cosmic Microwave Background radiation provides direct evidence that Matter, like at the proposed Big Bang, was at first tightly constrained while in a state of extremely hot, dense plasma. This state was attained within multiple closed dense plasma cores where particle and anti-particle annihilation was creating numerous photons and electrons. Primarily powered by the Gravity of a SMBH, Matter and Anti-matter annihilation energy now powers Matter creation cores where temperature rises to 10 trillion Kelvin and <u>mirrors the</u> <u>moment</u> at just one ten thousandth of a second after the said Big Bang.[005]

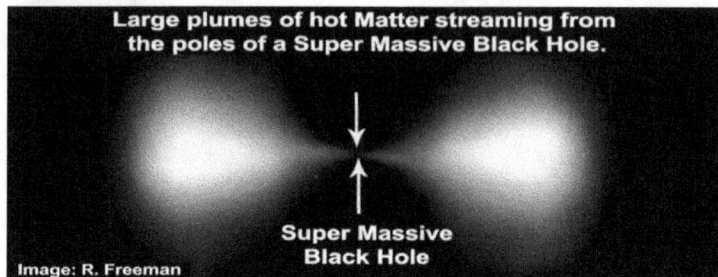

Briefly: Virtual protons are now being exposed to powerful, spiraling, magnetic fields which <u>forcibly separate</u> three negative anti-quarks from their <u>positive quark partners</u> which are momentarily tightly restrained by the <u>negative</u> energy density phase of virtual mass. This is the beginning of the <u>strong force</u> which by momentarily clinging on to three positive quarks actually <u>dictated</u> that the universe was made from Matter and not Anti-matter. The ripped away anti-quarks and the retained quarks were both born with <u>exactly</u> the same state of very little mass-energy. However, the ripped away anti-quarks are now residing without their COST and without a near interaction with a phase of <u>virtual mass</u> the anti-quarks act as massless waves which allows anti-quarks to be directed into the central Black Hole but contribute very little to its mass. **In sharp contrast;** because the three Matter quarks now cannot be annihilated by their anti-quark partners they <u>now live to enjoy a strong, near interaction</u> with a <u>complete</u> phase of virtual mass which by mimicking a tiny appearing and collapsing Black Hole <u>violently agitates</u> the three quarks at near <u>the speed of light</u> providing the quarks with newly found motion energy which provide the newly created regular protons with a comparatively <u>massive amount</u> of mass-energy, $E=mc^2$, given <u>free</u> by a tiny phase of <u>empty Space</u>. **Since, the newly created protons will get to <u>keep all of their inherited phase of virtual mass</u> a Super Massive Black Hole is not required to grow beyond its observed size.**

This correctly deals with the created anti-quarks which, unlike a Big Bang,

solves the Matter and Anti-matter asymmetry problem and saves our precious Matter from annihilation. The first batch of regular protons together with electrons are <u>explosively released</u>, providing a form of inflation, from being tightly constrained within dense, hot, plasma creation cores. If many of these events occurred almost simultaneously and close together than this time **closely resembles a period after the Big Bang.**

Now harboring three additional Matter quarks a positive proton continues to create from vacuum fluctuations virtual pairs of Matter and Anti-matter sea-quarks <u>as it did</u> when it was a virtual proton and <u>as it did</u> when it was once a simple Quantum vacuum fluctuation.

Ejected and exposed to a cooling environment protons grab an attracted electron to form hydrogen atoms which allows the plasma to become a gas and is when the <u>photons</u> of the CMB radiation are set free. The photons released from regions which had short spiraling magnetic fields and near stronger Gravity lost energy escaping the stronger Gravity so became slightly cooler than photons escaping from regions with long spiraling magnetic fields and weaker Gravity. **This provided the CMB radiation with its slightly cooler and slightly warmer regions.** Ejected Matter is <u>first directed to halo regions</u> where it quickly cools and begins forming many of the first stars before Matter falls to the Galactic plane. Cooling will allow Gravity to bring hydrogen atoms and molecular hydrogen together to form stars and adolescent Galaxies. Stars will <u>Gravitationally cling to</u> the already provided <u>essential</u>, Dark matter Gravitational scaffolds which are needed to quickly form and shape regular Galaxies, including Galaxies like our Milky Way Galaxy. By growing long spiraling magnetic fields the Matter creation system develops a Quasar which continues to eject <u>newly created Matter</u>. [105] <u>Because of the way Matter is released in long dense pulses</u> many young Galaxies may develop <u>correspondingly</u> bright bubble-like lumpy regions.

Unlike a Big Bang, a Natural Universe first creates Super Massive Black Holes so the first Galaxies <u>do not need to create</u> the hypothesized, super-sized stars <u>100,000 times</u> more massive than our Sun to form Semi Massive Stellar Black Holes at the hearts of first Galaxies. There is <u>no need</u> for first Galaxies to collide to combine their Semi Massive Stellar Black Holes to create Super Massive Black Holes (SMBH). Such hypothesized collisions would have <u>highly distorted the shape of the first Galaxies</u>. While some collisions do happen, Galaxies are generally being <u>moved apart</u> by Outer Gravity which makes such collisions <u>less common.</u> If the spiraling magnetic fields grow fairly quickly many of these first Galaxies are likely to gain early spiral arms. Rather than colliding and combining and creating peculiar shapes, Galaxies will grow themselves by forming stars from their self created Matter. Their central SMBHs and Dark matter scaffolds will normalize their appearance. Matter ejected far and sparsely will eventually form Irregular Galaxies which will naturally orbit their parent regular Galaxy. The Natural Model will explain how all Galaxies are formed, <u>including</u> their difficult to explain halo stars and Globular Clusters. Many halo stars are made first because **all Matter <u>first travelled</u> directly to the halo region of Galaxies** where it naturally formed many of the very first stars which are now the oldest, many of which reside in Clusters. Halo stars provide amazing, undeniable evidence of how the universe was truly made.

The universe is now said to be 13.8 billion years old. <u>As far back as can be directly observed,</u> which is now 13.62 billion years, practically all Matter is contained within numerous normal looking Galaxies like tiny islands within incredibly vast oceans of near empty Space, which is exactly what would be expected within a universe where Galaxies create their own Matter. The Natural Universe allows **the first Galaxies to form <u>before</u> the Big Bang.**

[005] The temperature at the core of Quasar 3C 273 has been discovered to be an astonishing <u>10 trillion Kelvin</u> which is said to be an astounding <u>100 times</u> hotter than the current theories for Quasars allow. To account for this Scientist are now scrambling for new theories. [038] Since this is where my Matter creation cores reside it is a very <u>unlikely coincidence</u> that this is <u>the</u>

exact same temperature which is calculated to have **only existed** during the creation of Matter and the annihilation of Anti-matter at just one ten thousandth of a second after the said Big Bang. There can now be no doubt that within these cores at the poles of Super Massive Black Holes, Matter exists in a state thought to have only existed at the proposed Big Bang and is amazing evidence that new Matter is being created here. The astonishing discovery will allow the evolution of Matter and the release of the photons of the Cosmic Microwave Background radiation to occur as Matter is exposed to a range of cooling events as theorized from a said Big Bang.

But is there any evidence that this is true? [006] NASA's Swift Satellite was launched in 2004. The Swift team says the jets from Quasars are *"made of protons and electrons"* (which make hydrogen atoms). [007] The ESO's Very Large Telescope has discovered 12 Regular Quasars located over 12.5 billion light-years from Earth were surrounded by huge halos of cool, dense hydrogen gas. [116] Astronomers have even discovered a Quasar, J2054-0005, powered by a SMBH and existing in the early universe which they say is *"spewing out molecular gas, the raw material needed to form new stars"*.

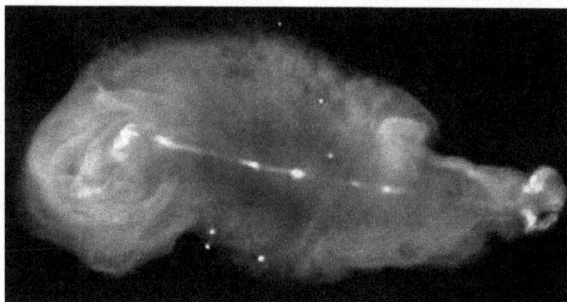

Image Credit: X-ray: NASA/CXC/Tokyo Institute of Technology/J.Kataoka et al, Radio: NRAO/VLA

[083] Above image; near the edge of the visible universe 12.4 billion light years away jets (narrow bar across center) are propelling material away from a central Super Massive Black Hole of Galaxy 3C353 (tiny bright spot at center). **Every computer simulation of Galaxy formation I have seen shows Matter, presumably created at the Big Bang, streaming in towards**

a forming Galaxy. While <u>actual images</u> from the early universe, like the image of 3C353, clearly show Matter and radiation streaming <u>away</u>!

[039] The Low-Frequency Array <u>radio</u> telescope is an array of 70,000 small antennae spread across nine European counties. The images <u>clearly</u> highlight massive plumes of hot Matter and radiation streaming from central Super Massive Black Holes (SMBH) of numerous early Galaxies. My Matter creation cores evolve into Quasars which provide an ongoing supply of Matter which cools and forms stars. The following illustration shows the number of Quasars since the beginning of the universe and the resulted rate of star forming, the correlation is clear and undeniable.

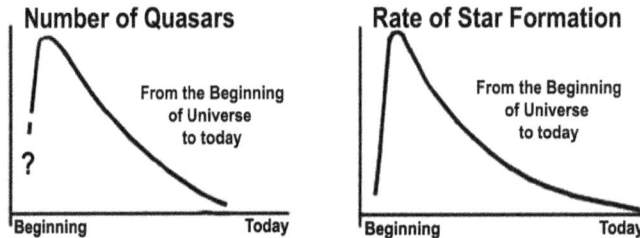

Number of Quasars

From the Beginning
of Universe
to today

?

Beginning   Today

Rate of Star Formation

From the Beginning
of Universe
to today

Beginning   Today

Within Matter creation cores at the poles of SMBHs my Natural Model uses powerful magnetic fields to separate Matter and Anti-matter. Although Scientists have discovered direct evidence of these powerful magnetic fields within these regions they say the process and the physics which generates the magnetic fields and associated plumes of hot Matter is not well understood. Observation and experimental evidence of every step of the Natural Model will be continually provided throughout this book.

[084A] **Length contraction and time dilation are predictions of Einstein's Relativity. Because (dimensional) Space and time cannot be separated,** length contraction and time dilation are naturally like two sides of the same coin. Length contraction contracts dimensional spatial distance and time dilation slows the passing of time. With a Natural Model length contraction and time dilation are induced by <u>both Gravity and speed</u>. [065] Length contraction and time dilation are real phenomenon. For example; if time dilation was not accounted for it is said your car's GPS navigation would be useless as your GPS would lose accuracy within about two minutes and errors would <u>accumulate</u> by more than 10 km per day.

The speed of light being a <u>constant</u> provides the <u>key</u> to understanding how time dilation and length contraction should operate. For a self-observer light must always travel the same dimensional distance over the same amount of passing time <u>regardless</u> of length contraction or time dilation factors. Dimensional distance is locked by Planck length and passing time is locked by Planck time and both Planck length and Planck time are provided by tiny Gravitational waves which can be stretched to provide both length contraction and time dilation. A stretched wave covers a greater distance which contracts to a Planck length which provides length contraction. A stretched wave takes longer to pass which <u>slows the passing of time</u> but because the passing wave maintains Planck time one's self time appears normal. **Because Planck values are maintained, length contraction and time dilation relating to one's self are masked from one's own entangled Dimensional Reality and are only revealed by an observer who has a significantly different entangled Dimensional Reality to one's own.**

Fast Mary's spaceship is travelling to an Earth like planet orbiting a star which for Stationary Bob on Earth is 100 light years away. If Fast Mary's spaceship travels at 99.99999999% the speed of light it will take her nearly 100 Earth years to arrive but only 7 seconds of her own self-observed time. Time dilation and length contraction shrinks Fast Mary's travel time from 100 years to just 7 seconds. Although not really possible for anything made from atoms, if Fast Mary was able to increase the speed of her spaceship to 100% of the speed of light, Mary's time dilation factor is now 100% and Mary will now arrive the very same instant she leaves and, like a photon of light, travel zero dimensional Space or time in doing so. Science today cannot explain the physical process which allows this to happen, other than say this is a ramification of the successful theory of Special Relativity.

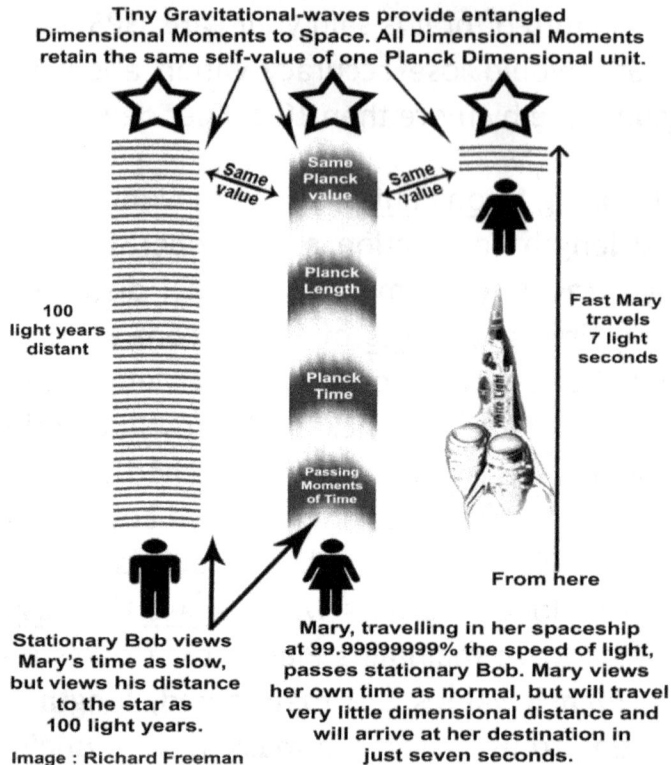

Tiny Gravitational-waves provide entangled Dimensional Moments to Space. All Dimensional Moments retain the same self-value of one Planck Dimensional unit.

Same value

Same Planck value

Same value

Planck Length

Planck Time

Passing Moments of Time

100 light years distant

Fast Mary travels 7 light seconds

From here

Stationary Bob views Mary's time as slow, but views his distance to the star as 100 light years.

Image : Richard Freeman

Mary, travelling in her spaceship at 99.99999999% the speed of light, passes stationary Bob. Mary views her own time as normal, but will travel very little dimensional distance and will arrive at her destination in just seven seconds.

Because Fast Mary's high speed is nearly the same as her tiny, dimensions providing Gravitational waves, the waves escape from her, in the direction of travel, <u>very slowly which effectively stretches the waves over a great distance</u>. This allows her to travel a greater distance during the escape of a wave, the greater distance of which now contracts to a Planck length. Since the passing of one of her tiny Gravitational waves continue to represent both a Planck length and a passing unit of Planck time her 'very slow' time <u>remains self-normal</u> but her distance to the star contracts. **This is how Fast Mary travels a great distance within very little passing of time.**

The Doppler Effect allows one to plainly see why Relativity says dimensions (length) are most affected in the direction of motion. The stretched waves <u>behind</u> Fast Mary also reflects the distance she has actually traveled and represents the distance which has severe length contraction due to the stretched Gravitational waves which contract distance to a Planck length.

The end result is Fast Mary will have travelled far fewer Planck lengths of distance and her travel time is reduced to just seven seconds.

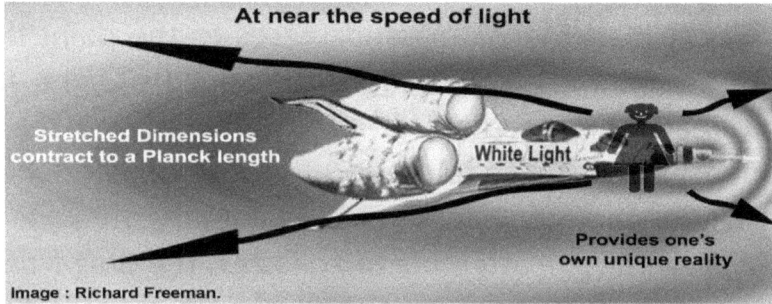

At near the speed of light
Stretched Dimensions contract to a Planck length
White Light
Provides one's own unique reality
Image : Richard Freeman.

At near the speed of light **the Doppler Effect** now causes intense length contraction which is contracting the distance behind Fast Mary. Attempting to exceed the speed of light this intense length contraction is now the same as applied by intense Gravity near the event horizon of a Black Hole. I will call this relativistic Gravity which, like attempting to escape a Black Hole, prevents Fast Mary exceeding the speed of light. Approaching the speed of light, due to Mary's relativistic mass, it now appears Mary is becoming her own Black Hole from which she cannot escape by going faster, since, exactly like attempting to escape from a Black Hole, it would require an impossible amount of energy to do so.

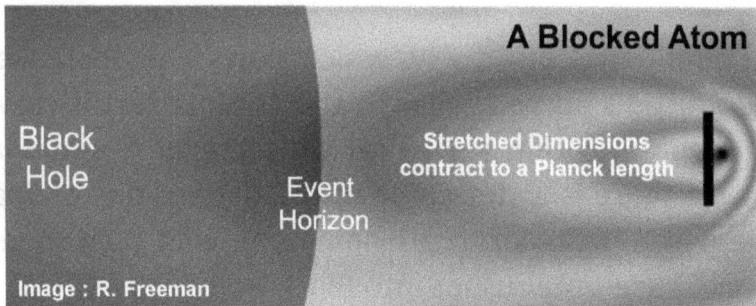

A Blocked Atom
Black Hole
Event Horizon
Stretched Dimensions contract to a Planck length
Image : R. Freeman

The length contraction from stretched dimensions behind Mary are clearly pulling against Mary in exactly the same way as if Fast Mary was attempting to escape the massive Gravity of the event horizon of a Black Hole. If Mary was stationary at the event horizon of a Black Hole she would effectively

23

have high speed into the flow of the dimensional Space which Scientists say flows into a Black Hole. Consequently, Fast Mary cannot exceed the speed of light for the exact same reason she cannot escape from a Black Hole. Special Relativity tells us Mary's mass increases to the point that it would require infinite energy for her to exceed the speed of light. If Fast Mary found a way to reach the speed of light she would discover she now has no entangled dimensional distance to exceed. Because her stretched and warp dimensions all normalize to Planck units Fast Mary views everything as normal. However, when she attempts to go faster, the drag from relativistic Gravity causes her to request more energy than her powerful twin Anti-matter stellar-drives can provide. Length contraction caused from near light speed is actually a very strong form of similar length contraction we feel as 1 G of Gravity holding one to the ground. My relativistic Gravity is most perceptible at speeds approaching light speed.

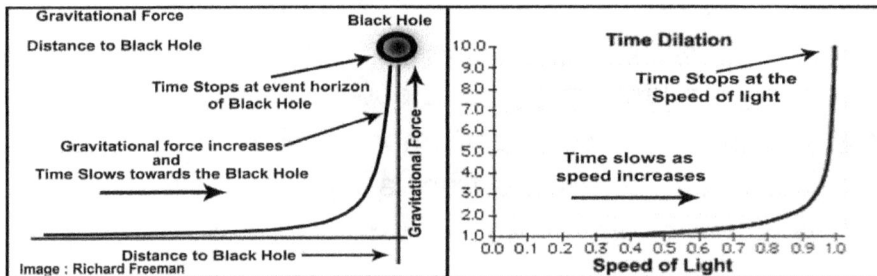

Image : Richard Freeman

Notice how the influence of Gravity, which slows time increases towards a Black Hole, produces a curve similar to the Lorentz factor curve for speed caused time dilation. Within a Natural Universe at a stationary position near a Black Hole there is a stretching of the dimensions of Space towards the Black Hole which is similar to providing one with high speed. Scientists say dimensional Space-time rushes into Black Holes.

Now imagine Fast Mary on board a super fast train travelling at <u>half the speed of light</u>. Mary places a mirror on the floor of the train and switches on an above light. For the same reason as onboard her spaceship, Fast Mary now has <u>fewer</u> passing dimensional moments of Planck length and

Planck time which now allows Fast Mary to observe the light, which speed is constant, travel a shorter distance than Stationary Bob at the station.

Because light from <u>all sources</u> is locked to a <u>dimensional</u> Planck moment, provided by a tiny Gravitational wave, light can only travel the same <u>dimensional distance</u> within the same dimensional moment regardless of whether a tiny Gravitational wave is stretched or normal. If the very <u>same event</u> has more moments of <u>passing time</u> in a different frame of reference, as with Stationary Bob, it also has more dimensional moments which allows light to be observed to have travelled a greater dimensional distance.

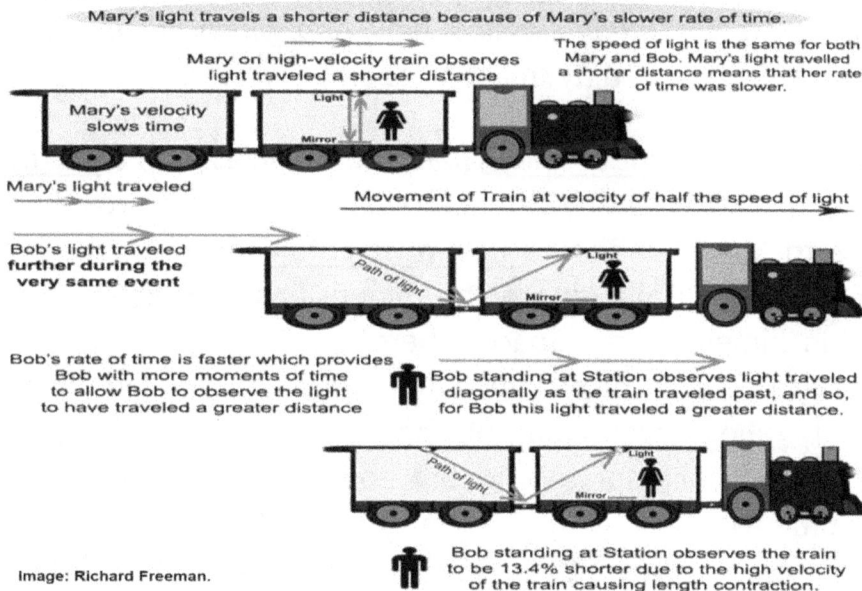

Mary's light travels a shorter distance because of Mary's slower rate of time.

Mary on high-velocity train observes light traveled a shorter distance

The speed of light is the same for both Mary and Bob. Mary's light travelled a shorter distance means that her rate of time was slower.

Mary's velocity slows time

Light
Mirror

Mary's light traveled

Movement of Train at velocity of half the speed of light

Bob's light traveled further during the very same event

Path of light
Light
Mirror

Bob's rate of time is faster which provides Bob with more moments of time to allow Bob to observe the light to have traveled a greater distance

Bob standing at Station observes light traveled diagonally as the train traveled past, and so, for Bob this light traveled a greater distance.

Path of light
Light
Mirror

Image: Richard Freeman.

Bob standing at Station observes the train to be 13.4% shorter due to the high velocity of the train causing length contraction.

This allows different observers such as Fast Mary and Stationary Bob to observe light <u>during the same event</u> to travel different dimensional distances at the <u>same</u> constant speed. Mary and Bob can see each other but each observes the light travel a different distance. Because Mary's fast train has less entangled, dimensional moments <u>Stationary Bob also observes</u> Mary's fast train as being shorter due to length contraction.

This is achievable because both Fast Mary and Stationary Bob reside at the center of their own but different <u>entangled</u>, Dimensional Realities which allows each to observe the same event differently. During the event <u>their actual dimensions of Space-time are able to be different</u> because both Mary and Bob reside within <u>their own</u> unique and entangled, Dimensional Realities. **Note: <u>Entanglement</u> prevents another observer's rate of time interfering with one's own. That is, <u>all other Dimensional Realities,</u> belonging to all other observers, are <u>totally ignored.</u>**

<u>Quantum Theory:</u> [071] Renowned for proving that Quantum behavior, though bizarre, really takes place, American Physicist Richard Feynman has said; *"We choose to examine a phenomenon which is impossible, absolutely impossible, to explain in any classical way, and which has in it the heart of quantum mechanics. In reality, it contains only mystery"*.

My Natural Model <u>cannot provide</u> Quantum particles with a <u>permanent</u> self mechanism to create the tiny Gravitational waves which provide <u>entangled</u> dimensions to Space. Without <u>their own entangled</u> dimensions to Space Quantum particles **<u>totally ignore</u>** the dimensions of Space and effectively reside within dimension<u>less</u> Space where everyplace is truly in the same dimensionally nowhere place and where time is reduced to an <u>instant</u>. Consequently, <u>unobserved</u> Quantum particles are <u>not self entangled</u> with dimensions and so completely ignore spatial dimensions, including the passing of time, which allow tiny Quantum particles to effectively reside as waves within dimension<u>less</u> Space and play by a completely different set of rules to objects provided with a permanent, <u>entangled</u> Dimensional Reality.

Naturally this means Quantum particles ignore the dimensions of Space-time and are <u>not influenced in the normal way by Gravity</u> which works by warping the curvature of the dimensions of Space-time. However, without being <u>restrained</u> at <u>stationary equilibrium</u> by the dimensions of Space-time a Quantum particle is free to have a **direct attraction to the strong attraction part of Gravity.** <u>Free of the shackles of being trapped</u> at stationary equilibrium within <u>entangled</u> dimensions allows Quantum particles, including photons of light, to <u>instantly</u> speed away in all directions due to the attraction part of Outer Gravity and for the attraction part of Inner Gravity to <u>redirect their path</u> when near an object of mass. Thus, Quantum particles now act exactly as they are observed to. **This provides Quantum particles with what can now be called <u>Quantum Gravity</u> which works with only the strong attraction part of Gravity and <u>without</u> the dimensions of Space-time.** <u>Instant speed</u> is a feature of Quantum Gravity and <u>accelerating speed</u> is a feature of Einstein's Gravity.

<u>**Solving wave and particle duality:**</u> My Natural Model will clearly show why Quantum particles effectively reside within dimension<u>less</u> Space as dimensionless waves. Observing or measuring from a Dimensional Reality causes wave contraction which <u>contracts a wave into a particle</u>. Wave contraction is observed or measured for the <u>exact same reason</u> as length contraction which is a prediction of Albert Einstein's Relativity. **Wave contraction, exactly like length contraction, is revealed when observing or measuring an object which is residing within a state of Space which is significantly dimensionally different to one's own.**

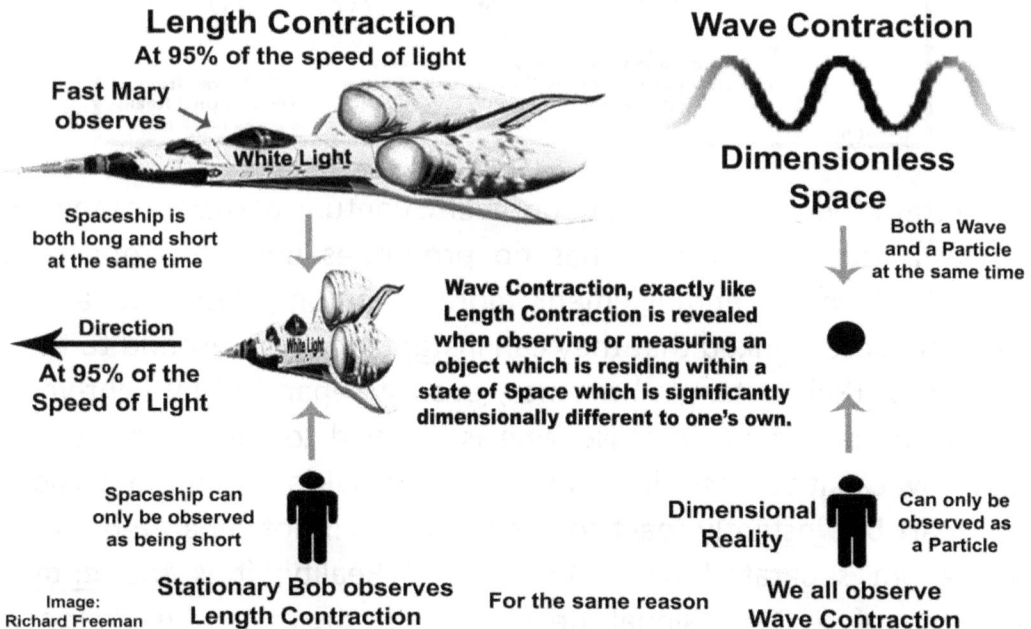

Length Contraction
At 95% of the speed of light

Fast Mary observes ↘

White Light

Spaceship is both long and short at the same time

← **Direction**
At 95% of the Speed of Light

Wave Contraction, exactly like Length Contraction is revealed when observing or measuring an object which is residing within a state of Space which is significantly dimensionally different to one's own.

Spaceship can only be observed as being short

Image: Richard Freeman

Stationary Bob observes Length Contraction

For the same reason

Wave Contraction

Dimensionless Space

Both a Wave and a Particle at the same time

Dimensional Reality

Can only be observed as a Particle

We all observe Wave Contraction

[076] My Natural Model clearly unravels all wave and particle duality mysteries including the <u>double slit experiment</u> where any attempt to observe or measure the wave passing through both slits will immediately contract the wave to a particle which can now only pass through one slit as a particle while an unobserved 'particle' clearly passes through both slits as a wave. Residing within dimension<u>less</u> Space <u>every part of the wave is in the same place</u>. When observed or measured from a Dimensional Reality, your own <u>Dimensional Reality correctly measures all facets of the wave as being in the same place which</u> collapses the wave and squeezes it into its <u>tightest possible configuration</u> of a particle. Within a Natural Universe one's own entangled <u>Dimensional Reality</u> acts as a <u>dimensional filter</u> from where the dimensionless wave can only be recorded as a dimensional particle.

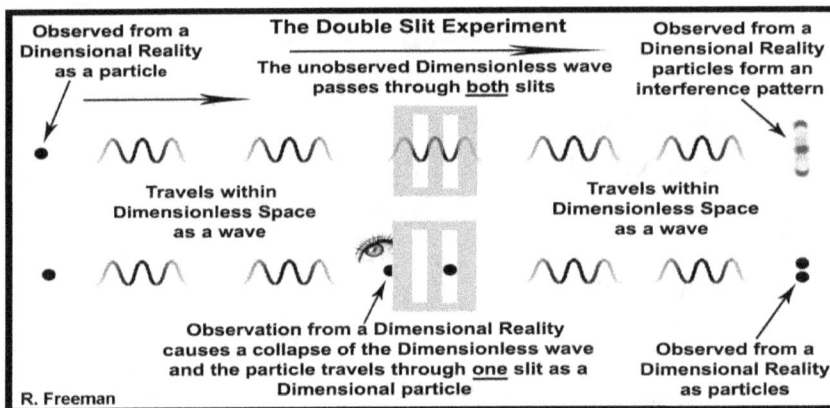

The Natural Model will rationally unravel Quantum particle entanglement and why a Quantum particle has no properties until it is observed or measured. Quantum entanglement more commonly refers to a pair of particles so closely linked that they can <u>instantaneously</u> respond to the spin or actions of their entangled partner, even if separated by light-years of dimensional Space and time. Nothing is allowed to exceed the speed of light so how can it be possible for an entangled pair of Quantum particles to be far apart but <u>instantly</u> react to each other as if not separated? Answer; they are not separated by a Dimensional Reality; it is the <u>entangled</u> dimensions of a Dimensional Reality which provides the dimensional separation not the dimensionless, nothingness of empty Space. My Natural Model clearly says unobserved Quantum particles are not naturally provided with a fluctuating mechanism to provide Space with entangled dimensions and the passing of moments of time. Consequently, Quantum particles effectively reside within dimension__less__ Space where dimensional distance and passing time are naturally, completely, absolutely irrelevant and so ignored. Two Quantum entangled particles can be entangled with each other's opposite entangled orientation properties but are not naturally entangled with a mechanism to provide entangled dimensions to Space. Residing as both a wave within dimensionless Space and a particle within an observer's entangled Dimensional Reality allows the particles to completely ignore the dimensions of Space and not be dimensionally separated and instantly react to each other's orientation properties, even if perceived to be separated by a vast dimensional distance when observed or measured as a particle from within an observer's entangled Dimensional Reality. **Because Space and (passing) time cannot be separated 'instant time' requires dimensionless Space, consequently, instant time is actually remarkable evidence of the true existence of a realm of dimensionless Space which only the modeling of my Natural Model can clearly provide.**

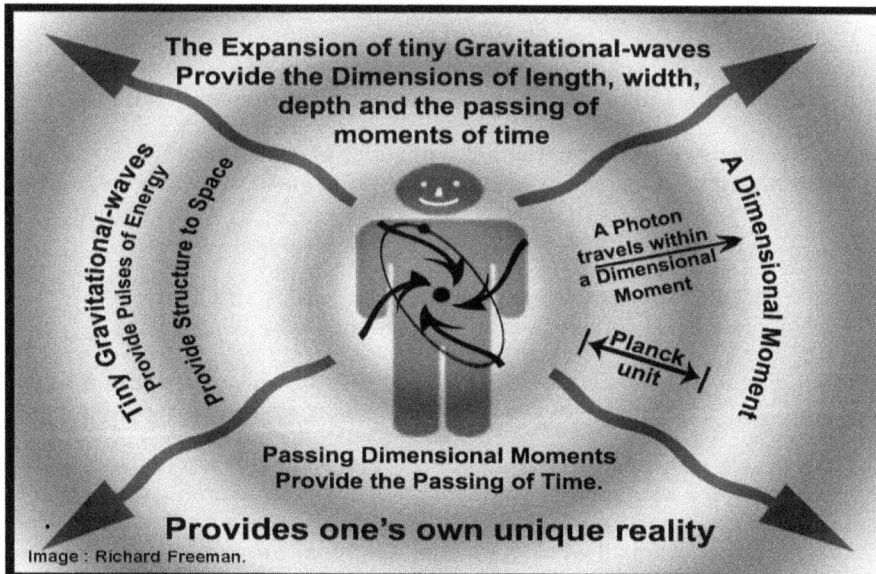

Image : Richard Freeman.

29

From the formation of Galaxies to the existence of you and me the evidence that the universe is made this way is truly everywhere. The following chapters will extensively detail all subjects covered in this brief chapter and provide links to an amazing amount of related evidence. The next chapter will begin by explaining why if one's wishes to obey the laws of physics; you and I could not have existed if there had been a Big Bang.

Briefly; Quantum vacuum fluctuations make virtual protons (Dark matter), virtual protons make regular protons, regular protons make hydrogen atoms, hydrogen atoms make stars and stars make everything else.

Time line of a Natural Universe.

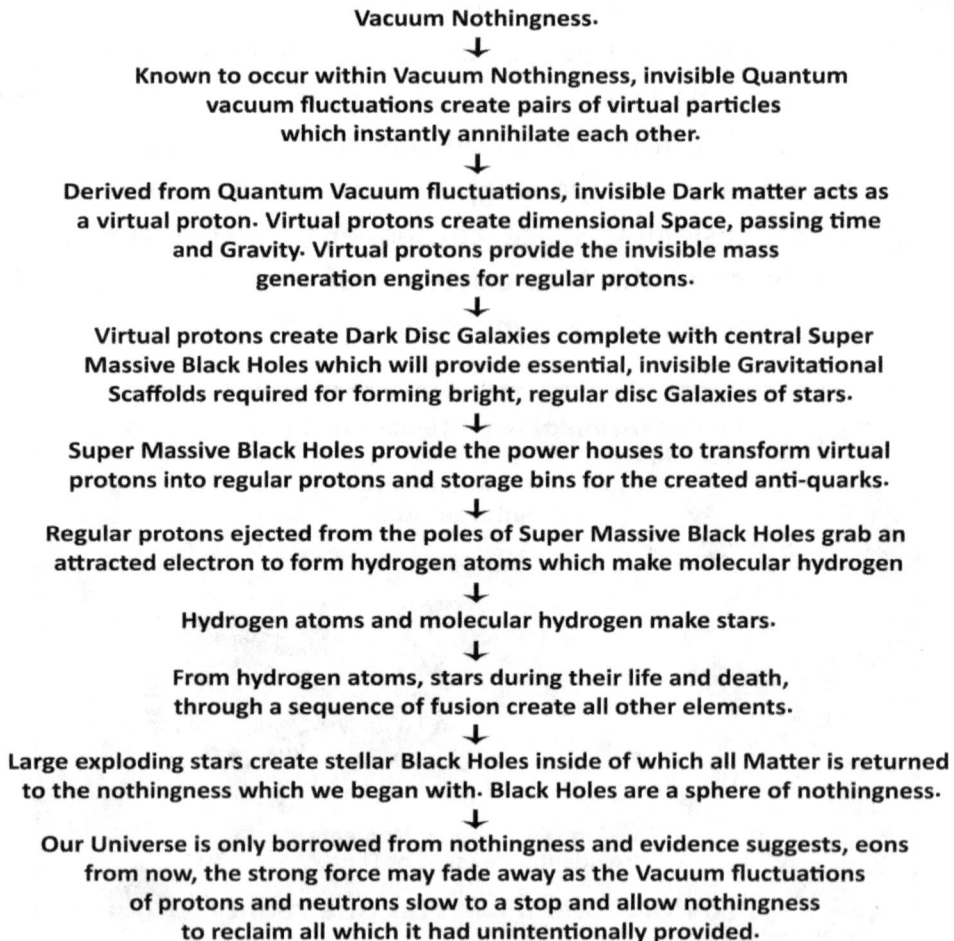

Vacuum Nothingness.

↓

Known to occur within Vacuum Nothingness, invisible Quantum vacuum fluctuations create pairs of virtual particles which instantly annihilate each other.

↓

Derived from Quantum Vacuum fluctuations, invisible Dark matter acts as a virtual proton. Virtual protons create dimensional Space, passing time and Gravity. Virtual protons provide the invisible mass generation engines for regular protons.

↓

Virtual protons create Dark Disc Galaxies complete with central Super Massive Black Holes which will provide essential, invisible Gravitational Scaffolds required for forming bright, regular disc Galaxies of stars.

↓

Super Massive Black Holes provide the power houses to transform virtual protons into regular protons and storage bins for the created anti-quarks.

↓

Regular protons ejected from the poles of Super Massive Black Holes grab an attracted electron to form hydrogen atoms which make molecular hydrogen

↓

Hydrogen atoms and molecular hydrogen make stars.

↓

From hydrogen atoms, stars during their life and death, through a sequence of fusion create all other elements.

↓

Large exploding stars create stellar Black Holes inside of which all Matter is returned to the nothingness which we began with. Black Holes are a sphere of nothingness.

↓

Our Universe is only borrowed from nothingness and evidence suggests, eons from now, the strong force may fade away as the Vacuum fluctuations of protons and neutrons slow to a stop and allow nothingness to reclaim all which it had unintentionally provided.

2 WHY THE BIG BANG EXPLAINS LITTLE.

The Big Bang Theory:

First propose in 1927 by Georges LeMaitre, the Big Bang is today commonly accepted by Scientists as the most probable beginning to the universe. Evidence that the universe is expanding provided support for a notion the universe began from an infinitely small point.

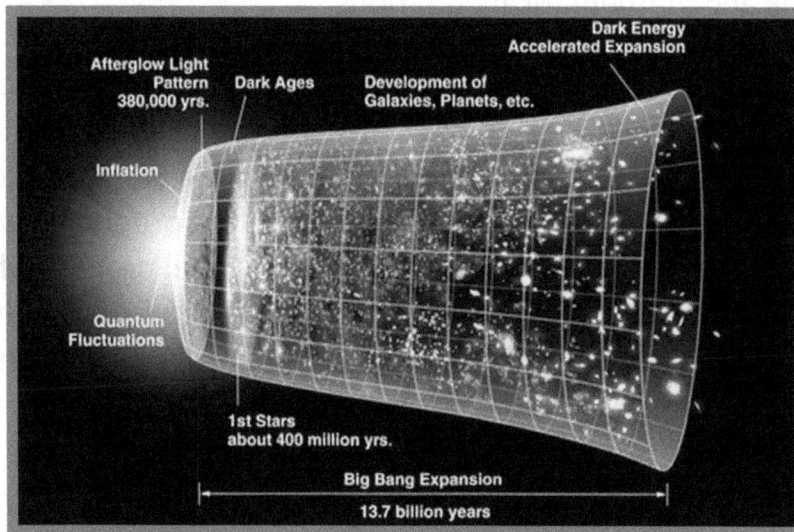

An artist's concept of the expansion of the universe. Credit NASA / WMAP science Team.

We really are very fortunate to have the many Scientists who are forever expanding our knowledge and enlightening us all on the many subjects dealing with the amazing mysteries of the night skies. I in truth have the greatest admiration for the same dedicated Scientists. I could not have created this model without their amazing achievements. I purchase and read their books, I thank those who have enabled me to so easily 'click' on my computer mouse and gain access to the latest science and knowledge available in these most fascinating subjects. **I am indeed <u>most indebted</u> to the Scientists and Physicists who first developed the theories and experiments which have now become common knowledge.**

Theoretical physics is indeed a valuable and powerful tool used to explain many of the mechanisms at play within our universe. Scientists spend many billions of dollars building an array of telescopes, Space craft and machines in an endeavor to 'see' if their equations are correct and to actually observe how events occur in a physical way. Scientists of course, do not expect to see their equations written in some faraway place deep within Space. They hope to observe the universe in a physical, traditional way which matches their theories and mathematical equations.

Today there remains many cosmic mysteries hidden within the mystic of theoretical physics, mysteries which some say are not even possible to ever be rationally revealed to our everyday, commonsense world of realism. Theoretical physics of the 'astro' kind can be good at predicting what occurs but are often poor at rationally explaining the physical process which allows it to occur. What <u>has been missing</u> is a physical reality model which clearly tells how the universe physically came into being and operates in a physical way which explains why the physics say what they say. My Natural Model is a physical reality model and is designed to explain in a physical, traditional way what the most successful physics models are telling with physics.

Often, the physical observations do not match mathematical models. For example, the modeling of the Big Bang told Scientists that by now, 13.8 billion years after the initial burst of powerful inflation at the said Big Bang, the expansion of the universe would be now slowing due to the effect of Gravity. Gravity brings objects together so for many years everyone believed the speeding apart of Galaxies was slowing due to Gravity. However, when Scientists built telescopes which could physically observe very distant supernovae and determine the distance to the very distant Galaxies, they discovered that the expansion of the universe was not slowing as their mathematical equations clearly predicted. Galaxies were in fact speeding up. Although this came as a startling, unexpected surprise this is not bad science; it is the way good science works, where the confirmation is always in the physical observation. The science, the physics, the ideas and the models must in the end **match with what is physically observed** and then ideally, what is observed can be explained in a rational manner and in plain words which we all can appreciate. One may say the proof is in the physical observation and the observation needs to match the theory.

Recently, CERN announced a series of unexpected, exciting measurements at the Large Hadron Collider. Scientists enthusiastically submitted over 500 papers, each proposing a novel way to explain the unexpected observed 'bump', which appeared to deviate from the standard model of particle physics. There was a lot of excitement as this may have been a new exotic particle which will lead to new physics? [101] However, later CERN confirmed that the 'bump' went away and the 'exciting measurements' were nothing more than statistical blips. The point here is that many scientific papers may be mathematically correct but none may be correctly telling the whole story of what is truly occurring. This is like your child leaving home with $100.00 in their pocket and returning home with $5.55. There could be any number of stories explaining exactly what happened to the missing $94.45, all may be mathematically correct but none may be correctly telling the whole story of what truly happened to the missing $94.45.

There have been numerous impassable roadblocks clearly <u>created</u> from beginning with a Big Bang. How did the universe arise from nothingness? The cause of the Big Bang itself, how did Matter prevail over Anti-matter, how protons and neutrons first formed, how did Super Massive Black Holes form, how did Galaxies get to be <u>inside</u> of their outer halos of Dark matter, how Matter was distributed to different regions of different types of Galaxies to match the different ages and momentum of their stars, Dark energy, Dark matter, the source of time, Galaxy formation, the origins of the <u>essential</u> fabric of Space-time and what it is made from, the origins of mass and even the actual source of Gravity <u>all harbor deep mysteries</u> which current theories are unable to totally unravel in a <u>physical way</u> which makes fully correlated, rational sense. The aim of the following Natural Model is to use rational and coherent methods which the lay person can understand. The methods will need to mirror the ramifications of complex scientific theories to sensibly configure a universe which physically acts as <u>one fully correlated</u>, rational entity which matches with what is observed.

The Big Bang theory has no explanation for its initial state; it was just conveniently there for us at the very beginning. Almost as bazaar and truly mind-boggling is that just one moment of time before the Big Bang it is said that there was absolutely nonexistence nothingness. Actually, one cannot even say, "one moment there was nothing and the next moment there was an instant Big Bang universe" because it is said that there was not even one moment of time or Space before the Big Bang. It is truly difficult to accept that there was absolute nothingness and then for no given reason what so ever, 'nothingness' exploded. Some say that it is wrong to even say "nothingness exploded" for the reason that even nothingness may be conjured as being something. Even when smaller than the full stop at the end of this sentence, the Big Bang universe contained all the ingredients for the makings of everything within the universe today. Although one must indeed be endowed with a grand imagination to do so, picture the energy of the entire contents of billions of Galaxies and all dimensional Space and time somehow contained within the following full stop.

Nothing it seems could have actually caused the Big Bang. The Big Bang was like the most astonishing magic trick of all time, similar to 'Abracadabra' and pulling a bunny out of a hat, except there was no bunny or even a hat, rather a whole universe instantly appeared from absolute, unconditional, non-dimensional, timeless, nothingness. Actually, the Big Bang apparently appeared from within itself and like a flash from the super-natural and in less than one astounding second there was our expanding universe complete with all dimensional Space and all Matter for all time.

Implausible as an event like this may appear, the Big Bang is now commonly accepted by Scientists as the most probable beginning to the universe. Corresponding scientific and associated satellite observations have helped confirm many aspects of the Big Bang theory. By surviving numerous attacks and without a plausible alternative, the Big Bang theory has surfaced as being the best overall fit to an observation that our universe is expanding. Nevertheless, to a great extent the very fundamentals of the greater part of our universe today remain a complete mystery.

Why is the Big Bang theory actually a poor theory?

In 1929 Edwin Hubble announced a remarkable discovery; the Galaxies and clusters of Galaxies were actually moving apart from each other. This naturally led science to the idea of extrapolating Hubble's discovery backwards to a time where Matter transformed from energy and back even farther to the said Big Bang. However this relatively simple reasoning has led science to many insurmountable problems. Scientists quickly realized that they had a very real dilemma because just a few parts of a second into the life of the Big Bang universe, energy is required to transform into mandatory, exactly equal parts of Matter and Anti-matter. When Scientists have managed to create small amounts of Anti-matter in the laboratory they have indeed always observed an exactly equal quantity of ordinary Matter. However, when Matter particles and Anti-matter particles are in the vicinity of each other, total annihilation takes place.

[008] At less than a full second old, a Big Bang universe really has what has

proved to be an insurmountable problem. <u>The laws of physics and as many experiments have undeniably confirmed</u> all Anti-matter and Matter should at this time come together and indisputably, <u>completely annihilate each other.</u> The obvious dilemma for obeying these mandatory laws of physics and the very precise results of many related experiments is that there would be no Matter (no you or me) in the universe today if there had been a Big Bang. This is called the Matter and Anti-matter asymmetry problem. I have read it many times; we shouldn't be here if there had been a Big Bang and the physics and many precise experiments have clearly confirmed it so.

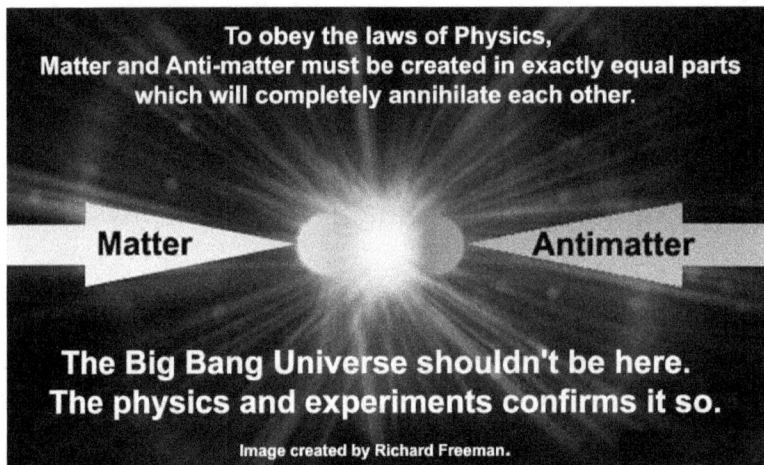

009 Confirming yet again that this is true, Scientists at CERN's Baryon Antibaryon Symmetry Experiment have now verified that Matter and Anti-matter particles are <u>perfect</u> mirror images of each other, with only their charges reversed. In an attempt to explain the missing Anti-matter it was theorized that anti-protons might decay faster than protons, however, Scientists have found within very strict limits that the charge-to-mass ratios are the same. **Because the laws of physics and <u>many experiments have confirmed</u> the theories, this really should have told Scientists that it is very unlikely that the universe could have begun by way of a Big Bang.**

Scientists believe only one Matter particle per billion survived annihilation at the said Big Bang. 010 Now new research suggests the Cosmic Microwave

Background radiation contains about 10 billion photons for every particle of Matter in today's Universe which propose for every particle of Matter which survived annihilation 10 billion particles of Matter were annihilated by anti-particles. Understanding that a lot of annihilation took place and in an attempt to solve this conundrum science has generally agreed to begin with a super, massive 1,000,000,000 times more Matter particles than what is required to account for all of the Matter in the universe then allow Anti-matter to annihilate all Matter but for one part in a billion. However, this theory pretends that all of the Matter in the whole universe inexplicably never had an Anti-matter counterpart! Some have suggested that anti-protons might be more inclined to decay than protons, resulting in a greater number of Matter particles than Anti-matter particles. [009] However, the resulting CP symmetry violation remains unsupported by any kind of empirical evidence and latest research all but rules this out.

The truth is the current laws of physics do not allow for even this tiny imbalance and modern science has to date failed to discover an appropriate solution to this dilemma created from beginning with a Big Bang. Today, science has still not solved this problem. The debt of missing Anti-matter is so very large since the debt of missing Anti-matter <u>will always</u>, in some way or another, directly mirror <u>all of the Matter in the whole universe</u>. The only true solution to this problem is that the missing Anti-matter is in fact somewhere but as many have already asked, where could it be?

The truth is there are far too many fundamental observations which the science of the Big Bang theory cannot explain. For example, every day we all observe Gravity holding us all to the ground, however the very source of Gravity has never been explainable by way of a Big Bang. How can it be that the very source of something which is so accessible cannot be fully explained in a physical way? The actual source of Gravity may never be fully explainable within a Big Bang universe. [059; 091] With a Big Bang the formation of Galaxies cannot be explained in a way to match the recently released startling images from the James Webb Space Telescope. It has not been possible to explain how Super Massive Black Holes with outer halos of Dark matter formed to be in place at precisely the right time to provide a Gravitational scaffolding to begin Galaxy formation. I have yet to read how Matter found a way to get inside of these vital Gravitational scaffolds.

The Big Bang theory says everything originated from the Big Bang. However, there is no evidence or modeling for the creation of Dark matter at the Big Bang. As everything is said to have originated from the Big Bang if the theory was complete it should fully explain the complete workings of the universe and all of its contents. The Big Bang theory is really a poor theory because it explains very little. A good theory for a universe should be able to explain all of the contents of this pie chart. Alarmingly, nearly all of the contents of our universe cannot be explained by way of a Big Bang.

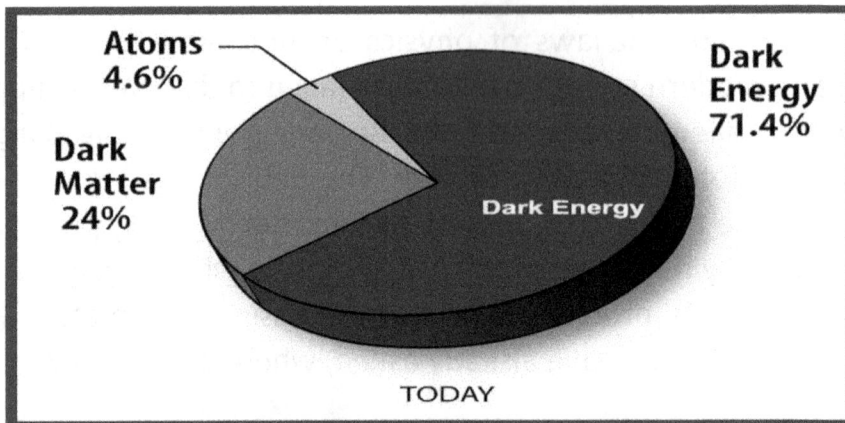

Image Credit: NASA.

Study this NASA pie chart of the contents of our universe; the Big Bang cannot explain the source of Dark energy or even what it is, cannot explain the source of Dark matter or even what it is, cannot explain the source of energy which is said to have transformed into ordinary Matter (atoms) or even what kind of energy this was and cannot explain the source of Gravity and so cannot explain exactly what Gravity is, how Gravity is reconciled with mass or how Gravity is reconciled with the tiny realm of Quantum. The

Big Bang cannot even explain what the fabric of dimensional Space-time is or how it was made. Yes, the vital substance which surrounds us all remains a complete mystery as to what it is. It is unwise to portray such a theory for our universe 'as fact', when it cannot fully explain even a small fraction of the contents of our universe. [001] Scientists today will readily admit that 95% of the entire contents of the universe are still unknown and so we are still *"staggeringly ignorant"* about almost everything in the universe.

Even the said understanding of the 'understood' 4.6% (atoms) is today in disarray. Amazingly, the mass of particles themselves, accounts for just 1% of an atoms mass. This 1% is said to be provided by an impossible to rationally physically describe interaction with the much publicized Higgs boson. [070] However, with the recent 'failure of Supersymmetry' Physicists are today faced with the reality of not having a fully workable Higgs model which comprehensibly agrees with the results of experiments and clearly explains the actual primary source of the mass of an atom's particles.

[011] No, the Higgs boson particle is in no way responsible for any of the remaining 99% of 'mass' which relates entirely to the binding energy of the strong nuclear force. This 99% of an atom's mass is loosely theorized to be created by tiny particles called gluons which create mass by strongly gluing the nucleus particles tightly together. However, exactly how gluons actually 'glue' to create this mass remains unknown. Strangely, the strong force created by gluing becomes weak and disappears when particles are closest and most tightly bound together and is strongest when particles are farthermost apart but disappears outside of the nucleus. It is a great unsolved mystery why this binding force works in this manner and how it can work back-to-front to all other forces which are strongest at closest range. Actually, this mass is created by madly agitating particles at near light speed and not by simply gluing the particles together.

The mass generation mechanism of protons has recently been described as *"the most urgent problem of modern particle physics"*. The theory for most of the regular mass of all Matter throughout the universe has yet to reveal its final answer. Consequently, exactly how atoms actually 'give rise' to their 'binding' mass-energy and <u>directly reconcile</u> this same amount of mass with an appropriate amount of Gravity remains an unsolved mystery.

Now that leaves none of the contents of the universe pie chart which is 'fully understood' and when one includes the flood of Gravity throughout the universe it truly leaves little which is fully rationally or physically explainable by way of a Big Bang. No this is not a 'tall story'; Scientists today cannot explain what Dark matter or Dark energy is, what the source of Gravity is, or exactly how an atom gives rise to 99% of its mass, leaving nearly all the contents of our home universe in a shroud of mystery. Don't get me wrong, even despite the fact that many of the very biggest questions remain unanswered science today has achieved a mind boggling amount of understanding and data on ever so much within our universe.

There have been other proposed alternatives to the Big Bang but like the Big Bang, none have explained the very basic fundamentals such as the source of Gravity, the source of dimensional Space including the passing of time, Dark matter, Dark energy or the true source of energy which transforms into Matter. Big Bang atoms have no known mechanism for Gravity or no known mechanism for the passing of time, so nobody knows exactly what Gravity or time is or how Gravity physically slows time. How completely empty Space was provided with the dimensions of length, width, depth and the passing of time to become the fabric of dimensional Space-time is a mystery. How the dimensions of Space-time were supplied in a way to be unique to the speed and Gravitational position of every atom in the universe and in doing so provide us all with our very own <u>absolutely unique</u>, Dimensional Reality is also a complete mystery. The Big Bang theory is said to explain the properties of our home universe but one can plainly see it explains very little, which is why it is actually a poor theory.

In an attempt to explain why Galaxies are now observed to be reaccelerating away science has attempted to add a new force or energy to the universe. Inserted by Scientists, this colossal, totally mystifying form of energy called Dark energy is said to now make up a massive 71.4% of the mass-energy of the whole universe. I say "Inserted by science" because, other than within mathematical equations designed for a hypothesized proposal, no one has any idea of where this overwhelming form of energy came from, how to directly detect it or even what it is. Scientists today do not know exactly what time is. While Special Relativity provides clues that time is variable <u>and cannot be separated</u> from dimensional Space, the exact mechanism which is able to physically provide us all with our <u>very own truly unique rate</u> of passing of time is a complete mystery. It is too easy, and not very scientific to say all of these things, including dimensional Space-time, were somehow created by a Big Bang and we should not ask where the Big Bang came from because there was no time before the Big Bang.

Scientists say there is no outside, center or outer edge of the Big Bang universe of dimensional Space and time. How can this be? How did numerous Super Massive Black Holes form so very early in the universe? [012] It is said that almost all theories containing hypothetical Gravitons suffer from severe problems and there is no experimental evidence supporting their existence so what could be the real physical source of Gravity?

How is it possible for 'particles' to be 'waves' which instantly become particles when simply observed or measured? How can simply 'looking at' physically change a wave into a particle? How are waves transformed into the particles to make up an atom so as you and I can exist? How can tiny particles completely disappear and reappear anywhere? When particles disappear where do they go? How can one rationally explain the mystic of the rhythm of time and time dilation? How can one rationally explain the mystic of length contraction? Why is a Planck length the smallest possible dimensional measurement? Why is Planck time the smallest possible duration of time? How can an entangled Quantum particle be 'instantly' aware of the actions of its entangled partner even if light years apart?

Despite decades of continuous attempts by many of the greatest minds in science, why have Physicists failed in all attempts, to merge the highly successful theories of Quantum Theory, Special Relativity and General Relativity into a grand theory of everything? Why have all attempts to combine Gravity with Quantum Theory failed? Why have all attempts to reconcile mass with Gravity failed? Why does Gravity appear to be much weaker than other forces? These are very basic but vital fundamentals. For models to be successful they must provide clear and correlated answers.

Probably the kindest way to put it is that these theories are not quite telling the whole true, rational story of what is <u>actually physically occurring</u>. If they were they would seamlessly and rationally integrate by flowing smoothly and seamlessly from scales of the smallest to scales of the largest in exactly the same way as the true universe must, without a doubt, operate.

The task ahead is to develop a <u>physical</u> way to piece together all known and proven science, as well as the known and observed contents of our universe, without having the many insurmountable mysteries which have proved not solvable by way of a Big Bang. However, without an instant Big Bang how could it be possible to produce the required gargantuan quantities of Dark matter and ordinary atoms from seemingly empty Space? Strangely, to do so, we will be frequently required to simply accept the proven laws of physics, the confirmed predictions of successful theories, the true results of scientific experiments and astronomical data (the true observation), rather than the belief of today's science. For example, by simply accepting the laws of physics and accepting the true results of scientific experiments, mysteries like the Anti-matter conundrum can be solved. In a similar way, this model will dig deep into seemingly empty Space to discover a strange Quantum realm which clearly exists but is truly hidden from our daily, traditional realm of Dimensional Reality.

3 WHERE ANSWERS LIE.

"Boldly go where no man has gone before".

Ask a Physicist or cosmologist what was there before the Big Bang or what is there outside of today's universe and you will most likely receive an answer somewhat along these lines *"yikes!, we do not go there, like nobody goes out there! Anyway it is completely pointless because today's theories tell us that there is no outside, no Space or time, not even nothingness".*

One may say it is nonsensical that the Big Bang universe of expanding dimensional Space and time does not expand into anything, while a Physicist will likely say it is nonsensical to even ask *"what is dimensional Space-time expanding into"*? The Big Bang universe of dimensional Space-time just expands from within, a concept which is always a little difficult for my humble human brain.

I have listened to Cosmologist tell us our universe is more like an ant on a ball who may think in a two dimensional world, where if the ant walked in any direction, the ant would eventually return to where he started. If one was to travel outside of our Home Universe what would one find? Just like that ant, it is said whichever direction one steered our spaceship one is likely to find that Space itself is curved so one would never find the outer edge of the universe. Funny thing, the ant kept jumping off the ball onto the ground before I could even focus the camera. The ant was not fooled into believing such a notion of a two dimensional world.

R. Freeman

But hang on, I'm not an ant, I know there is left and right, forward and backwards, up and down and before and after. One has little difficulty navigating the Space 'bent' by Gravity around the Earth, man has even gone to the moon, surely there is a direction which I can steer my spaceship to fool this most bothersome 'bent' Space at the edge of the universe.

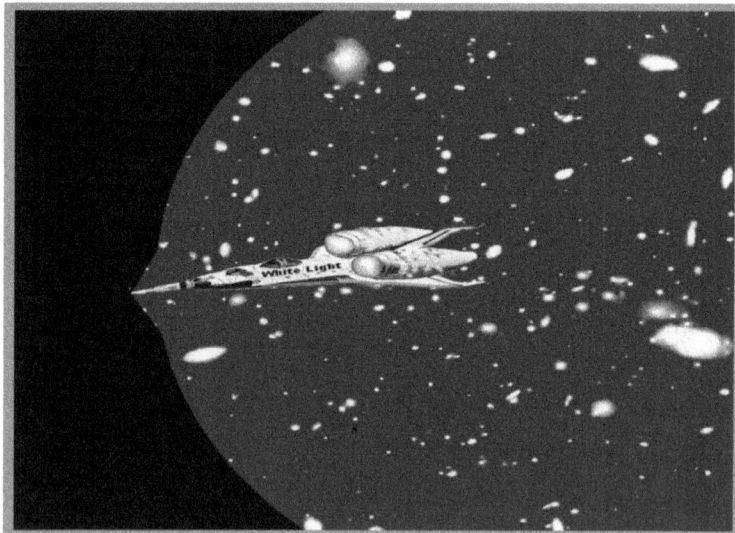

Many a time I have tried to envisage arriving at this edge of 'no place' in my Hybrid-super-modified Space-tunneling spaceship and what I may find way out there. Would it be a Domain wall of impenetrable blackness? Should it be called the big void, non-dimensional nothingness, emptiness, place of non existence and alien to modern science, dark Space, sub-Space, dimensionless Space, timeless Space, another dimension, place of nowhere, The super vacuum, The Dark side, Raw Space or Anti-Space?

'Dimensionless Space' may be the best true description and 'The Dark side' or sub-Space may be the most appropriate because this is where light is yet to shine and is what our dimensional Space-time will be expanding into. Star trekkers may prefer the term sub-Space, however, I do like the term 'Anti-Space'. It seems as though just by giving dimensionless, empty Space a name, one has now truly created something from nothing. It does not really matter whether Anti-Space is truly only a void of complete emptiness or whether it has true 'Anti' Space properties. **This model requires I give this dimensionless <u>void</u> a name to distinguish it from another dimensionless state which I will later refer to as the fifth dimension.**

An obvious description of our home universe would include this void now called Anti-Space as part of a whole universe. Now the whole true universe is endless and so has no physical or imaginable boundaries and <u>likely contains</u> many other universes like our own. A Big Bang universe is a closed universe of dimensional Space and time. Outside of dimensional Space and time is not allowed. A Natural Universe will be an open universe of dimensional Space and time which is <u>actively</u> expanding its dimensions into a seemingly never-ending, dimensionless state of Space called Anti-Space.

I simply love to gaze up at our mighty Milky Way Galaxy. Occasionally, on dark moonless nights I wander down to my local beach which is only a short leisurely walk away. I have a secluded special place where the comforting ocean waves drown out the noises of suburbia. I lie back on an easy slope with a She-oak tree on my left and a Banksia tree on my right. I feel a strange bond to the indigenous people who lived here long before my time and left behind their awesome stone tools for me to discover and marvel at. My mind drifts off to a wonderful place in complete awe of the grandness of the unknown. Millions of stars twinkle in the night sky and meteors frequently streak pass. I wonder how it can be possible to be sitting in complete comfort and seemingly completely stationary, when I am in fact actually spinning into the cosmos at around 1450 kilometers per hour, while at the same time travelling at 108,000 kilometers per hour, as our home Planet spins on its axis as it orbits the Sun. My ride becomes even more amazing when I realize that our solar system orbits the center of our mighty Milky Way Galaxy at a breathtaking 720,000 kilometers per hour, while our Milky Way Galaxy is travelling at an astonishing 2.5 million kilometers per hour on a magnificent journey where it will collide with the splendid Andromeda Galaxy in about 4.5 billion years from now.

I watch planets rise and make their way across the night sky. Sometimes I'm still there when a tiny slither of a waxing crescent moon rises over the ocean, which is always followed by the first faint aurora of a predawn. The kookaburras are laughing joyfully as they greet the very first soft colors of Sunrise which slowly erupts into an astonishing, vibrant display as our life-giving star peeps above the horizon. Finally, our nearest star the Sun, rises as a powerful fireball over the ocean. I am overwhelmed with a feeling that I am indeed very privileged to live on such a beautiful planet. I reflect if I could fly to the Sun in a jumbo jet the journey would take me 18 years. So far away, yet I can feel its fire burning my skin on a hot summer's day.

The point here is our home universe is vibrant and alive. Wherever one looks there are vast amounts of energy being expressed throughout our universe. There is one phenomenon more than any other responsible for maintaining all this activity and that is of course, Gravity. Be it the energy from our Sun, all the stars above and movements of all the planets or just the ability to walk on planet Earth, all would stop if Gravity was suddenly switched off. Gravity is delivered by the <u>mysterious</u> fabric of Space-time.

In order to solve the many unsolvable mysteries of the universe this model <u>has to be</u> different to the current understanding that everything was given from nonexistence nothingness at a Big Bang. When things are given this way there is no reason to explain how they were created. Many mysteries have to date been unsolvable by way of a Big Bang. My Natural Model is designed to begin, <u>not from nothingness</u>, but from seemingly empty Space and in a way which explains how things were physically created rather than simply given for no reason from nonexistence nothingness. The Big Bang fails to provide us with any understanding of what Space-time is made from. Einstein has made it clear that <u>Space and time cannot be separated</u> and is not nothingness but is *"like a fabric"* which has a curvature which can be warped or stretched to deliver Gravity. Hence, Space-time has to be some-<u>thing</u> which, like a woven fabric, can be created. **The actual dimensions of Space, because they cannot be separated from time, must like the arrow of time always advance.**

Surrounding us all, Space-time is the <u>little thought about</u> invisible stuff which makes everything work within Albert Einstein's world of classical reality. Thus, we are required to <u>create</u> dimensional Space-time, the truly <u>remarkable</u> invisible fabric which <u>actively warps to give the movement</u> to objects falling with Gravity, <u>provides dimensional</u> distance and dimensional measurements to all objects and <u>actually provides us all</u> with our very own <u>unique rate</u> of passing of time. We will also be required to <u>clearly provide</u> tiny Quantum particles with an entirely <u>different state of Space</u> which will finally <u>rationally solve</u> how a tiny particle <u>irrationally resides in all locations at the same time</u> and where tiny Quantum <u>particles</u> can naturally reside as <u>waves</u>. We need to provide a source for Dark matter which can provide it with <u>invisible properties</u> and eventually a source of particles of Matter and, somehow, provide both with the <u>very same source</u> of mass and Gravity.

The aim is to begin our universe with a <u>credible</u> but insignificant Quantum energy event which is known to spontaneously occur within seemingly empty Space. To construct almost anything and everything, Mother Nature always begins with the very smallest. Hence, the challenge is to begin with the smallest configurations of energy <u>known to spontaneously arise from empty Space</u> to construct the very largest and complexity of a universe.

4 AN AWESOME UNIVERSE.

*The Natural Model provides direct evidence and unprecedented correlation of all subjects. However, this superb correlation will require repeating many explanations in different chapters and associated subjects. For example; how Gravity works, how mass is created, what provides the passing of time and the creation of all Matter, dark or regular, are all **detailed in dedicated, stand alone chapters, like books within a book,** which repeat and <u>build on</u> common correlated themes by directly relating to text in previous chapters. This allows one to become familiar with common, fully correlated themes which <u>repeatedly solve</u> even the deepest, unsolved cosmic mysteries.*

The time is 1.45 am and the start of my work day. It is certainly not unusual for me to put to sea in the pitch dark; being a commercial fisherman this is my daily routine for as long as the weather allows. However, during this trip to sea I will have a somewhat unusual and captivating experience. I am about to launch my boat from its trailer at the boat ramp on the southern side of the Mooloolah River. Wading ankle deep in the water to unhook the boat from the trailer I can't help but notice an eerie glittering green glow as I move my feet around in the water. I find myself stirring the water with my feet just to stare at the phenomenon and I briefly become mesmerized by this very entertaining light show going on around my feet.

I know this to be the product of bioluminescent plankton and their light making chemiluminescence which is a light generating chemical reaction which occurs when the tiny creature is excited and serves as a defense system. This group of plankton is commonly referred to as 'Fire Plants'. Hence the term by fishermen "fire in the water".

I moor my boat to the nearby floating pontoon and park my land rover and boat trailer then proceed to slowly motor down to the entrance of the Mooloolah River. The first wash off the bow of the boat is glowing bright green as it permeates away from each side of the boat. The sea is calm and the swell is small so this morning there is no need to negotiate large, side on, breaking waves at the river's entrance allowing for an easy passage on into the Pacific Ocean. Once clear of the river break walls I power up to a lazy 16 knots. I am still aware of the amazing light show going on around me; however, I am busy setting my course. I turn on my auto pilot and move the cursor on my GPS plotter to where I wish to go and press the navigation button followed by the satellite button on my auto pilot.

Now my trusty 5.6 meter shark cat is being steered with the help of the GPS satellites out in Space to the exact place I wish to start my work day crabbing. I turn off all but my navigation lights and sit back to enjoy the brilliant bioluminescent light show. There is no moon and the sea is glassy calm. The sky is clear but for a bit of cloud on the horizon and the air has a little chill to it. I love these mornings they are great to be alive mornings. I have the two Pointers and the Southern Cross off to my starboard quarter and the Milky Way is so bright that I can clearly see the dust lanes which mingle among the millions of visible stars.

But the awe of the Milky Way is being overshadowed by the fire show going on around me. Every droplet of spray on each side of the boat is alight with fire and as tiny droplets fall back to the water they are lit up and appear to float around inside my boat like cinders from a true fire, I even have sparks coming into the boat through a tiny slit to the side door. The wash behind the boat starts around the propellers as a ball of brilliantly bright green light which slowly dims over a hundred meters astern. I am thinking what a spectacular sight this is. Unexpectedly, an immense flash so bright that it floodlights the whole boat then another and another which are like flashes of lightning and I am blind for a few seconds until my eyes adjust to the darkness again. The boat has obviously encountered a series of small dense and concentrated pockets of the bioluminescent plankton.

The light bombs stop and I settle back again. Now out passed the Gneering Blinker the shower of sparks continue to flood off the chines of the boat. All of a sudden I notice a long trail of bubbly light approaching at high speed on my starboard side, then another and another. Four high speed torpedoes of light approaching on an intercepting course on the starboard side! My heart rate rises as the speeding trails of light barely miss the boat and disappear under the starboard sponson. The Dolphins are having a game with me and one leaps out high above the bow rail from between the two sponsons of my shark cat. I instantly grab the hand rail on the dash and duck for cover. Peering above the dash, with my eyes wide open, my field of vision is filled with an unbelievable image of a dolphin drenched and dripping with liquid bioluminescent light silhouetted against the starry sky. With my heart racing, it appears as if time stands still as the dolphin literally hangs there high in the night sky for a few amazing seconds. Though there for only a brief moment of time this image of a dolphin dripping with bioluminescent light and silhouetted against the awe of the Milky Way will remain with me forever.

Although I have studied cosmology for well over five decades, I am not a Scientist or a Physicist, I am but a humble fisherman who catches spanner crabs with dillies and catches fish with a rod and reel or more frequently with the use of simple hand lines. Consequently, I write this with the words and phases of a fisherman and not with the words and phases of a Scientist. I have accumulated a lifetime working alone at sea. Countless night hours alone, well out to sea and far away from the glow of the on shore lights.

On moonless cloudless nights at sea our mighty Milky Way Galaxy is breath taking beautiful with its darker dust lanes silhouetted against a dazzling display of stars. I think to myself what a splendidly fitting name for our home Galaxy. Meteorite showers and even the occasional comet over the years have added even more to the delight of viewing the brilliant night sky while at sea. Grandstand views over the ocean of alluring moon rises and moon sets are pure magic. When alone at sea witnessing these spectacular night skies one feels a surreal sense of self being.

Almost all of my working life has been alone at sea and urged on with the sounds of the ocean my mind is always motivated and free to wander and speculate about the mysteries of the night sky and what more amazing mysteries may lie beyond. I wondered what could really be going on far out there, how did it all first begin and will it ever end? I wondered how it can be physically possible that I'm fixed to my current location while a tiny Quantum particle is truly at all locations at the very same instant of time. I searched repeatedly for rational answers but found only more mysteries. The deeper I searched the deeper the mysteries became. Decades went past but frustratingly the very same mysteries remained; for me it was clearly time for a new approach and time to search out realistic answers.

I think of how it would take me 18 years at the speed of a jumbo jet to travel to the closest star our Sun but just over eight minutes at the fastest speed allowed by the laws of the universe, the speed of light. My mind wanders even further away; if I had commenced a journey at the amazing speed of light, at about the time our calendar began, to travel across the width of our Milky Way Galaxy, by now 2024 years later I would have only completed barely more than 2 percent of my journey. At least a hundred thousand years at the speed of light is the unimaginable distance across our mighty Milky Way Galaxy. Unimaginably large as it is the Milky Way is but a tiny pin prick of light among billions of other Galaxies, all accelerating while speeding through the vastness of the universe itself.

Credit: NASA, ESA and S. Beckwith (STSci) and the HUDF Team. Galaxies, Galaxies everywhere.

5 DARK MATTER - THE COST.

Within a Natural Universe Dark matter is a virtual proton and the vacuum fluctuation system which powers it is the COST.

[103] **85% of all mass and 90% of our Milky Way Galaxy's mass is in the form of invisible and not directly detectable Dark matter. Consequently, it is surely common sense to begin a universe with Dark matter, rather than a Big Bang which <u>only accounts</u> for the remaining 15% of all Matter. This is why my Natural Universe begins by first creating Dark matter from a <u>known source</u> which is capable of <u>providing</u> it with its invisible properties.**

Why are stars and Galaxies <u>only found</u> in regions which are richest with Dark matter? Answer; Dark matter is a <u>virtual proton</u> which provides the crucial rudimentary seed for creating regular protons which create hydrogen atoms and molecular hydrogen to make stars. Naturally, this is why stars and Galaxies are <u>created</u> within regions which are richest with Dark matter. The simple configuration of Dark matter will lay the essential foundations for the <u>unwavering laws</u> which will <u>rule a Naturally Forming Universe</u>. Unlike any other model the Natural Model will tell exactly what Dark matter is. The <u>known, invisible, properties</u> of Dark matter will be used to expose Dark matter's <u>heritage</u> to the known to exist virtual realm of vacuum fluctuations and virtual particles. A **virtual proton's Gravity** will power a path from the virtual realm to the creation of all regular Matter.

<u>Why the first 'Particles' are not 'Particles' at all:</u> I found <u>past</u> theories which say Dark matter originated from before the Big Bang and others which say Dark matter originated after the Big Bang. Today it is now said Dark matter originated alongside ordinary Matter at the Big Bang. The Natural Universe begins by first creating <u>Dark matter and Gravity</u> at a time <u>long before</u> the Big Bang and within seemingly empty Space. Thus, Dark matter must be constructed from the same 'empty Space' which is why the COST is not an actual 'particle' but rather a **C**onfiguration **O**f **S**pace-**T**ime. To create a universe a vital step is to provide empty Space with dimensions which will transform empty Space into dimensional Space-time. **Einstein clearly realized Space-time is not nothingness** but rather a flexible fabric which can be stretched and warped to provide Gravity, time dilation and length contraction. What physically changed seemingly empty Space into <u>an essential</u> woven fabric of Space-time? A physical reality model is required to create the physical components which allows us to observe our wonderful universe 'as it exists' within one's very own Dimensional Reality.

Accounting for at least 85% of all mass and Gravity, invisible Dark matter is only indirectly exposed by its Gravity. Consequently, whatever it is which provides and delivers Gravity has to also be totally invisible. Wherever one finds Gravity one also finds mass. Because Gravity slows time we now require a totally invisible and not directly detectable kind of mass which provides Gravity in a way which somehow slows the passing of time. Gravity is why the universe and all within it, is constructed the way it is. Therefore to understand how the universe <u>really works</u>, one must <u>first</u> thoroughly unmask the source of Gravity. To understand the origins of Gravity and since Dark matter accounts for **<u>at least 85%</u> of all mass and Gravity,** one must first understand the origins of Dark matter. When one understands the origins of Dark matter, one will also be able to sensibly understand its said characteristics, the very same characteristics which have been derived from many years of extensive scientific research.

<u>There really has to be</u> a very good and logical reason to why Dark matter

completely dominates the mass and Gravity of all the Matter in the universe. The Natural model will expose exactly how Dark matter first creates Super Massive Black Holes which will provide the power houses to transform Dark matter into regular Matter to make a Galaxy's stars which is exactly why Dark matter <u>needs to</u> dominate all the Matter in the universe.

There are no theories today which even attempt to <u>physically</u> and rationally explain <u>exactly how</u> Gravity or Space-time was <u>physically</u> created. Because Gravity is coupled to mass, solving the source of Gravity needs to solve the origins of mass. Because Gravity works by warping the curvature of dimensional Space and time, solving the source of Gravity needs to solve what physically provides this <u>curvature</u> of the dimensions of Space and time. Gravity slows time, thus, one will need to physically unmask how Gravity slows the passing of time and because nobody knows what causes the passing of time one will need to unmask exactly what passing time is and why passing time <u>cannot be separated</u> from the dimensions of Space.

Gravitational Attraction is said to be a **negative energy** state which cancels the positive state of all Matter, dark or regular, so as the sum-total mass-energy of the universe remains zero. One must show why Gravitational Attraction makes this so. This is fitting as we begin the Natural Model within sum-total zero mass-energy of empty Space and because we are obligated to respect the conservation of energy law, one will need to show how a universe full of Matter maintains this sum-total zero mass-energy.

Today it is believed that for some <u>unknown reason</u> the unidentified ingredients for Space-time and Gravity to develop were just astonishingly, instantaneously 'there' **as part of the Big Bang.** Now, by somehow popping into existence and becoming a part of atoms a <u>purely hypothetical</u> messenger particle called a Graviton somehow acts, by somehow reaching out over huge distances to somehow stretch or warp a strange **curvature of Space-time which somehow affects** all dimensions of Space, thus somehow delivering Gravity to all objects with mass. [012] In actual fact almost <u>all theories</u> containing Gravitons are said to <u>suffer from severe problems</u>. Gravitons have never been directly detected and there is no experimental evidence supporting their existence and it is never said that Dark matter has Gravitons. So to solve the many ramifications of Gravity one will need to solve exactly how Gravity stretches or warps the curved dimensions of Space and time. However, nobody knows how empty Space acquired <u>dimensions</u> to become dimensional Space-time which has a **curvature** which can be stretched or warped. As one cannot warp completely empty Space, **one will <u>need to first provide</u> whatever it is that gives empty Space 'flexible dimensions' which Gravity can stretch or warp.**

Dark matter has Gravity and since our universe today cannot exist without Gravity, any theory for the universe which cannot explain the exact <u>physical</u> source of Gravity and <u>exactly how</u> Matter first acquired Gravity should not be too highly prized. Within a Big Bang universe if Gravity had not 'somehow appeared' from some unidentifiable source there could only have been atoms of mostly hydrogen, floating aimlessly throughout the entire universe. The probable truth is that <u>Gravity</u> is so entwined with the <u>mass</u> of atoms that even these first hydrogen atoms could not have existed.

Gravity undeniably plays such an overwhelming role in creating a universe for without Gravity there could be no Galaxies, Stars or Planets. Truly, anything and <u>everything which we can visually observe</u> and even the very atoms of the type which we are all made from, would not have ever existed without Gravity. <u>Gravity created</u> and sculptured the dimensional universe to be as we know it today. One could argue that without an <u>entire physical understanding of Gravity</u>, one cannot with complete confidence explain another thing. **Given that nothing meaningful 'can exist' within our universe without Gravity, it is not only common sense but is in fact 'absolutely imperative', that in creating or understanding a universe or almost anything within a universe, one should <u>first begin</u> by <u>laying the</u>**

essential foundations responsible for supplying a universe with Gravity.

If one simply accepts all research and observation, Dark matter appears to be nothing more than raw Gravity itself, devoid of any true Matter. Dark matter has proved to be devoid of any detectable true or known forms of Matter but how could this possibly be so? [022] Despite many years of very determined searches and the fact that Dark matter is far more plentiful than ordinary Matter, sophisticated Dark matter detectors designed by science have truly detected nothing. For example, the Large Underground Xenon experiment (LUX) which cost in the region of $10 million to construct and employed over 100 Scientists and engineers from 18 institutions found no signs of their 'proposed' Dark matter contender; WIMPs. Scientists were eager to discover a hypothetical '**W**eakly **I**nteracting **M**assive **Particle**', or 'WIMP', which may mirror the properties of Dark matter and somehow came into existence not long after the Big Bang but such a **particle** would not explain the source of Gravity. The exact nature of these proposed WIMP **particles** is not presently known and WIMPs are purely hypothetical and are not even predicted by the standard model of **particle** physics. [032] video at time 42.00. Today Physicists say, in their hunt for Dark matter, they require *"something totally and completely new"*. [022] See: The 'WIMP Miracle' Hope For Dark Matter Is Dead. *All direct detection experiments have (now) thoroughly ruled out the WIMPs Scientists were hoping for.*

[026] The Large Hadron Collider has also failed to find even a trace of **particle** based evidence of Dark matter. By eliminating all known and detectable **particles** Scientists are more certain what Dark matter is not than what it is. Even though said to be all around us, to the point that we do not have to reach out to touch it, Scientists have not been able to find or directly detect even one tiny **particle** of Dark matter or a trace of Dark energy. [001] Consequently, science today cannot directly locate or understand what it is that makes up at least 95% of the said contents within the whole universe. Even the standard model of **particle** physics doesn't explain what Dark

matter or Dark energy is. **Interesting to note that although Dark matter behaves as if it is not an actual particle all research has been centered on discovering a Dark matter <u>particle</u>.** More often than not, when you do not know exactly what it is that you are looking for, it is most difficult to find.

Scientists have now labeled Dark matter as a 'ghost particle' which is fitting because the COST will clearly operate as a ghost proton. While I really like the term ghost proton the more correct term will obviously be 'virtual' proton. CERN researchers say *"that unlike normal Matter, Dark matter does not interact with the electromagnetic force which means it does not absorb, reflect or emit light".* [004] Being totally invisible and able to pass through ordinary Matter as if it was not there permits Dark matter to truly act as a ghost-like particle. Dark matter has only been revealed by the tell tale <u>Gravitational influence</u> it has on ordinary Matter.

For instance, our Milky Way Galaxy would simply fly apart without the Gravitational influence of its Dark matter. The influence of Dark matter has also been observed by light from far away sources which has been distorted or bent by Dark matter's Gravity. [013] The ghost-like characteristics of Dark matter have been exposed by Scientists studying the colliding of Galaxies within the 'Bullet Cluster'. They discovered colliding gas clouds interacted as expected and were slowed down by the collision. However, <u>detected by its Gravitational influence</u> and in a ghost-like fashion, some form of totally invisible Matter had passed all the way through and out the other side of the colliding Galaxies. Apart from its Gravitational influence, this ghostly form of 'Matter' had merely coasted through the gas clouds as if they were not even there. It truly is like Dark matter is not composed of anything meaningful, allowing Dark matter to act in a <u>true ghost-like manner</u> and freely coast through ordinary Matter as if it was not even there. How could this be? **How is it possible to create a 'true' ghost particle which truly <u>acts</u> like there is nothing really there but somehow has Gravity? Answer; you create a <u>virtual proton</u> based on a Quantum vacuum fluctuation.**

Quantum vacuum fluctuations are said to spontaneously arise from every tiny point of <u>empty Space</u> and <u>exactly like Dark matter</u> the vacuum fluctuations are totally invisible, do not absorb, reflect or emit light and ordinary Matter pass through as if they are not there. All is <u>reason and evidence</u> why the COST in its Dark matter state will not be an actual particle but rather a permanently active, Quantum 'configuration' of Space-time and therefore cannot be directly detected by the methods in which true particles may be detected. Like all vacuum fluctuations the COST is a fluctuation mechanism which <u>creates pairs of virtual particles</u> which come and go. The COST will provide this active mechanism for regular protons as well as a <u>documented phase</u> which will provide mass and Gravity. However, <u>without</u> three Matter valence quarks the COST is not yet a regular proton.

Sometimes the truth is found by simply accepting the actual results of research and experiments. While others did not accept experiments which implied that the speed of light did not change with regards to one's speed, Albert Einstein accepted the results and deduced that <u>instead</u> of the speed of light changing the speed or rate of <u>time</u> varied in regards to one's speed, which allowed him to develop his brilliant theory of Special Relativity.

Since Scientists today cannot explain what this truly ghost-like Dark matter substance, which produces most of the Gravity within the universe is and cannot explain the very source of Gravity or the primary source of mass; one may easily reason that the answers to these things are probably closely related. Since Gravity and mass <u>are like one</u>, solving the source of Gravity and what Dark matter is, should also solve the source or origins of mass. By eliminating all known and detectable particles, science has eliminated all of what Dark matter cannot be and has been left with, well, nothing.

[103] Consequently, despite many years of immense research, all that is known about the properties of Dark matter is it looks like and behaves like invisible nothingness and it possesses Gravity. **Being based on a Quantum vacuum fluctuation is clearly what will provide Dark matter with its**

invisible, ghost like, virtual properties and is why nobody has been able to directly detect a Dark matter particle. No known actual 'particle' has such properties. By actually embracing these features, the COST will have a tiny beating heart of absolute empty Space which will provide a tiny phase of negative energy for the source of both Gravity and mass for <u>all Matter</u>, dark or regular.

<u>The beginning of everything without a Big Bang:</u> The task now is to lean heavily on known and proven science and theories to produce gargantuan quantities of Dark matter and ordinary atoms from the nothingness of empty Space. Because our universe contains far more Dark matter than ordinary Matter, this model will first focus on a means which will be truly capable of producing huge quantities of Dark matter from the nothingness of empty Space. In doing so, the model will begin with the smallest of the small to unravel one of the biggest mysteries of today's astrophysics; the most plausible composition of truly, invisible, ghost like Dark matter. One of our main objectives is to base this concept on known and proven science and physics. So how could it be possible to begin the very first expansion of dimensional Space and supply all Matter without a once only Big Bang? Our Natural Universe will begin with an insignificant, most minuscule, Quantum vacuum fluctuation. This is like the smallest possible, insignificant energy event <u>known to spontaneously arise</u> within the vacuum of <u>empty Space</u>.

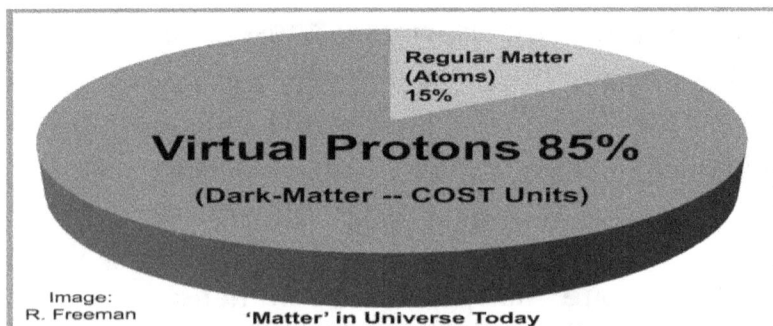

Regular Matter (Atoms) 15%

Virtual Protons 85%

(Dark-Matter -- COST Units)

Image: R. Freeman

'Matter' in Universe Today

[014] Undeniably, the Heisenberg's Uncertainty Principle, related theories, mathematical calculations and experimental evidence imply that 'empty'

Space contains an enormous amount of 'hidden' mass-energy relating to the materialization of Quantum vacuum fluctuations. Vacuum fluctuations at every tiny point of empty Space can add up to an absurd, unrealistic <u>amount of mass and energy</u> which will be fully addressed in a later chapter.

The COST will be based on these same Quantum vacuum fluctuations. The energy of Quantum vacuum fluctuations has been explained by the <u>uncertainty principle</u> first put forward in 1927 by German Physicist Werner Heisenberg. It says that <u>at any observed tiny point</u> of empty Space there are temporary changes in energy. Frequently this energy found in empty Space is changed into <u>mass</u>, by pair production of <u>virtual</u> particle and anti-particle pairs. Usually these pairs of virtual particles combine and then mutually annihilate. **However, due to <u>external energy forcing them apart</u> they can sometimes avoid annihilation and <u>become actual, real particles</u>. The physics and experimental evidence tell us that this is true which provides a vital clue to how our universe came into being from empty Space.**

Long before the proposed Big Bang, a Natural Universe begins building a Dark Universe complete with Dark Galaxies. To do this we first begin with an insignificant, most minuscule, Quantum vacuum fluctuation like which is found at every tiny point of empty Space. This provides a <u>known source</u> for the COST (Dark matter) to be sourced in the massive quantities required to build a universe. The COST is an <u>enhanced</u> Quantum vacuum fluctuation which becomes cyclic and amplified with <u>energy</u> when <u>squeezed</u> by Gravity's warped Space-time. <u>Like a regular proton</u> this energy will be produced by pairs of virtual particles which immediately annihilate to produce the tiniest splash of energy. Physics require this energy is <u>immediately returned to Space</u>; so we need to provide for this to happen.

Virtual particles are less like actual particles and more like minuscule wave like (+) energy and (-) anti-energy configurations within Space. This is why the COST <u>is invisible</u>. To complement the phase of virtual particles the COST will fluctuate to a documented phase of virtual mass which because it is a tiny state of <u>less energy</u> than 'the vacuum' of dimensional Space is a phase of <u>negative energy.</u> [111] **Note: Negative energy is simply pure empty Space.** [109] **Quantum field theory permits the existence of regions with <u>negative energy densities.</u>** The COST acting as a virtual proton will create a bond between mass, energy, Gravity, dimensions, time and the speed of light.

<u>The deep vacuum of a tiny phase of virtual mass and the roles it will play:</u>
Virtual mass is unique to my Natural Model and is a powerful, deep-vacuum phase of a Quantum vacuum fluctuation which because of the task it will be required to perform <u>mimics</u> a minuscule, rapidly collapsing and reappearing Black Hole. [021] **Theories allow a Planck scale *virtual* Black Hole such as this to appear spontaneously but only very briefly.** I wrote in my first book <u>published in 2014</u>: *'Very little is known about the makeup of Dark matter other than it looks like invisible nothingness and it possesses Gravity. Our 'Dark matter' will, in fact, have a heart of <u>invisible emptiness</u> and not only possess Gravity but provide the mechanism for Gravity for all Matter and will, cleverly, do this without the use of (hypothesized) Gravitons'.* Thus, Dark-matter will fluctuate to a phase of <u>empty Space</u> which will provide strong, negative energy for the <u>source of both Gravity and mass</u> for all Matter, both dark and regular. Although I first modeled this more than three decades ago, several more recent scientific experiments have now remarkably discovered this phase of a deeper vacuum of empty Space.

[015] Astonishingly, a team of Scientists headed by Professor Alfred Leitenstorfer at the University of Konstanz have discovered the best evidence which aligns most closely with this modeling. They say they've ***"Manipulated Pure Nothingness* and observed the fallout".** The Scientists say they have directly detected Quantum vacuum fluctuations. Their experiments also clearly detected something strange when they *'<u>squeezed the vacuum</u>'* the fluctuations became <u>louder</u> and so amplified and were observed to fluctuate to a level which is *'**lower than the ground state (the vacuum) of empty Space'.*** The Scientists called this unexpected discovery

of a lower state of Space an **"*astonishing phenomenon*"**.

The Natural Model very clearly identifies the *"astonishing phenomenon"* as a very important tiny phase of virtual mass which will provide Gravity which provides motion. **By causing a Quantum vacuum fluctuation to become louder with energy and to phase to a lower state than the ground state of empty Space the Scientists appeared to have, without realizing it, actually coaxed a simple Quantum vacuum fluctuation into becoming a virtual proton (Dark matter).** [011] **The most important lower state phase of vacuum, exactly like found within a regular proton, provides excellent evidence of the creation of a virtual proton.** [119] While intentionally manipulating the Quantum vacuum fluctuations, the Scientist reference they can bring time to a stop, temporally increase the speed of light, and (dimensional) Space and time behave absolutely equivalently. Ask; why would the dimensions of Space be affected in this way? The Natural Model uses the Quantum vacuum fluctuations of virtual protons to actively create the dimensions of Space and time. A manipulation of the same Quantum vacuum fluctuations would now be expected to cause a corresponding manipulation of the same newly created dimensions of Space and time.

I created the following illustration for my book which I first published in 2014 where my illustrations of my Dark matter showed a tiny state of completely empty Space at the center from where the dimensions of Space-time expanded away. Anti-space is a state of complete nothingness.

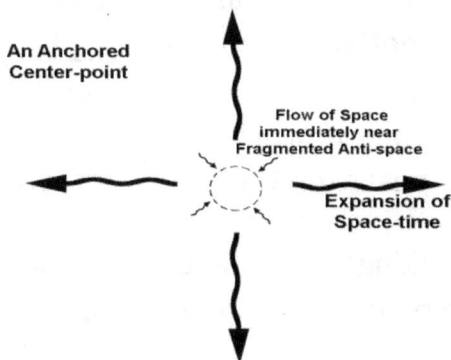

An Anchored
Center-point

Flow of Space
immediately near
Fragmented Anti-space

Expansion of
Space-time

016 false vacuum. <u>Further evidence</u> of this same tiny phase of a lower state of the vacuum now comes from an entirely different source; the energy state of the vacuum can be calculated from the potential energy of the Higgs field and the masses of the Higgs and top quark; refined measurements and calculations reveals evidence of the existence of <u>a lower state of vacuum</u> than the vacuum of dimensional Space. The discovery is said to be <u>one of the biggest unsolved mysteries in physics as it means we live in a false vacuum</u>. This mystery is easy solved; there is nothing in the phase of virtual mass; it is a tiny phase of true vacuum nothingness while what is now called 'the vacuum' is not a true vacuum but a false vacuum since it contains the positive energy of the dimensions of Space and time. **Einstein clearly realized that dimensional Space was not nothingness but like a fabric the curvature of which can be stretched or warped to deliver Gravity.**

002 Thomas Jefferson National Accelerator Facility has discovered that inside every proton the pressure peaks <u>ten times greater</u> than at the **heart of a neutron star**. Most interesting, the researchers based their pressure findings on the interactions of a proton's quarks, but not its gluons. Phiala Shanahan an Assistant Professor of Physics at MIT has via a more recent public lecture webcast stated that ongoing super computer modeling confirms; *"protons and neutrons are made from quarks and gluons – that's it. There is no deeper structure to be found. Quarks and gluons are the fundamental building blocks of the universe".* 032 excellent video: Phiala Shanahan also revealed the pressure inside of a proton peaks higher than inside of a neutron star. My phase of virtual mass is a tiny phase of the negative energy of a deep-vacuum nothingness which <u>needs</u> to mimic a minuscule, rapidly collapsing and reappearing Black Hole so as it will be able to provide the **<u>only likely source capable</u>** of providing the immense pressure inside of a proton which amazingly peaks ten times higher than inside a neutron star.

003 Scientists at CERN working with the Large Hadron Collider have now discovered protons contain numerous quarks, anti-quarks and gluons which

are wildly, erratically moving about at near the speed of light creating *"unimaginable, internal pandemonium"*. A minuscule, rapidly collapsing and reappearing Black Hole is likely the only possible object <u>clearly capable</u> of providing particles with velocities very near the speed of light and would certainly cause the described *"unimaginable, internal pandemonium"*. The energy obtained from particles being erratically moving about at near the speed of light is now said to be responsible for at least 99% <u>of all mass.</u>

[011] This extraordinary video **actually reveals empty Space is the source of most of our mass.** For me this video helped to yet again confirm my source of all mass. In the video **'Your Mass is Not from the Higgs Boson'**, Prof. Derek Leinweber speaks about the findings of their research at the University of Adelaide. The research focuses on a super computer simulation of the theory of Quantum Chromodynamics which describes the action of the strong force within a proton. The video clearly shows how quarks gather around an area of <u>empty Space</u> which is <u>clearly identified as being the source of the strong force</u> which creates regular mass. The conclusion is that it is *"extraordinary, because what we think of ordinary empty Space, that turns out to be the thing that gives us all <u>most of our mass</u>"* (99%). Apparently, this (phase of) empty vacuum actually costs an enormous amount of energy to create. A tiny appearing and collapsing Black Hole would certainly provide an enormous amount of energy.

With this Model the **'extraordinary'** empty Space <u>exposed</u> by this computer simulation is <u>derived</u> from the same [015] **"astonishing phenomenon"** <u>discovered</u> by Prof. Dr. Alfred Leitenstorfer and team from the University of Konstanz in Germany. [016] This is also the same <u>lower state of deep vacuum</u> which Scientists have <u>discovered</u> by a calculation from the potential energy of the Higgs field and the masses of the Higgs and top quark which has been described as one of the **biggest unsolved mysteries in physics**. That is, the extraordinary, astonishing phenomenon which is one of the biggest unsolved mysteries in physics is actually a phase of <u>virtual mass</u> provided by

Unmasking the Natural Universe

the vacuum fluctuations of my COST. **The three amazing discoveries of a lower state of the vacuum are actually providing <u>amazing evidence</u> of a thick web of tiny, positive Gravitational waves which are providing dimensions to Space and are <u>increasing</u> the ground state of Space to the ground state of dimensional Space which includes the passing of time.**

While this model will unravel the mystery of exactly what this lower state phase is and the role it plays in creating mass and Gravity, <u>I credit</u> the many Scientists, from the many different avenues of research, that have now discovered direct evidence of this phase of a lower state of the vacuum. When I first published this in 2014 I believed it would be impossible to provide direct evidence of a tiny phase of completely empty Space, I wrote in my first book; *'We are referring to a weird piece of emptiness which if we were to use a particle collider it may not be detectable because you simply cannot collide any particle with emptiness'*. However, Scientists regularly carry out brilliant research which sometimes delivers unexpected results.

Mirroring the Anti-space of the outer empty void I at first called this piece of empty Space 'fragmented Anti-space'. But as it plays a direct role in the way this model creates regular mass I have now suitably called this tiny phase of lower state of Space 'virtual' mass which is a state of negative energy which mirrors the positive energy created by pairs of 'virtual' particles. **There is nothing in the tiny phase of virtual mass; it is a tiny phase of pristine Space and a state of <u>true vacuum nothingness</u> while what is now called 'the vacuum' is not a true vacuum but a false vacuum since it contains the positive energy of the dimensions of Space and time.**

Quantum vacuum fluctuations are often described as; *particle and anti-particle pairs of virtual particles materializing in empty Space and annihilating which produces a Quantum fluctuation.* For the COST, it is <u>important</u> that I am far more comprehensive; at a tiny point of empty ground-state Space the ground-state fluctuates to a lower state of vacuum. Ground-state dimensional Space rushes in, at the speed of light, into the

66

tiny hole of a phase of a lower state of the vacuum and <u>clashes</u> at the very center. The all important '<u>clash</u>' produces a tiny splash of energy in the form of a particle and anti-particle pair of virtual particles which annihilate. Annihilation energy radiates into Space in the form of tiny <u>Gravitational waves</u> created by the 'fluctuations' of the event of what is effectively a minuscule, rapidly collapsing and reappearing Black Hole. <u>The actual annihilation event</u> causes a new tiny lower state hole in the vacuum of dimensional Space and so the process has become cyclic.

With a Natural Model the actual '<u>clash</u>' of Space which produces the virtual particle and anti-particle pair will eventually be the actual source of all regular Matter particles responsible for the creation of all Stars, Planets and objects made from Matter within the universe. With this model, at this very early stage all virtual particles are nothing more than a temporary, uncertain, energy wave of potentials. Due to its energy, the energy phase protrudes above the ground-state of dimensional Space and is a phase of <u>positive state</u> while the phase of a lower state of the vacuum sits below the ground state of dimensional Space so is a phase of <u>negative energy (density)</u> [111.] The negative state cancels the positive state so as the whole event remains sum-total zero mass-energy. This is required because it preserves the conservation of energy law. The energy phase can become louder by producing extra pairs of virtual particles causing the annihilation event to be more powerful and the lower state phase of virtual mass to likewise increase its volume below the ground-state of dimensional Space.

This is a simple '<u>for every action, there is an equal and opposite reaction</u>' scenario. The whole process becomes permanently cyclic and sustainable. In a <u>cyclic, amplified state</u> a common Quantum vacuum fluctuation becomes a COST unit which mimics all of the known properties of Dark matter. The COST is now working much like a proton but without three primary <u>valence</u> quarks it has no regular particles or regular mass and in this Dark matter state can be best called a <u>virtual proton</u>.

The virtual quarks and anti-quarks of the COST <u>annihilate before they interact or are held strongly together</u> by the strong force which will be created by the tiny phase of virtual mass. Because the quarks of the COST are yet to interact similar to when the COST becomes a regular proton they are yet to be assigned different roles identified by colors or flavors. Although the quarks of the COST remain in their primordial colorless, flavorless and virtual state they can be said to have <u>unassigned potentials</u>. All is why the COST, residing in its Dark matter state as a virtual proton, is more related to a Quantum vacuum fluctuation than a baryon or a meson.

As it is a requirement of physics that energy of the energy phase is immediately returned to Space, modeling will accordingly explore a seemingly, classic notion of the annihilation energy from Quantum vacuum <u>fluctuations</u> of the COST being immediately returned to Space in the form of expanding waves which are produced due to the cyclic fluctuations.

The model will require the <u>extraordinary tiny,</u> expanding waves to speed away at the speed of light, <u>carry away a portion of the Quantum fluctuation's energy</u>, actually provide the universe with dimensions and provide the delivery part of Gravity. Amazingly, there is one kind of wave which appears to be perfectly capable of performing these tasks and that is an appropriately called a Gravitational wave. Gravity makes Gravitational waves but in this case the source of Gravity creates extraordinary tiny Gravitational waves. Gravitational waves travel through Space at the speed of light and are <u>fittingly a prediction of Albert Einstein's General theory of Relativity which is a theory of Gravitation</u>. Whenever an object emits Gravitational waves the waves radiate away a portion of its energy. Although the tiny Gravitational waves will carry a portion of a vacuum fluctuation's energy, the waves carry no state of the deep vacuum phase of virtual mass. Consequently, the waves carry <u>no mass</u> which will allow the tiny Gravitational waves to travel in a massless state at the speed of light.

[017] A Quantum fluctuation or oscillation of energy to a lower state alone can

be expected to produce <u>extraordinary tiny</u>, invisible Gravitational waves in Space. For instance, simply waving your hand back-and-forth, like a fluctuation, is said to produce tiny Gravitational waves. These very tiny, expanding Gravitational waves will be caused by the Quantum vacuum fluctuations of the COST fluctuating to a phase of virtual mass. The virtual mass phase acts as a very minuscule Black Hole which is in a state of rapidly <u>collapsing and reappearing, a process which is ideal for producing extraordinary tiny Gravitational waves</u> which are <u>spontaneously</u> entangled from birth with the energy from which they were created. Although incredibly tiny, these will be the universe's first Gravitational waves which radiate away <u>entangled energy</u> from the event which made them.

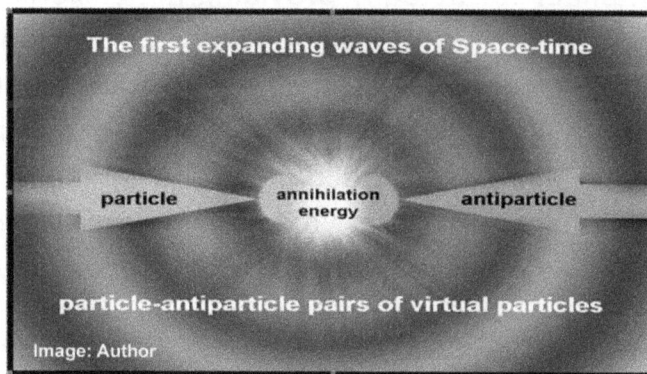

The first expanding waves of Space-time

particle   annihilation energy   antiparticle

particle-antiparticle pairs of virtual particles

Image: Author

[017] The largest Gravitational waves are produced by ultra powerful events such as colliding regular Black Holes. On September 14, 2015, LIGO, a Gravitational wave interferometer detected passing Gravitational waves generated by such an event. LIGO consists of two interferometers, each with two 4 km long arms arranged in the shape of an "L". LIGO is designed to detect tiny <u>changes in the dimensions</u> of the distance between mirrors.

Passing Gravitational waves are predicted to stretch, squeeze and distort, so as to actually interfere with the dimensions of Space-time and so change the physical dimensions of anything in their path which is why they are ideally the most logical source of the actual dimensions to all Space. Gravitational waves passing through an object are less like an earth quake but more like looking into a distorted mirror. Only instead of just a reflected distorted image the actual dimensions of Space are squeezed and stretched or distorted so as the dimensions of an object change, usually by a <u>very tiny amount</u>, as a Gravitational wave distort the dimensions of Space when a wave passes. The passing of powerful Gravitational waves are actually predicted to <u>distort the dimensions of Space-time</u> between objects similar to overriding existing dimensions by <u>providing</u> new, different dimensions to Space. If Gravitational waves <u>have the extraordinary ability to physically interfere</u> with the dimensions of Space-time, <u>so as to actually change the dimensions, of any object or Space which they are passing through,</u> than very tiny Gravitational waves have to surely be the <u>most obvious, and the only possible contender</u> for actually providing dimensionless empty Space with dimensions. Although extraordinary tiny these positive state waves produced by the COST will build their influence in huge numbers. The tiny <u>positive</u> Gravitational waves are driven outwards in all directions at the speed of light by a relentless attraction to the <u>negative energy density (negative energy)</u> state of a far outer void of completely empty Space called Anti-space. This is also a positive pressure to negative pressure scenario.

Because the tiny Gravitational waves will be required to provide single units of dimensions they are required to be so extraordinary tiny that the distance across one wave, termed as the wavelength, will provide <u>the smallest possible dimensional measurement within a Dimensional Reality.</u> This should, of course, ideally be a <u>Planck length</u> and because the passing of these tiny Gravitational waves will provide the passing of time each passing wave will provide one unit of Planck time. In order to be so extraordinary tiny the Gravitational waves are provided by the tiniest fluctuating energy events which are the Quantum vacuum fluctuations of the COST. I will continually provide correlated evidence of this throughout the modeling.

Planck length will be the distance trough to trough or the <u>wavelength</u> of a tiny Gravitational wave which will naturally be the smallest measurable unit within a <u>Dimensional Reality</u>. Smaller than the Planck constant can only be characterized as a <u>dimensionless state</u> and is where Quantum Theory allows actions not allowed within an entangled, Dimensional Reality. Where dimensions are <u>not an entangled property</u> is where Quantum Theory excels and is where Einstein's Relativity will clearly give way to Quantum Theory.

Limits of General Relativity: Gravity and the dimensions of Space and time will be delivered in tiny but <u>continually produced waves</u>. Thus, dimensions of Space-time will not be infinitely divisible. Although dimensions can be *quantized* into extraordinary tiny segments, just a Planck length across a wave, the tiny Gravitational waves are continually produced by every COST to provide a '<u>seamless' succession of waves</u> from their source COST unit.

A Planck length is only 0.000 000 000 000 000 000 000 000 000 000 000 016 of a meter and a hydrogen atom is about 10 trillion, trillion Planck lengths across. One wave, effectively just a Planck length thick, naturally provides a COST with an entangled dimension of length and the passing of which provides the passing of a moment of time. The significance of the dimension of length and the passing of time has been realized with the predictions of Albert Einstein's Relativity which reveal both length contraction and time dilation are induced by both speed and Gravity.

The COST will be transformed into the protons and neutrons of all atoms. Because the tiny Gravitational waves expand on a flat plane it will require many sets from <u>many atoms to come together</u> to build an all around smooth Dimensional Reality. It is said there are generally more atoms in a grain of sand than there are grains of sand on all of Earth's beaches. Thus, even a grain of sand will be given numerous sets of tiny Gravitational waves all of which <u>work in unison</u> to provide a robust all-around <u>four Dimensional Reality</u> which includes the passing of time and is effectively built from <u>just one dimension</u>; the dimension of length. These are the dimensions which will come together with the amalgamation of millions of atoms to build an object's complete Dimensional Reality from which there is no escape. I will later provide direct evidence that a Dimensional Reality is <u>truly built by the amalgamation of many atoms</u>. [018] Physicists do not understand why this is so and say there must be an objective threshold between the tiny scale of quantum and the scale of Einstein's Relativity. My Natural Model will clearly explain why it takes an amalgamation of many atoms to build a completed all around Dimensional Reality from where there is no escape.

[019] General Relativity expresses Gravity as a warping of the <u>curvature</u> of Space-time; <u>so how is dimensional Space somehow curved?</u> Gravitational waves act much like the <u>circular</u>, expanding ringlets from raindrops on a pond which is why dimensions are actually provided as a <u>curvature</u> of dimensional Space-time. Naturally, this is further evidence that tiny Gravitational waves which can clearly provide this <u>curvature</u> are the source of dimensions to Space. Consequently, it is this <u>curvature</u> of the dimensions of Space which can be stretched or warped, similar to the <u>curvature</u> of Space-time as described by Albert Einstein's General Relativity. All of which will provide <u>the delivery part of Gravity</u> which allows a grain of sand to fall to Earth. Gravity will also require a source for <u>an attraction part</u> which will be provided by the negative energy phase of deep, pristine vacuum of the COST. Thus, although it has no regular particles the COST in its Dark matter state provides both parts of Gravity and both parts are distinctly different.

Planck length and Planck time are indeed extraordinary tiny and are the <u>smallest possible</u> such values that Scientists say surface naturally in the <u>traditional laws</u> applying to our universe which means Planck values are universal, even for an alien on a far away Planet. Special Relativity requires that the laws are the same for all observers regardless of length contraction or time dilation induced by speed or Gravity.

[020, 21] Although Scientists say *"the Planck scale defines where Gravity meets Quantum Theory, (dimensional) Space and time"* the Scientists also say ***"currently, we don't know much about this interaction"***. A Natural Model unmasks tiny Gravitational waves which provide the <u>essential key</u> required

to <u>unlock</u> <u>exactly how</u> this interaction occurs: Tiny, entangled Gravitational waves originating from the hearts of virtual protons (Dark matter), protons and neutrons provide <u>entangled</u>, dimensional moments or Planck units to dimension<u>less</u> Space. Gravitational waves are a prediction of General Relativity and length contraction is a prediction of Special Relativity. The distance trough to trough or the wavelength of a tiny Gravitational wave provides a Planck length which will be the smallest possible dimensional distance within a Dimensional Reality. The passing of one of the same tiny Gravitational waves provide a unit of Planck time which will be the shortest possible duration of passing time within a Dimensional Reality. This changes dimensionless Space into dimensional Space-time where the dimensions of Space cannot be separated from the passing of time, hence, Space-time.

Gravity is delivered by a 'stretching' causing a warping of the very same tiny Gravitational waves. A stretched Gravitational wave retains the same Planck value as a non-stretched wave. Because Planck value is always maintained, the distance across a <u>stretched</u> wave contracts to a Planck length which contracts distance causing an object to fall, this is Gravity. A stretched wave takes longer to pass which is why Gravity slows the passing of time. Planck values are retained because a stretched wave retains the same vital <u>'single pulse' of energy</u> from the event which created it. Because Gravity is delivered in Planck lengths, General Relativity breaks down and fails at <u>smaller</u> than Planck scale. This is where <u>Quantum Gravity</u> takes over which works <u>with only</u> the strong attraction part of Gravity and in a state of Space which is, in effect, <u>dimensionless</u> (Page 347). Within a Dimensional Reality it is not possible to determine a position to smaller than a Planck length and <u>is where General Gravity meets</u> the indefinable scales of Quantum Theory. **This is how a Natural Universe allows** *"Planck scale to define where Gravity meets Quantum Theory, dimensional Space and time".*

Because the tiny Gravitational waves retain <u>the same vital pulse of energy</u>, their Planck value is retained regardless of the waves being stretched or passing slowly. Thus the laws of the universe remain the same for all observers regardless of length contraction or time dilation.

Smaller than a dimensional moment or when dimensional moments provided by the tiny, entangled Gravitational waves are not recognized will be where Quantum Theory and the Heisenberg Uncertainty Principle takes over from the realm of Relativity. Different observers may rightly observe objects and events differently but because the value of a Planck unit remains constant all observers are really observing the same. This is like two different observers each drinking a measured cup full of water, drink it slow or drink it fast there is only a full cup of water to drink.

Planck length is the distance light travels in Planck time. The passing of a tiny Gravitational wave provides Planck time. [021] Because Gravitational waves travel at the speed of light, this will allow light to always travel a Planck length (the width of the wave) in Planck time (the passing of the wave) within a Dimensional Reality regardless of length contraction or time dilation. This locks the speed of light to always travel a Planck length in Planck time within one's entangled, Dimensional Reality and is why you will always observe the speed of light as a constant.

Expanding from the fluctuations of COST units, the extraordinary tiny, dimensional Gravitational waves are fittingly closely modeled on a prediction of Albert Einstein's General theory of Relativity. [017] General Relativity delivers Gravity by warping the curvature of the dimensions of Space which is also why tiny Gravitational waves, their curvature of which can easily be stretched or warped, are ideal for delivering Gravity. The tiny expanding dimensional waves will for the first time provide dimensionless Space with a source of dimensions. This is why a Natural Universe is required to begin with these tiny Quantum fluctuations. Before this time there was nothing within seemingly empty Space to provide Space with dimensions to become the fabric of dimensional Space-time as described by Albert Einstein and there was nothing in Space for Einstein's General

Relativity to bend or warp the 'curvature' of to deliver Gravity.

Entanglement is clearly recognizable by the way it allows one to always, regardless of your speed, observe the speed of light as a constant and observe objects as they are within one's own, unique, Dimensional Reality. Quantum particles which have no such mechanism to provide entangled, tiny Gravitational waves are naturally free of being locked to entangled dimensions and so totally ignore the restrictions of entangled dimensions.

COST units (Dark matter) are Quantum vacuum fluctuations which have become permanently amplified to the extent of being permanently self-active. This will allow a tiny configuration of invisible Space-time to display energy and so mass, entangled dimensions to Space and, last but not least, the simple, invisible, permanent configuration of Space-time will express one other all important property, which is of course Gravity. **Being based on a Quantum vacuum fluctuation is clearly what will give Dark matter its invisible, ghost like, virtual properties and is why nobody has been able to directly detect Dark matter.** The COST, acting as a virtual proton, is designed to exactly match the known 'properties' of Dark matter.

The size of the COST: If one measures the COST at its fluctuating heart the COST is very, very tiny. However, as an entangled element of the COST, tiny Gravitational waves extend its entangled influence to a universal scale.

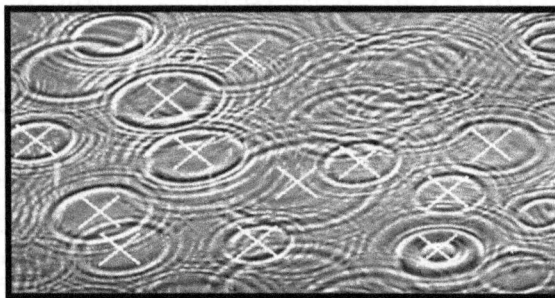

Why entanglement? Tiny, expanding Gravitational waves rippling through dimensionless Space from our first minuscule Quantum vacuum fluctuation events now naturally provides dimensionless Space with entangled dimensions. The tiny Gravitational waves act much like the expanding ringlets from raindrops on a pond which, like Gravitational waves, also <u>carry the energy</u> from the event which created them.

The energy of the happening event is naturally entangled with the event. Each wave is instantly born with and carries the energy of the event. By carrying energy from the event which created them, the tiny Gravitational waves which provide dimensions to Space are <u>spontaneously entangled</u> from birth with the energy of the event which created them. Entanglement will allow us all to reside within our very own unique, self entangled Dimensional Reality which <u>cannot interfere</u> with, <u>and so ignores</u>, all other self entangled Dimensional Realities. This allows your <u>time dilation factor</u>, which slows your passing of time and is induced by your unique speed and Gravitational position, to <u>not interfere</u> with and so **totally ignore** all other self entangled Dimensional Realities provided to all other <u>observers.</u>

Before this time there was no physical mechanism within empty Space to provide Space with dimensions and provide a coordinated, spatial system. <u>Entangled from birth</u> with the event which creates them, the first tiny Gravitational waves provide the first expanding dimensions to Space and provide Space with an <u>all important</u> <u>invisible structure which can be stretched, warped or bent</u> which delivers the first ripple of the delivery part of Gravity. An attraction to an absolute dimensionless far outer void now relentlessly drives the expansion of the tiny dimensional waves at the speed of light. **At the speed of light <u>distance to the void is meaningless.</u>**

The dimensions of Space, including the passing of time will remain directly associated and so <u>entangled</u> with the COST from which they originated. <u>Thus, dimensions will be more of the property of the COST and less a direct property of Space.</u> Science is inclined to portray the dimensions of Space as a common property of all Space which we all observe uniquely differently. Within my Natural Universe the dimensions of Space are personal and so are more a part of us, provided by the vacuum fluctuations of the protons and neutrons at the heart of our very own atoms.

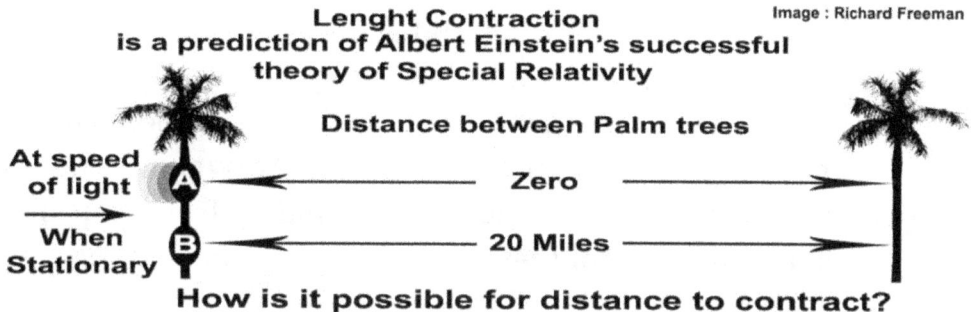

Lenght Contraction is a prediction of Albert Einstein's successful theory of Special Relativity

Image : Richard Freeman

Distance between Palm trees

At speed of light — A — Zero

When Stationary — B — 20 Miles

How is it possible for distance to contract?

Why the dimensions of Space <u>need to be</u> an entangled property: Einstein's Special Relativity infers the speed of an object contracts the dimension of length and slows the passing of time. At the speed of light, dimensions contract to a single point and the passing of time stops. 'Light speed' of one object does not cause all dimensions of all Space for all objects to close out and the passing of time to stop for all objects as if dimensions were a universal, common property of all of Space. For this to occur in a <u>physical way</u> that dimensions are unique to the object which has speed, dimensions are more likely to be <u>an entangled property</u> of an object and <u>not likely</u> to be a common property of all Space. For example, a photon (light) may travel for 100,000 light years across an entire Galaxy; however, in its own time and perspective, the same instant a photon begins its journey it also arrives at its destination and does so <u>without travelling any dimensional distance</u>. At the speed of light there is no dimensional distant or passing of time.

How can photons of light effectively <u>reside within</u> dimension<u>less</u> Space, but you and I reside within dimensional Space? Scientists say wherever one is, one is at the center of the dimensional universe, how is this so? How can both <u>movement and Gravity</u> slow your own rate of time and slow your wrist watch but <u>always appear</u> to you to be running at the same <u>unchanging</u> rate of time? How can the length of an object contract in the direction of motion? How can your passing time run faster, so as you are actually ageing faster, when you are standing still at the top of a tall building? These are predictions of Albert Einstein which have been repeatedly proven beyond any doubt by today's atomic clocks. According to Einstein we all carry our own personal rate of time which can run at very different rates. So how can your own rate of passing time be totally unique to yourself and instantly change with every move you make? Why does not another observer's rate of time interfere with you own rate of time? **Answer:** having your own **entangled** dimensions of Space and time expanding from your own atoms is what will allow every one of these things to physically and logically occur.

With this model a photon of light, like all tiny Quantum particles, have no fluctuation mechanism to provide entangled dimension to their Space. Without entangled dimensions, the particles effectively reside within dimensionless Space from where a photon travels instantly to anyplace.

Ultimately, there is very little choice other than to accept that the universe had to begin within empty Space. A Natural Universe begins by first creating Gravity powered by nothing more than empty Space.

[011] This extraordinary video **actually reveals empty Space is the source of most of our mass**. In the video **'Your Mass is Not from the Higgs Boson'**, Prof. Dr. Derek Leinweber speaks about the findings of their research at the University of Adelaide. The research focuses on a super computer simulation of the theory of Quantum Chromodynamics which describes the action of the strong force within a proton. At time 3.57 one will find an image of quarks sitting on 'lumps' in the gluon field where the quarks are positioned around a central area of empty Space. It is this actual empty Space, vacuum nothingness, which the research **clearly identifies** as being the source of the strong force which agitates and binds quarks to create a proton's mass. **The conclusion of the super computer simulation is that it is *"extraordinary, because what we think of ordinary empty Space* that**

turns out to be the thing that gives us all, <u>most of our mass</u>" (99%). Thanks to this extraordinary research we can now <u>confidently say</u> the source of mass is a tiny phase of empty Space supplied by a vacuum fluctuation.

Mass and Gravity go together like two peas in the same pod; <u>exactly where you find mass you will find Gravity</u>. Since the source of a proton's mass is empty Space, and to reconcile Gravity with mass, Gravity should <u>ideally</u> share the same source of empty Space. Dark matter is responsible for 85% of all mass and <u>Gravity</u>. To produce all this Gravity, Dark matter likely has a similar mechanism to a regular proton; a <u>virtual</u> proton will provide this.

Quantum vacuum fluctuations <u>spontaneously arise</u> from every tiny point of empty Space which <u>provides a known source</u> for virtual protons (Dark matter) to be sourced in the massive quantities required to create a universe. Like Dark matter, Quantum vacuum fluctuations are invisible. Within a Natural Universe, Dark Matter is a virtual proton which effectively provides the missing link between a common Quantum vacuum fluctuation and the <u>most similar</u> Quantum vacuum fluctuations powering a regular proton. **A virtual proton (Dark matter) is effectively a proton without the three additional Matter valence quarks which <u>give a proton its properties</u>.**

This model shows why a proton is in essence an energy amplified **Quantum vacuum fluctuation** which has acquired three additional Matter valence quarks and developed a phase of negative energy density (empty Space) which is the source of the strong force which creates our mass.

Within the universe, natural progression creates many things. For example; protons attracted an electron to progress to forming atoms of hydrogen which progressed to form molecular hydrogen gas which progressed, with the <u>help of its own Gravity</u>, to make stars.

Because common Quantum vacuum fluctuations will progress to being regular protons it is <u>helpful</u> to understand the <u>very similar properties</u> of a common Quantum vacuum fluctuation and a regular proton. Both have Quantum vacuum fluctuations which create pairs of virtual particles which can be referred to as quarks and anti-quarks. Both have been documented to express phases of a lower state of empty Space, a state of <u>negative energy density</u>. Thus, the relationship which will allow common Quantum vacuum fluctuations to progress to becoming regular protons is clear. Virtual protons (Dark matter) provide a natural missing link between the two which allows this <u>progression</u> to occur. See exactly how at chapter 8.

How could a universe begin from nothing more than empty Space? The *astonishing* results of the following experiment will be used to tell us how.

How to make Dark matter: [015] Scientists may have made virtual protons (Dark matter) without knowing it. Astonishingly, a team of Scientists headed by Prof. Dr. Alfred Leitenstorfer at the University of Konstanz have discovered the best evidence which aligns most closely with this modeling. <u>Please note</u> the Scientists actually say they've **"Manipulated Pure Nothingness and observed the fallout"**; this is important since the Natural Model actually begins within the nothingness of empty Space. The Scientists say they have directly detected Quantum vacuum fluctuations. Their experiments also clearly detected something strange when they *'squeezed the vacuum'* the fluctuations became <u>louder</u> and so amplified and were observed to fluctuate to a level which is **'lower than the ground state (the vacuum) of empty Space'**. Not knowing what to call this unexpected discovery of a <u>phase of vacuum</u> in a lower state than empty Space the Scientists have simply called it an **"astonishing phenomenon"**.

So what could have caused a common Quantum vacuum fluctuation to become a virtual proton (Dark Matter)? [015] The above experiment tells us; all that it is required is to squeeze the vacuum of Space around Quantum vacuum fluctuations which will cause the vacuum fluctuations to become very amplified which increases their energy phase and significantly reveals their phase of negative energy. The most likely thing which can do this is Gravity which happens to be the only known property of Dark-matter.

But can Gravity, by squeezing the Space around Matter, really alter Matter? It certainly can; in the Sun, warped Space-time squeezes atoms so tightly together that they fuse into completely new elements. In neutron stars, warped Space-time squeezes atoms so tightly that all that is left of atoms are neutrons. In Black Holes, warped Space-time squeezes all atoms so tightly that they are crushed out of existence. To change common Quantum vacuum fluctuations into virtual protons will require just a gentle squeeze. **Thus, the same procedure which makes stars and planets also began the Natural Universe.** See page 80 the experimental evidence used for this [015].

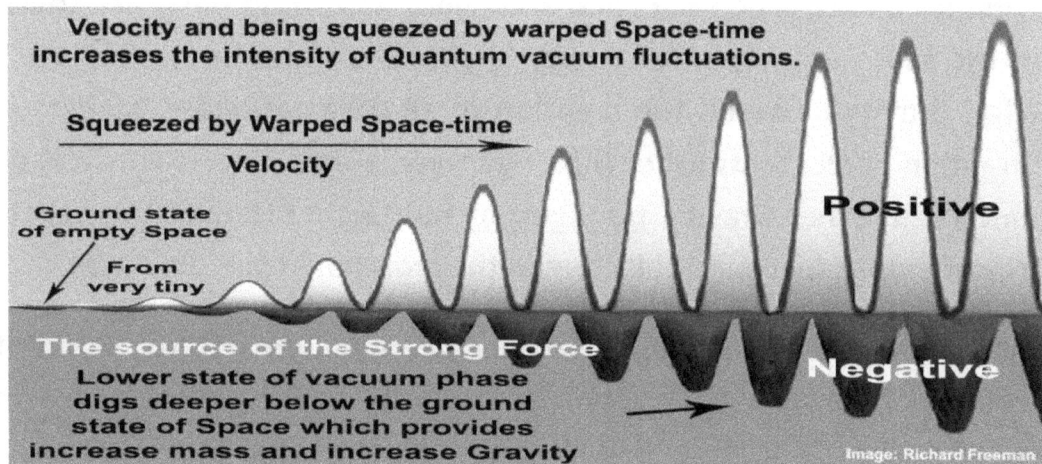

Velocity and being squeezed by warped Space-time increases the intensity of Quantum vacuum fluctuations.

Squeezed by Warped Space-time

Velocity

Ground state of empty Space

From very tiny

Positive

The source of the Strong Force
Lower state of vacuum phase digs deeper below the ground state of Space which provides increase mass and increase Gravity

Negative

Image: Richard Freeman

At the very beginning, very weak Quantum vacuum fluctuations filled seemingly empty Space. Their feeble breath of Gravity began to slowly move them closer to each other. Gently squeezed by surrounding warped Space-time created by their own combined Gravity, Quantum vacuum fluctuations progressively gain energy, **like observed in experiments,**

causing their <u>negative energy phase</u> of virtual mass to significantly increase in volume which significantly increases their <u>Gravitational attraction phase</u> and gave Gravity a more <u>purposeful delivery part</u>. The more the vacuum fluctuations are squeezed by Gravity, the more Gravity they produce which squeezes and <u>powers up</u> the vacuum fluctuations even more. It seems to survive virtual protons (Dark matter) need to be surrounded by their own kind. **This is how** <u>insignificant</u> Quantum vacuum fluctuations <u>powered up</u> to become seemingly mobile <u>virtual protons</u> which <u>will eventually power</u> the <u>more significant</u> vacuum fluctuations <u>required to power the regular protons and neutrons of all atoms</u>. **This is also how Gravity, powered by nothing more than <u>empty Space</u>, slowly arose from squeezing numerous Quantum vacuum fluctuations and began building a <u>Naturally Forming Universe</u>.**

Thus, Gravity gave birth to the Natural Universe. A Natural Model's Gravity has an attraction part and a delivery part; both parts are sourced from the same tiny phase of <u>empty Space.</u> The universe has thrived on Gravity. Gravity powered the creation of all Galaxies, stars, planets and the creation of all atoms other than hydrogen atoms. This is why it is only common sense that Gravity, <u>powered by</u> nothing more than the negative energy of <u>empty space</u>, actually began the creation of the universe.

All virtual protons now fluctuate to a <u>negative energy density</u> phase of a lower state of the vacuum (empty Space) which, like the dimensionless void state of empty Space we began with, <u>also attracts</u> the tiny positive Gravitational waves from all other Virtual protons. The <u>fluctuating</u> lower state phase of vacuum provides <u>strong negative energy</u> for the <u>attraction part</u> of Gravity which is the source of Gravity. The <u>fluctuating</u> lower state phase of vacuum also creates tiny Gravitational waves, which provide <u>entangled</u> dimensions to Space. The <u>energy carrying</u> waves, which can be stretched and warped, will provide the <u>variable</u> delivery part of Gravity which <u>provides movement</u> and <u>mirrors</u> Einstein's General Relativity. Virtual protons being **trapped and restrained** at <u>stationary equilibrium</u> **by the variable delivery part of Gravity** will allow Gravity to be <u>weakly delivered</u> here on Earth and <u>strongly delivered</u> at the event horizons of Black Holes.

So how can a common Quantum vacuum fluctuation become a seemingly <u>mobile</u> virtual proton? Gravity moves virtual protons while they remain restrained, at <u>stationary equilibrium</u>, <u>at the center</u> of their own entangled dimensions of Space. **Thus, to a 'moving' and so mobile virtual proton it is not actually moving, since it <u>always remains stationary, at rest, at the center</u> of its own unique, <u>entangled</u> dimensions of Space and time. Because of <u>entanglement</u> all other dimensions of Space are <u>totally ignored</u>. <u>Entanglement</u> is what will allow us all to carry our own <u>unique</u> rate of passing time based on our movement and Gravitational position.**

The lower state phase of vacuum, because of the role it plays in this modeling, I have fittingly labeled as a phase of '<u>virtual mass</u>'. The COST now has a phase of <u>virtual</u> particles fluctuating to a phase of <u>virtual</u> mass, all of which happens within a time period defined by Werner Heisenberg's uncertainty principle. <u>Virtual</u> particles will later provide the seed for the creation of regular particles and <u>virtual</u> mass will interact, in a fitting and understandable manner, with regular particles to provide regular mass.

The aim now is to maintain the symmetry of empty Space. Symmetry is related to the conservation of energy law where total energy remains constant over time. Energy can neither be created nor destroyed. Energy can be transformed or transferred from one form to another. Consequently, if the universe begins from a sum-total <u>zero</u> energy empty Space one must show how this sum-total zero energy is maintained when creating a universe. Symmetry and the conservation of energy law are so important that they <u>cannot be violated</u> which means, because we begin with zero mass-energy, the total mass-energy of the universe when everything is taken into account must add up to and remain sum zero.

Although significantly different, since I have also included the source of mass, I have partly based this sum-total, zero mass-energy model on related modeling as expressed on page 136 of my copy of Stephen Hawking's book '*A Brief History of Time*' in which the modeling relates negative energy to <u>Gravitational Attraction</u> where negative energy is required to separate objects <u>held together</u> by Gravity and positive energy relates to all Matter. My model also includes modeling of <u>the zero-energy universe hypothesis</u> which was first proposed in 1973 by Edward Tryon, who <u>proposed</u> the universe is the result of a <u>Quantum fluctuation</u>. Albert Einstein said mass and energy are equivalent and with Edward Tryon's model, the positive energy associated with mass is counterbalanced by <u>Gravitational potential energy</u>, which is negative. My Cost is aptly also based on a <u>Quantum fluctuation</u> which includes a negative energy phase of virtual mass which by providing the source of both <u>Gravitational Attraction</u> and <u>Gravitational potential energy</u> as well as being the negative source of regular, positive mass-energy includes the negative facets of both theories which have been designed to cancel the positive mass-energy of all Matter.

It is important to note that Gravitational Attraction in this way can be regarded as being a negative state. This is important because a negative state attracts both a positive state and a neutral state. [111] With this modeling the source of Gravitational Attraction is a lower state of Space which we can now understand is a state of negative energy. This will enhance the effectiveness of both the delivery part of Gravity which is a positive state and the binding energy of the strong force which is required to bind the particles of the positive proton and the neutral neutron.

I do like such a scenario for the reason that my Natural Model lends itself beautifully to such concepts. The big advantage my Natural Model has for such a concept is that it has an 'actual source' for both Gravity and mass which allows one to be <u>very explicit</u> when fine tuning such modeling.

So how is it possible to extract energy from empty Space? One exposes negative energy from 'less than Space' which <u>action</u> naturally gives an opposite amount of positive energy from Space. Naturally if this positive energy is given back to repair the violation in Space it will leave empty Space again, this is symmetry. Think of this like building a small sand castle near water's edge at low tide. There are no mounds of sand to fill your bucket so you dig a little hole in the flat sand and use the sand from the hole to fill your bucket. Now you turn the bucket upside down to make your little sand castle. The sand in the sand castle has actually gained Gravitational Potential Energy because it is higher than the ground-state which it is sitting on. Now the tide comes in and flattens the sand castle and fills the hole up with sand again. Everything is now status quo again.

With a scenario like this the energy from the energy phase of the COST is positive energy in relation to the negative energy density state of the phase of virtual mass. The proposed modeling here is that the <u>positive</u> energy of all regular mass-energy of all Matter, dark or regular, and the <u>positive</u> dimensions of Space-time should be cancelled out by the phase of virtual <u>mass</u> of the COST which is the source of both the <u>attraction and the delivery part of Gravity</u> and the <u>primary source of regular mass</u>. Meaning the sum-total mass-energy of our universe should begin with and always remain exactly zero. The same modeling will advance from the virtual particles created by the Quantum vacuum fluctuations of the COST (Dark matter) to creating regular particles (protons and atoms). This model will show how it can account for 100% of all regular, positive mass-energy being directly contributed to the nothingness of the lower state of vacuum phase of the COST. All of which will allow a universe of dimensional Space and time containing Matter, mass and Gravity to be created from empty Space.

With such modeling what is today considered ground-state of empty Space (the vacuum) is actually dimensional Space-time which is positive because it contains a thick web of tiny, positive Gravitational waves which provide dimensions to Space. Negative (-) energy sits below the ground-state of dimensional Space. Virtual mass punches below positive dimensional Space-time so virtual mass is of negative energy. As the phase of virtual mass becomes stronger it increases its volume below dimensional Space-time. Positive dimensional Space-time immediately rushes into the negative energy of virtual mass which naturally momentarily repairs the abnormality in dimensional Space-time before a new phase of virtual mass reappears.

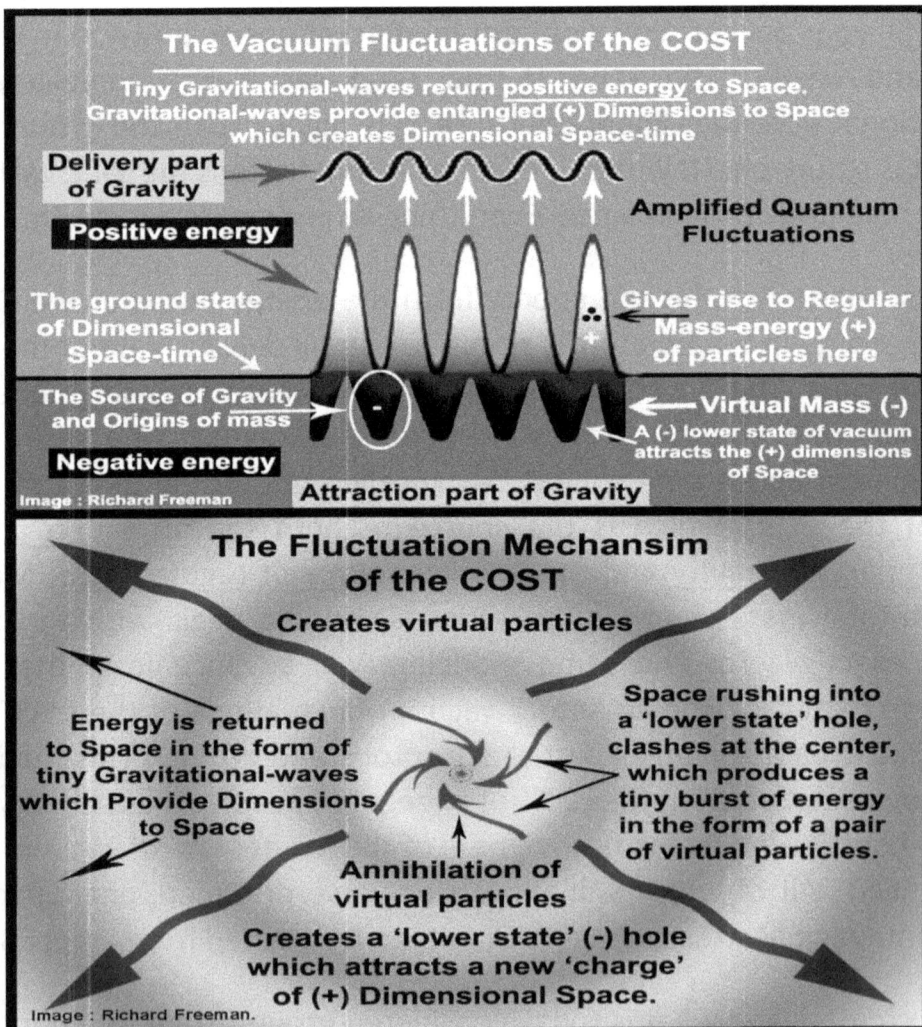

The Vacuum Fluctuations of the COST

Tiny Gravitational-waves return positive energy to Space. Gravitational-waves provide entangled (+) Dimensions to Space which creates Dimensional Space-time

Delivery part of Gravity

Positive energy

Amplified Quantum Fluctuations

The ground state of Dimensional Space-time

Gives rise to Regular Mass-energy (+) of particles here

The Source of Gravity and Origins of mass

Virtual Mass (-)
A (-) lower state of vacuum attracts the (+) dimensions of Space

Negative energy

Image : Richard Freeman

Attraction part of Gravity

The Fluctuation Mechansim of the COST

Creates virtual particles

Energy is returned to Space in the form of tiny Gravitational-waves which Provide Dimensions to Space

Space rushing into a 'lower state' hole, clashes at the center, which produces a tiny burst of energy in the form of a pair of virtual particles.

Annihilation of virtual particles

Creates a 'lower state' (-) hole which attracts a new 'charge' of (+) Dimensional Space.

Image : Richard Freeman.

The previous two illustrations of the COST (Dark matter): The first shows its vacuum fluctuations in relation to the ground-state of dimensional Space and the second is how the COST (Dark matter) may look like if one was able to observe it. Notice when the positive energy phase peaks the lower state phase of the vacuum of negative energy density is least and when the lower state phase peaks the positive energy phase is least. The Gravitational waves are many times larger for illustration purpose.

The virtual <u>mass</u> phase is negative energy (-) and the annihilation of virtual particles creates an energy <u>phase</u> which is positive (+) energy. As one cancels the other the COST, acting as a virtual proton, is neutral. The <u>origin</u> of regular mass-energy is virtual mass. Regular <u>mass and energy cannot be separated</u> and is positive (+) energy. Because virtual mass will create regular mass-energy and provides Gravitational Attraction it provides the negative energy which cancels the positive energy of all Matter, dark or regular, for creating a sum-total, zero mass-energy universe.

The <u>attraction part of Gravity</u> is virtual mass which is negative (-) energy. The <u>delivery part of Gravity</u> only applies Gravity by the amount the positive Gravitational waves (+) are stretched by the attraction of (-) virtual mass. The delivery part of Gravity is positive energy (+). Note Gravity <u>requires</u> both a negative part and a positive part. Opposites charges attract which contributes to the negative pressure attraction of the tiny (+) Gravitational waves to (-) virtual mass. The Ground-state of expanding <u>dimensional</u> Space-time is positive and is positive pressure in relation to virtual mass.

Because it is dimension<u>less</u>, the outer void which I have called Anti-Space is negative energy. This is because the tiny Gravitational waves which are positive (+) energy have yet to reach this seemingly endless void. This also means that this void is of a lower state of vacuum than the ground-state vacuum of dimensional Space-time. Consequently, the positive state tiny Gravitational waves which provide the dimensions of Space-time are also attracted to the negative energy of the outer void which is why they expand in all directions at light speed and in doing so deliver Outer Gravity.

[100] **Experimental studies today reveal** negative energy relating to Gravitational Attraction accounts for <u>only about a third</u> of the negative energy required to cancel the positive energy of all Matter. This is why the negative energy of virtual mass which by being responsible for Gravitational Attraction <u>plus</u> all regular, positive mass-energy <u>plus</u> the positive dimensions of Space-time, is required to balance the scales to achieve a sum-total zero mass-energy universe. Consequently, despite creating positive Matter with mass-energy and the positive dimensions of Space-time the total energy of a Natural Universe is designed to remain zero. <u>Gluons can't achieve this</u> which is why **virtual mass has to be responsible** for creating all regular mass-energy and is revealing evidence that the strong phase of virtual mass is indeed the source of all mass-energy created by the strong force. By always complying with these rules any amount of Matter complete with dimensional Space-time can be <u>continually</u> created.

This is how my Natural Universe begins from empty Space. By continually obeying these rules this model progresses from the virtual realm to regular Matter particles endowed with regular mass-energy and Gravity. Thus, Mother Nature found a way to create a universe from a <u>virtual realm</u>. The empty Space of virtual mass can provide Matter; however, it can also fittingly take it back. One may ask; should not the negative energy relating to virtual mass destroy the positive energy of Matter and so destroy all Matter? Yes it should and the evidence of this is a Black Hole where all Matter is crushed out of existence. If the variable delivery part of Gravity **did not have a restraining element** and applied the full strength of the attraction part here on the surface of the Earth, the Earth could be immediately crushed down to a Black Hole the size of a small coin and all of Earth's Matter would be fittingly crushed out of existence. The Earth would now be a small sphere of empty Space where all of Earth's Matter has been returned to a status quo of the <u>empty Space we began with</u>. In this way the negative energy of Gravitational Attraction which provides Gravitational Pressure can indeed destroy and so cancel out the positive energy of Matter and provides evidence of how all Matter was created from empty Space. **This model begins within empty Space and all Matter can be returned to empty Space.** I will later show why all Matter, eons from now, may degrade and return symmetry to the natural nothingness of Space.

What about the Higgs Boson? The total mass-energy of our universe is <u>required to remain zero</u>. This complies with the laws of physics and the conservation of energy law. However, the method the Higgs provides mass has no understanding to comply. At the very best the Higgs is said to be responsible for only 1% of all mass, however, related modeling still has <u>many</u> worrying problems which the world's smartest particle Physicists have yet to solve. I will address some of these problems at chapter 7.

A universe which creates all Matter from the properties of Dark matter will clearly, permanently lock Gravity to mass. One superb thing with this modeling is it will clearly explain how and why one <u>cannot source</u> regular mass without having Gravity and cannot deliver Gravity without slowing the passing of time; the direct relationship is not only most compelling, it is also easily explained and easily understood.

The positive dimensions of Space time are attracted to the <u>negative energy density</u> **(from now mostly abbreviated to <u>negative energy</u>)** phase of virtual mass. Virtual mass is found at the heart of virtual protons (Dark matter), regular protons and neutrons. Consequently, the dimensions of Space-time also flow into atoms. This is also why dimensional Space rushes into Black Holes. Black Holes are made by the Gravity from the virtual mass phase of Matter; consequently, dimensional Space also rushes into atoms of Matter. The negative energy of the deep vacuum phase of virtual mass has to be <u>continually reproduced</u> by the <u>vacuum fluctuations</u> because if they were not <u>continually reproduced</u> the phase of virtual mass would immediately fill with ground-state of dimensional Space and the attraction would <u>stop.</u>

No, all regular Matter was not likely derived from an extraordinary, unknown-to-man form of energy at a super-natural-like event of a Big Bang. If one was attempting to create regular Matter and was searching for a very plentiful resource which may provide the properties of both mass and Gravity than surely Dark matter would be your prime contender. Because Gravity physically slows time, in a way which physically slows clocks, one must physically reconcile the source of time with the physical source of Gravity which clearly physically moves objects. A ramification of beginning with an <u>unknown-to-man</u> form of energy at a Big Bang is that Scientists today cannot <u>completely</u> explain the raw physical source (origin) of Gravity or the actual, raw physical source (origin) of at least 99% of an atom's mass in a way which physically and clearly reconciles the source of mass with the source of Gravity. Neither can the source of mass and Gravity be rationally and physically reconciled with an actual source for the dimensions of <u>Space and time which is what physically gives the movement to a falling object</u> and <u>advances</u> time so as the future physically becomes the present. Thus, Gravity <u>actively</u> moves objects, passing time <u>actively</u> passes by and the strong force <u>actively</u> gives mass by <u>actively</u> agitating tiny particles at near the speed of light at the nucleus of atoms. **None of these crucial, physical like phenomena are likely achievable with a particle but are <u>rationally</u> achievable when linked to a fluctuating mechanism which clearly <u>actively</u> continually advances</u> from a fluctuation to a successive fluctuation.**

<u>Uniqueness</u> is provided by a COST unit's <u>unique</u> speed and Gravitational position within the universe. Vacuum fluctuations at the hearts of all protons and neutrons of all of our atoms provide tiny Gravitational waves which provide entangled dimensions to Space. This is a little like tuning your radio to a unique frequency and listening to your favorite song where all other stations on other frequencies are ignored. You may later listen to your favorite song on a different frequency and to you it sounds exactly the same. In the same way, regardless of one's own Dimensional Reality being <u>totally unique</u>, every unique Dimensional Reality appears self-same.

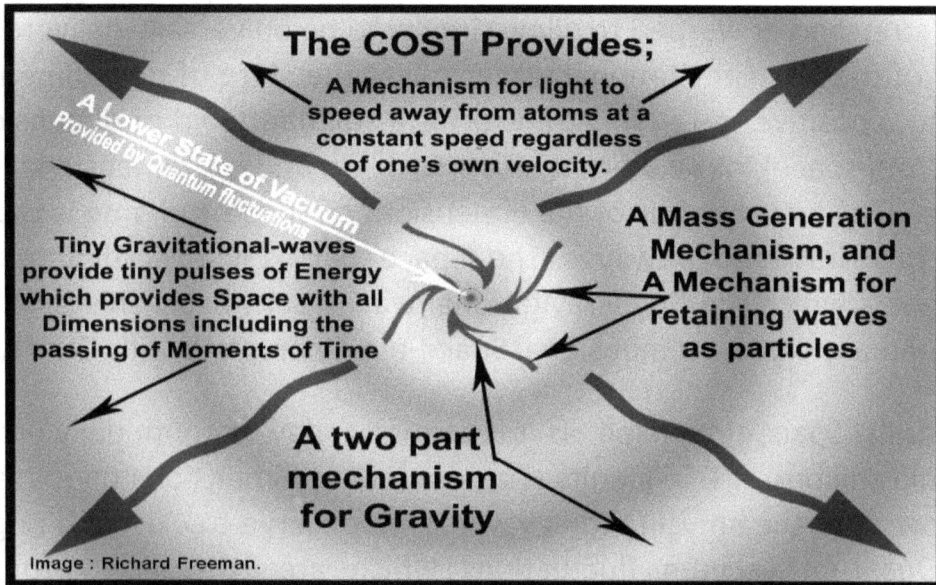

The COST Provides;

A Mechanism for light to speed away from atoms at a constant speed regardless of one's own velocity.

A Lower State of Vacuum Provided by Quantum fluctuations

Tiny Gravitational-waves provide tiny pulses of Energy which provides Space with all Dimensions including the passing of Moments of Time

A Mass Generation Mechanism, and A Mechanism for retaining waves as particles

A two part mechanism for Gravity

Image : Richard Freeman.

With my Natural Model your own unique passing rate of time is protected by entanglement. Entanglement prevents any other rate of time interfering with one's own. We will all reside at the center of our own personal universe of dimensional Space and time where, though uniquely different, time dilation will allow all to appear self-same for all. One truly lives within one's own unique rate of time. Scientist know it is true, they tell us the sums confirm we all, at all times, reside at the center of the universe which we all observe uniquely different based on our speed and Gravitational position in the universe. Within a Natural Universe we all reside at the center of our very own unique, entangled, dimensional universe of Space and time. Within a Big Bang universe the expansion of the universe is measured at the rate the Galaxies are accelerating away. Within a Natural Universe the dimensions of Space-time speed away from all atoms at the speed of light and Galaxies are being accelerated away by Outer Gravity.

Tiny Gravitational waves expanding from trillions upon trillions of new COST units now permeate all of reachable Space, eventually crisscrossing each other, weaving an expanding, invisible, woven fabric of dimensional Space and time, a fabric which now has an all important structure which can be stretched or warped to allow Gravity to be universally delivered. The energy of the tiny Gravitational waves raises the state of now dimensional Space to a positive state. This transforms a stagnant void of dimensionless Space (Anti-space) into a vibrant universe of expanding, dimensional Space and time. The woven fabric of the dimensions of Space and time is provided by the tiny Gravitational waves. The 'Space' part is just the dimension<u>less</u> nothingness which the tiny Gravitational waves permeate. This is how this model actually creates dimensional Space-time, complete with Gravity.

Dimensional Space-time itself, as it is termed today, is continually provided by untold trillions of COST units. The expanding dimensions of Space-time will allow all of the amazing things which we observe to exist and originate from Quantum vacuum fluctuations which are consistently found in theories and mathematical calculations. Evidence of Quantum vacuum fluctuations can be observed with the Casimir effect where the effect from Quantum vacuum fluctuations on two parallel plates causes a slight but measurable force pushing the plates together. [014] The Casimir effect: a force from nothing. Even a <u>perfect vacuum</u> at absolute zero has fluctuating fields known as "vacuum fluctuations", the mean energy of which is said to correspond to half the energy of a photon. The Cost in its Dark matter state acts as a virtual proton. A virtual proton is a regular proton <u>without</u> the <u>three</u> main valence quarks which provide a regular proton <u>its properties</u>.

It is as if a COST unit's size is infinitely large, as its working, beating heart remains at the center of its self-created, unique universe of dimensional Space and time. One can now argue that the true size of a single COST can be measured on a universal scale, however, this model has focused on the tiny, Quantum scale at its very center. A new Natural Universe created from an expanding fabric of dimensional Space and time has now begun.

Being based on a Quantum vacuum fluctuation is what clearly provides our Dark matter with its invisible, ghost like, virtual properties and is why nobody has been able to directly detect Dark matter. One only needs to accept the known properties of Dark matter. No known, true particle can mirror these properties. However, without the vital mechanism of the COST, an apple would not fall from a tree, moments of passing time would be reduced to an instant, the atom would have no mass or energy, atoms would be unable to supply the fundamental forces and an atom's particles may reside as waves without displaying particle properties of any kind.

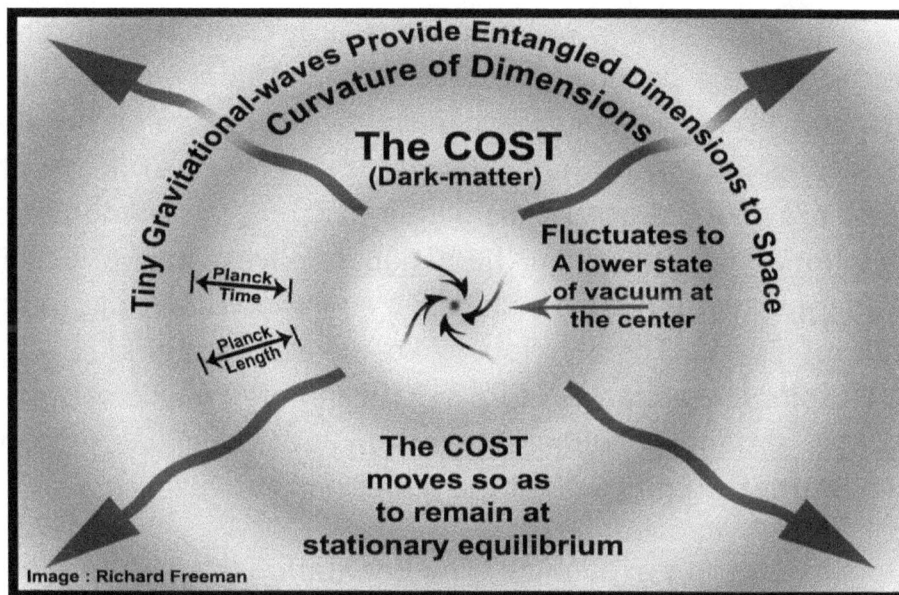

The Gravity of the COST: (Gravity is explained in depth at chapter 11)

How can Dark matter be totally invisible but account for the largest portion of the universe's Gravity? Answer; Gravity is provided by a tiny phase of invisible empty Space. The vacuum fluctuations of the COST fluctuate to a deep vacuum phase of <u>negative</u> virtual mass (empty Space) which mimics a rapidly appearing and collapsing Black Hole which is ideal for creating very tiny, <u>positive</u> Gravitational waves which provide entangled dimensions to Space. For Gravity, the <u>same phase</u> of <u>negative</u> energy of <u>other COST units attracts</u> the massless, tiny, <u>positive</u> Gravitational waves which are providing entangled dimensions to Space. The ground state of the outer void of Anti-space is Space without dimensions and is both negative energy and negative pressure in relation to the expanding dimensions of Space-time. The outer void also attracts the dimensions of Space-time and in doing so expands the dimensions of all Space-time at the speed of light. **At the speed of light <u>distance to the void is meaningless</u>.**

The COST in its Dark matter state is a <u>virtual proton</u>. Tiny positive Gravitational waves expanding from a virtual proton <u>provide entangled dimensions</u> to Space. The wavelength of a Gravitational wave provides a Planck length and the passing of a wave provides Planck time. The Gravity of a virtual proton is a two part mechanism consisting of a strong attraction part and a <u>variable</u> delivery part. The <u>negative energy</u> of a tiny phase of virtual mass provides the <u>strong</u> attraction part of Gravity and the <u>positive</u> tiny Gravitational waves provide the <u>variable</u> delivery part of Gravity. The <u>negative energy</u> of <u>virtual mass</u> attracts, in a whirlpool fashion, the <u>positive</u> dimensions of Space. The whirlpool like attraction provides the virtual proton with spin. The attraction to a negative energy phase of virtual mass <u>stretches the expanding dimensions</u> from other virtual protons. The 'stretching' warps the curvature of the dimensions of Space-time.

When a virtual proton's expanding, tiny Gravitational waves which provide dimensional units of Planck length are stretched, their Planck value is maintained. [017a] The amount of <u>energy that Gravitational waves carry is fixed</u> as they travel perpetually through space. Planck value is maintained because the Gravitational waves maintain a single, <u>passing pulse of energy</u> from the <u>fluctuation</u> event which created them. This causes a <u>contraction of the stretched distance</u> covered by a stretched wave to a Planck length which causes Gravitational length contraction. Gravitational length contraction now causes a virtual proton to <u>move</u> to maintain a position of <u>stationary equilibrium</u> where it remains **restrained** at the center of <u>very tiny units of expanding energy.</u> This movement is Gravity. Since dimensional Planck values and a position at <u>stationary equilibrium</u> are maintained the virtual proton moves <u>without feeling movement</u> or acceleration. **The <u>variable delivery part</u> of Gravity has a <u>restraining part</u> which operates in the <u>opposite direction of a falling object</u> causing Gravity to be <u>very weakly delivered</u>. This solves a long time mystery of why Gravity is <u>far weaker</u> than other forces.**

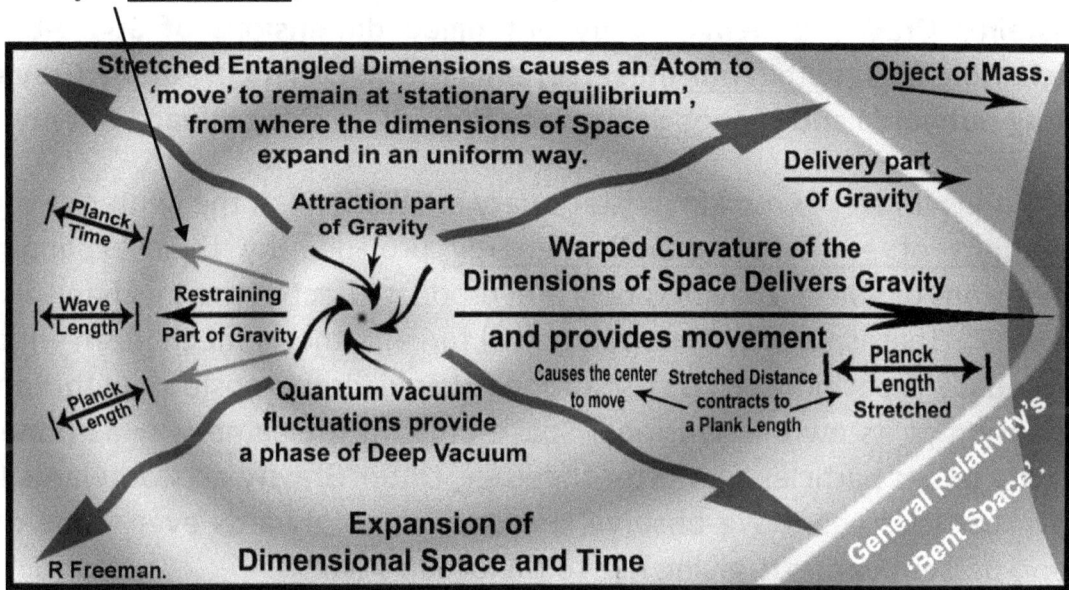

A stretched Gravitational wave takes longer to pass which slows the passing of time and is why delivering Gravity slows time. An object now moves in the same direction which the dimensions have become most stretched. A stretched Gravitational wave maintains Planck value because it retains the same 'single pulse of energy' from the event from which it was created.

The far away outer <u>dimensionless</u> void of Anti-space is naturally void of the energy of Gravitational waves so is a lower state of vacuum than that of the ground-state of <u>dimensional</u> Space. This means the void provides a massive resource of negative energy. At the event horizon of a Black Hole Gravity drives all dimensions, at the speed of light, <u>one way towards</u> the massive, Inner Gravity of the Black hole. <u>Away from a Black Hole</u> the infinitely large far away outer void <u>similarly</u> attracts the tiny dimensions which is <u>why</u> dimensions generally expand in all directions at the speed of light. However, the expanding <u>energy carrying</u> Gravitational waves which are providing dimensions to Space naturally become slightly more stretched in the direction closest to the attraction of the outer void. This provides Outer Gravity which acts in the direction of expansion. **Note: As this is an outer void of empty Space, it <u>contains nothing</u> which may crush our universe. Einstein's Gravity is delivered by entangled dimensions of Space-time originating from <u>within our universe</u>, thus the void can only apply Gravity acting outwards and <u>cannot deliver</u> any form of crushing inward Gravity.**

COST units 'move' to remain at 'stationary equilibrium' at the center of <u>very tiny units of eternally expanding energy</u> provided by their expanding dimensions of Space. Gravity is only a byproduct effect of the mechanism of the COST. That is, Gravity is not an actual force, rather a mechanism which continually repositions a COST's stationary equilibrium position within the expansion of its own dimensional Space-time. In the same way as (Inner) Gravity brings particles of Matter together, Outer Gravity acts outwards to move particles apart. We observe Gravity acting outwards every time the moon passes overhead, giving rise to the ocean's tides.

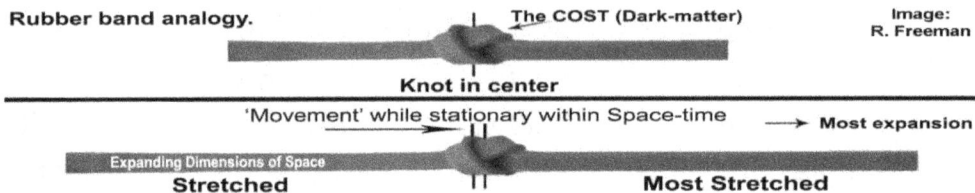

Above image; notice with this strip of rubber band analogy for Gravity, how there may be a large amount of expansion and a small movement of the knot. The knot resides at 'stationary equilibrium', that is, the knot has <u>not moved</u> through the rubber. Stretching (expanding) the strip of rubber band slightly more in one direction has both given the knot movement and, at the same time, actually allowed the knot to remain completely stationary, at a center of 'uniform' expansion within the expanding rubber.

Because the knot (a COST unit / virtual proton) has <u>not moved 'through'</u> the rubber (dimensional Space), one may call this 'stationary movement'. That is, the knot remains <u>stationary within the rubber</u> in the same way as a virtual proton remains stationary within its expanding dimensions of Space-time. Without expansion (the stretching of the rubber band) the process stops. Now release the least stretched end of the rubber band. Now the knot seemingly <u>instantly</u> speeds away; welcome to <u>Quantum Gravity</u> (see chapter 13).

The <u>first created</u> virtual protons, which began to accelerate first, may catch up with the newly created virtual protons ahead. This may occur because the <u>light speed</u> expansion of the warped dimensions of Space-time is activating new COST units, while a virtual proton lags far behind because it is trapped at 'stationary equilibrium' at the center of its own expansion of the dimensions of Space-time. I will call this process 'Jump Start'. Jump-start may possibly cause a hole at the center which, to a certain extent, Inner Gravity may regularize. However, this hole at 'cosmic center' may possibly grow with expansion and evidence of this hole would be a very large, cold super-void. **The existence of a mysterious, unexplainable, very large, cold void has now actually been detected within the Cosmic Microwave Background image.** The imprint of this super-void has grown to now be at least 1.8 billion light years across. A Natural Universe may have begun at the center of this enormous void where one lone virtual proton, the one which began it all, remains at the very center while all other virtual protons have been driven outwards by Gravity acting outwards.

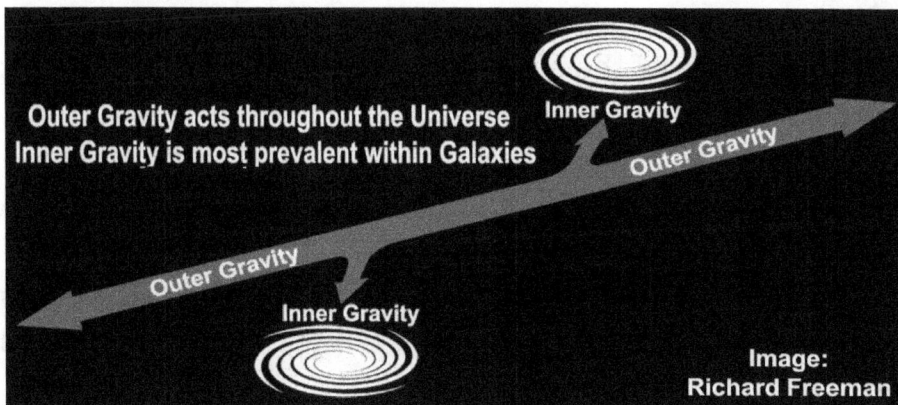

Outer Gravity acts throughout the Universe
Inner Gravity is most prevalent within Galaxies

Image: Richard Freeman

A battle between Inner Gravity and Outer Gravity creates a process a little like pulling partially melted cheese apart. Virtual protons desperately cling together, while Outer Gravity attempts to rip them apart, creating a honey comb of cosmic web like filaments which are <u>bulging at intersections</u> with accumulating super large clouds of virtual protons.

Within the largest clouds a forming dense center begins to rotate where Gravity will eventually cause a Gravitational collapse which creates a Super Massive Black Hole. The COST in this Dark matter state is a virtual proton.

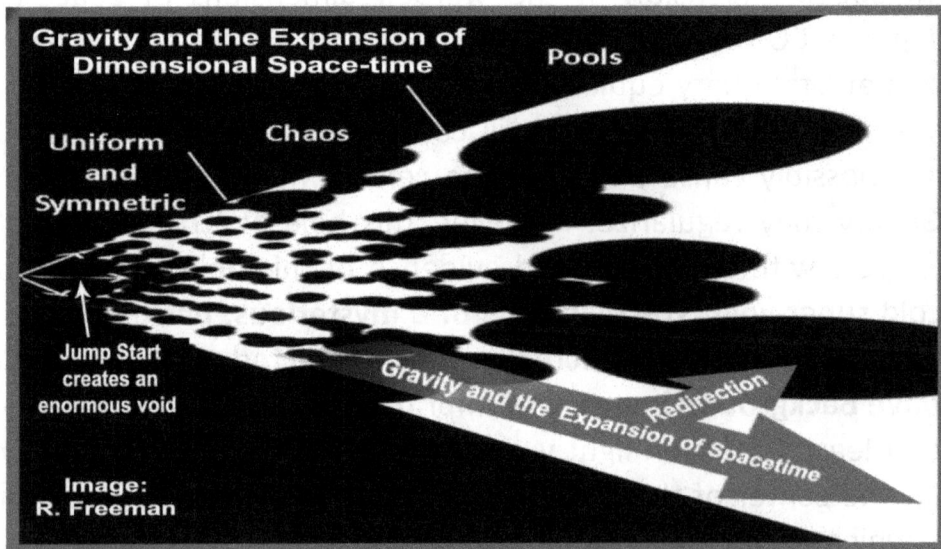

The dimensions of Space being stretched from all directions towards the accumulating clouds of virtual protons now prevents these clouds from being split apart. Some of the effect of Outer Gravity, which first separated virtual protons into web like filaments, is now locally redirected towards the cloud like pools of virtual protons (Dark matter). Both Outer and Inner Gravity are <u>driven in unison</u> by the very same exact mechanism. Actually, modeling has now created an uninteresting dark universe consisting of only invisible virtual protons (Dark matter), invisible dimensional Space and time and Gravity. However, modeling has now laid the essential foundations for the <u>unwavering laws which will rule a Naturally Forming Universe</u>.

6 CREATING SUPER MASSIVE BLACK HOLES.

The COST in its Dark matter state is a virtual proton. Virtual protons are first brought together by Gravity to create super large clouds. Driven by Gravity the center of the clouds becomes dense. Squeezed by Gravity, virtual protons are further <u>amplified</u> to carry their <u>strongest Gravity which prevents</u> their tiny Gravitational waves escaping. This allows virtual protons to come tightly together which causes a central Gravitational collapse which forms a Super Massive Black Hole. Consequently, Super Massive Black Holes are made from the virtual mass of virtual protons.

We will explore known physics while applying the concepts of a Natural Model. The reality is <u>present day physics cannot cope</u> with many of the ramifications of Black Holes. When it comes to Black Holes, <u>nobody knows</u> what the final accepted answer will be. The model will avoid the use of the notion of infinity and its resulting, erroneous singularity. In doing so, the modeling will expose some <u>unexpected properties</u> of Black Holes. At 2.7 million light years away M33 X-7, with its <u>15.65</u> solar masses, is the largest known stellar Black Hole and at 10.4 billion light years away TON 618, with its <u>66 billion</u> solar masses, is the largest known Super Massive Black Hole.

Scientists have a good understanding of how <u>stellar</u> Black Holes are formed, however, Astrophysicists will readily express that they do not understand how Super Massive Black Holes were first formed and say that this remains one of the biggest mysteries of today's astrophysics. Because the existence of Super Massive Black Holes at the center of Galaxies and at a time when the universe was too young to form such monsters in accordance with the current understanding of Matter created from a Big Bang, it has been suggested that they were created by *"strange unseen forces"*. Stellar Black Holes are formed when the <u>inner</u> core of a very large star exhausts most of its fuel supply and begins to make iron which costs energy. Without excess energy to fight its crushing Gravity the mass of the core suffers a sudden, inward collapse of a Gravitational implosion. The collapse triggers a type 2 supernova, the term used for an exploding star that blasts the outer parts of the star into Space and, if massive enough, the star's central <u>dense core</u> suffers a Gravitational collapse which produces a central stellar Black Hole.

There is evidence that Super Massive Black Holes formed 'before' the first stars but strangely science seems to be fixated on beginning the process by first forming stars. Because of their brightness Quasars are among the farthest 'back in time' observable objects within the universe. Because Super Massive Black Holes are said to 'power' Quasars, Super Massive Black Holes must have existed before Quasars. A Super Massive Black Hole can power a Quasar but a Quasar is not known to make a Super Massive Black Hole. Quasars are most prolific towards the beginning of the universe which means Super Massive Black Holes were <u>abundantly forming</u> even before the observable Quasars from near the very beginning of the universe. **This is exactly what is observed and is exactly what this model requires.**

[092] Quasar (J1342+0928) is over 13 billion light years away, placing its existence at just 690 million years after the said Big Bang. This Quasar is said to be especially interesting because it comes from a time when the universe was just beginning to emerge from its dark ages. Scientists say they are puzzled to how the mass of the Quasar's Super Massive Black Hole is comparable to the mass of Super Massive Black Holes that are around today and say it challenges their theories of Black Hole formation. Black Holes are theorized to feed on Matter and grow over billions of years. How can it be that these relatively very young Super Massive Black Holes from

near the beginning of the universe are also some of the most massive? Scientists suggest Super Massive Black Holes may have begun as stellar Black Holes and grow over time by consuming Matter.

Even more difficult is to mathematically understand a Black Hole's said singularity where the 'ramifications of infinity' of the seemingly less than zero dimensionless state of an <u>endlessly deeper</u> singularity are truly incalculable. A model which is capable of completely understanding the baffling 'zero dimensionless state' of a singularity will be imperative if one is to fully explain Black Holes. An <u>infinitely small</u> singularity is mystifying because today's related theories of physics are said to be incomplete.

The Gravity of mass makes Black Holes. When both the source of Gravity and mass are today bound in mystery it becomes most difficult for Scientists to fathom how Super Massive Black Holes originally formed at the beginning of the universe. However, this model has workable modeling for the source of both <u>Gravity and mass</u>. This will allow one to use this same modeling to show how these first Super Massive Black Holes were most likely formed. Modeling will sensibly show how the same state of Space which provides a Black Hole's mass <u>also sensibly provides an atom's mass.</u>

Two most important and obvious questions to ask are why Super Massive Black Holes are prevalent within our earliest universe and why are Super Massive Black Holes <u>only observed to be at the center of Galaxies</u>? Unlike a Big Bang beginning, the answer is again straightforward, simple logic for a Physical Reality Model. Super Massive Black Holes were created first and then the Super Massive Black Holes provided the means to create a Galaxy's own Matter which is clearly why all regular Galaxies, including our home Milky Way Galaxy, have Super Massive Black Holes at their center.

How to create a Super Massive Black Hole: Although Super Massive Black Holes are prevalent within our very earliest universe, very little is known about how Super Massive Black Holes first came into existence and so this is an ongoing topic of intense investigation by today's science. How and why did the universe produce these invisible monsters at a time when the universe was in its infancy? Science agrees that the first Super Massive Black Holes must have played a large part in the early evolution of Galaxies. However, the path of Matter from a Big Bang has made the early evolution of Galaxies difficult and complex to explain, so exactly how this early evolution took place remains surrounded by mystery. Extreme Gravity can produce a Black Hole but the actual source of Gravity is today also a mystery, so precisely what it is at a <u>fundamental level</u> that makes a Black Hole can also be said to be another unsolved mystery.

Astronomers have long been mystified by how such massive, complex structures as Galaxies, with centers harboring Super Massive Black Holes, which are often powering Quasars, could have evolved so bewildering quickly, from <u>nothing more than atoms of mostly hydrogen</u>, at a very early time within the Big Bang universe. Scientists know the ferocious expansion (inflation) of the instant Big Bang universe would override the weak Gravitational influence of individual atoms of hydrogen.

There are several theories to choose from all of which describe different scenarios for the formation of Super Massive Black Holes, which have been commonly detected by their Gravitational influence at the hearts of Galaxies. The most common theories proposed by science suggest Matter first clumped together. The only mechanism which can possibly bring copious amounts of Matter together is Gravity. Gravity and Matter has a snow ball effect. As Gravity increases it attracts even more Matter until Matter has concentrated enough to form the first stars. **Today, there is <u>no known or observed process</u> for Matter, initially comprised of mostly hydrogen, to avoid the star making process in order to make a Black Hole.**

Once a large star is formed strong solar winds blow out from the star, preventing the star from obtaining more hydrogen and becoming more massive. In an attempt to explain the existence of large Black Holes within our earliest universe Scientists have hypothesized the first stars may have somehow been super-sized at <u>100,000 times</u> more massive than our Sun. Scientists are hoping the recently launched James Webb Space Telescope will directly reveal these super-size stars. Such stars are far larger than any star ever observed and drastically challenge current theories of stellar evolution. R136a1 is the most massive known star in the universe. With its 315 solar mass R136a1 is plainly puny compared to the super-sized 100,000 solar mass of the first stars hypothesized by Scientists as a possible way to progress to the Super Massive Black Holes found at the hearts of Galaxies. Such a proposal suggests that these very large, super-sized stars will go hypernova at the end of a short lifetime, which results in the star's core collapsing and forming a super stellar-size 'seed' Black Hole. The Natural Model has <u>no need</u> for these hypothesized, super-sized stars.

Within a Big Bang universe the largest Stellar Black Holes known to exist would be required to somehow grow over a billion times in order to match some of the large Super Massive Black Holes found only at the hearts of Galaxies. Consequently, even the largest proposed 'seed' Black Holes are required to either swallow an enormous amount of Matter or collide and combine with numerous other seed Black Holes. Today's theories portray Galaxy collisions, which combine Black Holes at the center of evolving Galaxies to finally create Super Massive Black Holes. Collisions do happen, however, within my Natural Model Galaxies are generally moving away from each other due to Outer Gravity. New research reveals very large Black Holes are likely to just orbit each other rather than combine. Other proposed theories include massive dark stars and short lived stars made from hypothetical 'weakly interacting massive particles', or 'WIMPs', which may mirror the properties of Dark matter and somehow came into existence not long after the Big Bang but such a particle would not explain the source of Gravity. The exact nature of a proposed WIMP is not presently known and WIMPs are not even predicted by the standard model of particle physics. [022] The 'WIMP Miracle' Hope For Dark Matter Is Dead.

[089] **The big problem is such theories generally require many billions of years to grow and evolve Super Massive Black Holes and the relatively very short time frame from the proposed Big Bang to the first observable Galaxies means that such theorized evolution is not feasible.** Millions of Stellar Black Holes are said to exist within the Galactic Disk of our Milky Way Galaxy alone but none have apparently grown beyond the size of a stellar Black Hole. There is likely a good reason why Stellar Black Holes do not readily combine to create intermediate or Massive Black Holes.

Very few intermediate Black Holes have been authenticated and there is commonly a huge gap between the largest stellar Black Holes and the smallest Super Massive Black Holes. Such a gap naturally suggests that Super Massive Black Holes were probably formed by a different method. The existence of Super Massive Black Holes at the hearts of first Galaxies indicates that they formed at a time before the evolution of first stars. As science observes deeper into Space and so farther back in time issues of events happening too quickly for them to occur from a Big Bang beginning is becoming more problematical and more common. So one should ask; why do these objects appear to be forming too quickly to agree or to be even possible with the currently proposed models for our earliest universe?

The facts point to numerous Super Massive Black Holes forming very quickly at around the beginning of the universe which has proved to date to be unexplainable from a Big Bang beginning. Instead of many millions or billions of years a Natural Model can create a Super Massive Black Hole in probably weeks or possibly a few days and also avoid the intricate modeling currently proposed by Scientists. Simply allowing Gravity to suppress the tiny Gravitational waves, which are providing entangled dimensions to Space, will allow COST units to come tightly together and cause a near speed-of-light collapse which creates Super Massive Black Holes. I love clever and simple. An efficient and problem free process of COST units simply coming tightly together with Gravity completely avoids the intricate, very time consuming process of forming theoretical and never observed super large stars and exploding them as 'super nova' and somehow growing Super Massive Black Holes from combining many 'seed' stellar Black Holes.

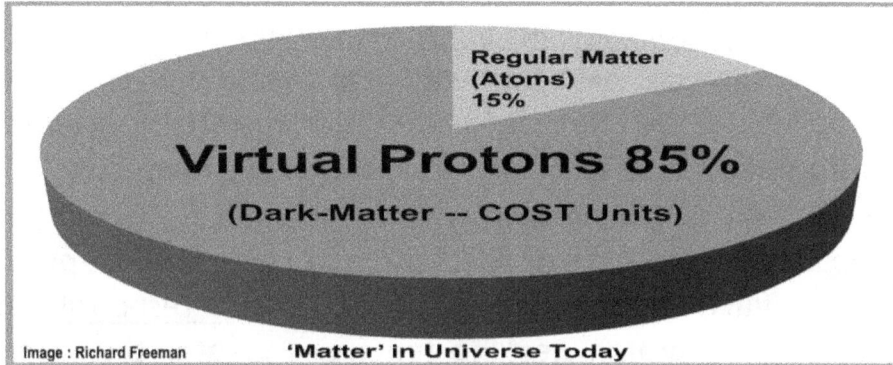

Image : Richard Freeman   'Matter' in Universe Today

Although COST units, in their Dark matter state, cannot form stars or any useful observable object they can easily create Super Massive Black Holes. There are several very good reasons why COST units are excellent for forming Super Massive Black Holes. The universe today contains far more COST units (Dark matter) than all other Matter combined. Unlike hydrogen atoms, the process is not severely limited in size or time by a star making process. The COST fluctuates to a phase of virtual mass which is the source of Gravity. The tiny phase of virtual mass mirrors a minuscule, rapidly collapsing and reappearing, Black Hole. The COST in its Dark matter state acts as a virtual proton which will provide protons and neutrons with a phase of deep vacuum which effectively mimics a rapidly 'fluctuating' minuscule, rapidly collapsing and reappearing Black Hole and is a phase which I have called virtual mass. The (negative) energy of a 'fluctuating' minuscule Black Hole will be required so as it has the clout to provide the near light speed agitation of a proton's particles, the energy of which represents a proton's regular positive mass (mass-energy) which is different to virtual mass. Virtual mass supplies Gravity and will create regular mass, consequently, wherever one finds regular mass, one will also find Gravity.

Consider; a neutron star is so dense that one teaspoon of neutron star material weighs around four billion tons which is like more than 26,500 ocean liners as large as the Queen Mary 2 all in one teaspoon. The Gravitational Pressure is so great that a neutron star is just a small step away from becoming a Black Hole. [002] Scientists have discovered that the pressure <u>inside of a proton peaks ten times greater</u> than at the <u>heart of a neutron star</u>. To provide such tremendous pressure one obviously needs something which is a step further than a neutron star. The very next step from a neutron star is a Black Hole or as in this case a minuscule fluctuating Black Hole. By providing a phase of virtual mass the COST will effectively provide a minuscule 'fluctuating' Black Hole at the center of protons and neutrons. The 'fluctuating' <u>minuscule</u> Black Hole <u>is responsible</u> for the pressure inside of a proton peaking ten times greater than <u>inside of a neutron star</u>. It undoubtedly requires something very special to apply these incredible 'forces' stronger than at the <u>heart of a neutron star</u> and there can hardly be anything more capable and <u>fitting</u> than an exceptionally tiny 'fluctuating' Black Hole. However, before we create regular protons and neutrons to make atoms we first need to create Super Massive Black Holes and the COST by fluctuating to a negative energy <u>phase</u> of deep vacuum which mimics a very <u>tiny</u> Black Hole will obviously provide the perfect tool.

Today it is recognized regular mass is a mirror of the <u>energy</u> acquired by tiny <u>particles</u> being both agitated at near light speed and held together by gluons <u>within</u> protons and neutrons. However, inside of Black Holes <u>all particles</u>, for all sense and reasoning, <u>crush themselves out of existence</u> at an infinitely diminishing point of <u>zero volume,</u> called a singularity. So, how can this same mass remain inside of Black Holes? Short answer; it cannot which is why the source of regular mass is required to be something which can both agitate and constrain the particles within protons and neutrons but cannot be crushed out of existence inside of a Black Hole. Thus, <u>gluons cannot be used as a source of mass in this way</u>. This is why the virtual mass phase of a virtual proton is ideal for creating regular mass and also an ideal source of Gravity. Virtual mass is a deep vacuum phase of empty Space which <u>cannot be crushed out of existence</u> and is what will <u>remain behind</u> to provide mass and Gravity when all particles are crushed out of existence in a Black Hole. [023] **The fact that inside a Black Hole all particles are crushed out of existence leaving only a sphere of pure empty Space, which is also a sphere of pure mass and Gravity, is all the evidence one needs to know that empty Space is indeed truly the source of both mass and Gravity.**

Inside of a Black Hole all time stops, consequently, whatever collapses to a Black Hole must show why the collapse caused the passing of time to stop.

The only item required to make a massive Black Hole is Gravity, which happens to be <u>the only known property of Dark matter</u>. Given the fact that the universe contains so much more Dark matter (virtual protons) than ordinary Matter it appears to be only rational, common sense that the first Black Holes must have formed directly from virtual protons and many were Super Massive for the reason that there was an immense supply of virtual protons confined to an early compacted universe. The true fact that Super Massive Black Holes may be billions of times more massive than a stellar Black Hole which formed from ordinary Matter, surely reinforces this concept. When forming Super Massive Black Holes, our <u>nebula like, super large clouds of virtual protons</u> can completely avoid the restrictive and very time consuming procedure of Star making, a procedure which is only known to form Black Holes millions of times <u>smaller</u> than the kind of <u>Super Massive Black Holes</u> required to power the hearts of Galaxies. Unlike the regular protons of hydrogen <u>atoms,</u> dense clouds of virtual protons <u>cannot</u> fuse together to make helium nuclei so as to begin fusion to create stars, which actually provide virtual protons an easy and passive <u>direct short-cut</u> to forming Super Massive Black Holes. Because virtual protons can so easily form SMBHs at a time when the universe requires Super Massive Black Holes provides evidence that this modeling for Dark matter is correct.

Virtual protons can become regular protons, and their <u>same</u> vacuum fluctuations will then provide mass and Gravity within all atoms, stars, neutron stars and <u>finally within stellar Black holes</u>. This model will show how virtual protons can also <u>directly collapse</u> to make Super Massive Black Holes. With a Natural Model the first Super Massive Black Holes are constructed, like their smaller, stellar Black Hole cousins, by a Gravitational collapse. **Consequently, one only needs to clearly show what happens to the mechanism of a virtual proton to cause a Gravitational collapse.**

For example, if one could collapse the Earth down to the size of a small coin, planet Earth would form a Black Hole. [023] Now a Black Hole, the only trace left of Earth is its mass and Gravity. A Gravitational singularity is a position in Space-time where a Gravitational field of a massive body becomes <u>infinitely</u> small and where reliance on coordinates is loss. At such a place dimensions are meaningless and the passing of time stops. Since all ordinary Matter is crushed out of existence, for all sense and reasoning, all Black Holes are no more than <u>a sphere of empty Space</u> which attracts massive volumes of ground-state dimensional Space and provides Gravity.

Why does dimensional Space naturally flow into Black Holes? The flow of dimensional Space is driven by the amplified vacuum fluctuations of virtual protons, which phase from positive energy to a tiny deep vacuum state of virtual mass (-). The negative pressure of virtual mass is a negative energy state which attracts the positive dimensions of Space-time. The virtual mass phase acts as <u>a hole</u> or abnormality in the ground-state of dimensional Space which the positive dimensions of Space are naturally attracted to.

How virtual protons first form a Galaxy's Super Massive Black Hole:

To understand how a very large cloud of virtual protons collapses to form a Super Massive Black Hole I will model the following on the known way a large, cold cloud of gas collapses to form a star. Because virtual protons <u>cannot make a star is exactly why</u> the process continues until it forms a very Massive Black Hole. Expanding from virtual protons 'entangled dimensions' in the form of tiny Gravitational waves provide tiny pulses of energy which naturally provide cushioning dimensions or distance between virtual protons. This <u>gently prevents</u> virtual protons from crashing tightly together. At first virtual protons, due to their own Gravity, generally just hang around their own kind more like bubbles creating foam. Atoms have the <u>regular mass</u> of regular particles which provide objects with the regular '<u>momentum</u>' to crash tightly into each other. Virtual protons have a phase of virtual mass but this form of mass remains virtual. To give virtual protons the clout to aggressively come tightly together to create a Super Massive Black Hole virtual protons <u>will require another boost of Gravity</u>.

[015] **The research by Prof. Alfred Leitenstorfer and team at the University of Konstanz is <u>again most valuable</u>.** Their <u>experiments revealed</u> when they '<u>*squeezed the vacuum*</u>' of Space around Quantum vacuum fluctuations their energy phase became louder and a phase '<u>*lower than the ground state of empty space*</u>' **greatly increased its volume into deep vacuum** which for this model is a tiny, negative energy state of deep vacuum I have called virtual mass which **provides the strong attraction part of Gravity.**

Gravity has now restrained virtual protons (Dark matter) to massive web-like filaments where <u>at intersections</u> virtual protons <u>gather and accumulate</u> into **<u>extraordinary large</u> clouds**. Driven by Gravity, virtual protons <u>migrate towards</u> the center of a very large, super cloud which causes the center to become denser and <u>begin to rotate</u>. Rotation now flattens the central area 'inside' of a cloud <u>into a central, thick 'dense' disk</u>. Exposed to increased motion and further <u>squeezed by warped Space-time</u> virtual protons continue to gain energy causing their phase of **virtual mass, as <u>observed in experiments</u>** [015], to <u>increase in volume</u>, which increases their Gravitational attraction. **This is now a far more aggressive form of Dark matter armed with much stronger Gravity which may reach as high as a regular proton.** Surrounded by other virtual protons their <u>combined</u> phases of virtual mass, which are providing the attraction part of Gravity, begins to suppress, slow and eventually stop the expansion of their <u>tiny Gravitational waves,</u> which are providing tiny <u>pulses of energy</u>, that provide entangled dimensions, including the passing of time to Space. This slows and eventually <u>stops their passing of time</u> which happens because <u>Gravitational waves</u> travel at the speed of light and the escape speed of anything inside of a Black Hole <u>exceeds the speed of light</u>. Now the tiny Gravitational waves, which were providing entangled dimensions and passing time to Space, cannot escape from virtual protons. **Now without the <u>energy</u> of their expanding, <u>entangled</u> dimensions to keep them separated, virtual protons are now finally free to use their deep vacuum phase of <u>Gravitational attraction</u> to come tightly together which causes a central Gravitational collapse.**

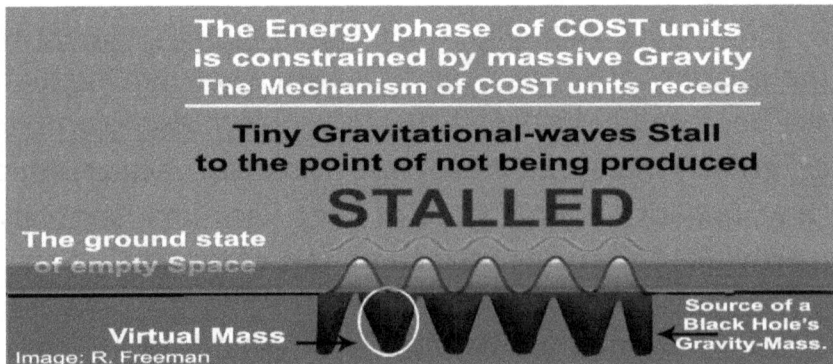

The collapsed <u>most dense</u> central regions of the super clouds of virtual protons now forms a Super Massive Black Hole (SMBH). The Gravitational collapse happens at near the speed of light and naturally leaves behind a configuration of a central SMBH, a thick 'dense' disk of Dark matter completely surrounded by a <u>less dense</u> outer halo of Dark matter. This first creates a <u>disk shaped Dark Galaxy</u>. Virtual protons from the dense, dark disk <u>continue to migrate</u> to the center of the Galaxy where virtual protons will be <u>used</u> within Matter creation cores at the poles of the central SMBH causing the dense, dark disk to <u>lose density</u> and mass as much of the dark disk is <u>replaced</u> with a disk shaped, bright Galaxy of stars. **The outer <u>halo</u> of <u>original density</u> Dark matter <u>remains today as obvious evidence</u> of how virtual protons were <u>sourced</u>, so removed, from a central area of a large cloud of virtual protons to create a Galaxy of stars and its central SMBH.**

**How COST Units (virtual protons) form
Super Massive Black Holes.**

The dimensions of
Space-time expanding
away normally prevents
COST Units from
combining like atoms.

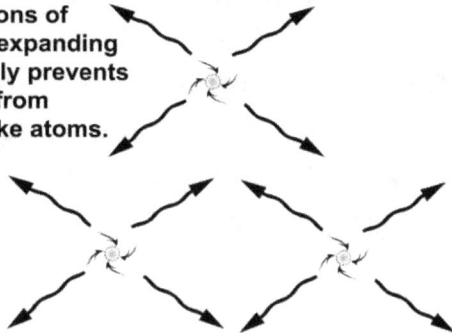

The transition from COST Units
to units of only virtual mass
causes a Gravitational Collapse

The expanding
dimensions of
Space-time slow
which allows
COST Units to come
closer togerther.

Slows time.

Massive Gravity prevents
the excape of the tiny
Gravitational-waves which
are providing dimensions
to Space. Space is now
dimensionless.

Creates Fluctuating Units
of virtual mass which are
a lower state of Deeper
vacuum than that of the
ground state of empty space.

Without dimensions
units of virtual mass are
free to combine to
provide the strongest
form of Gravity

The passing of
Time stops

A resulting Gravitational
Collapse forms a
Super Massive Black Hole

Trillions x trillions of units of virtual mass
combine to form a 'dimensionless area' which mirrors
a singularity and creates a Super Massive Black Hole

Image created by
Richard Freeman

But would it be possible to find evidence that Galaxies like our Milky Way now reside within a <u>depleted</u> Dark matter disk surrounded by an outer halo? [114] See: A dark matter disk in our Galaxy. *An international team of scientists say their supercomputer simulation has exposed the presence of a disk of Dark matter within our own Milky Way Galaxy but the <u>rotating</u> dark disk is (today) only about <u>half</u> of the density of the Dark matter of the outer, <u>non-</u>*

rotating halo. Ask why the dark disk appears to have lost at least half its density? Answer; originally denser than the halo, the lost Dark matter has been used to create the Galaxy's regular Matter. **As this closely matches the known configuration of a Galaxy's <u>essential</u> Dark matter Gravitational Scaffold, it is unlikely that these Gravitational Scaffolds could have been created by any other method. Note;** to finally solve how Super Massive Black Holes formed, modeling <u>requires</u> Dark matter to be a virtual proton.

The original <u>size</u> of the cloud of Dark matter (virtual protons) naturally provides a <u>relationship</u> between the mass of the <u>newly</u> formed Super Massive Black Hole and the mass of the Dark matter halo. [See 024] **Scientists have <u>actually discovered</u> this distinct relationship but are mystified to why it should exist. Obviously, for a Natural Universe this is no mystery.**

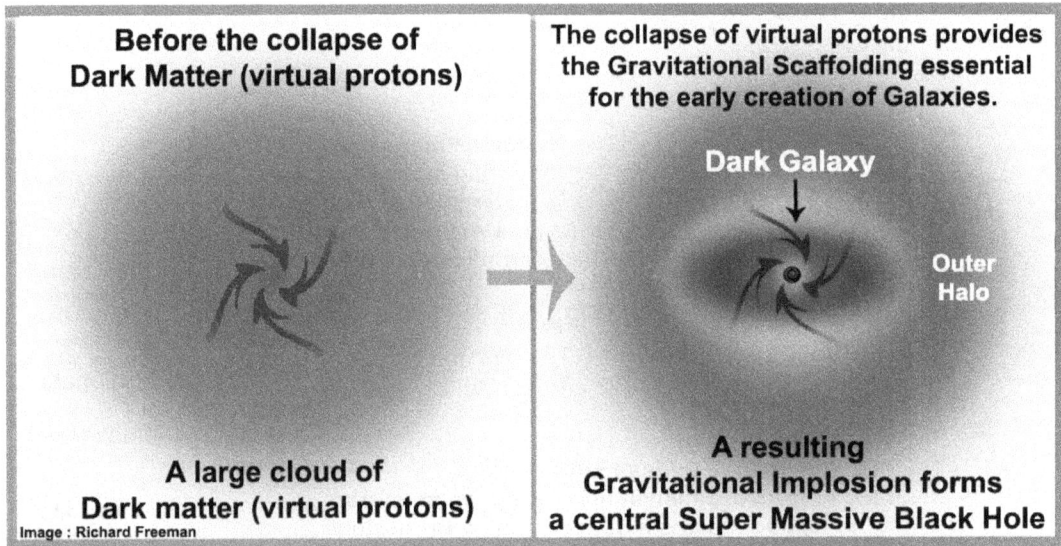

Before the collapse of Dark Matter (virtual protons)	The collapse of virtual protons provides the Gravitational Scaffolding essential for the early creation of Galaxies.
A large cloud of Dark matter (virtual protons)	A resulting Gravitational Implosion forms a central Super Massive Black Hole

Image : Richard Freeman

For a Natural Universe the reasoning is straight forward; larger clouds of virtual protons naturally created larger central **Super Massive Black Holes** (SMBHs) surrounded by larger amounts of virtual protons which will create larger amounts of regular protons for creating larger amounts of hydrogen atoms which <u>eventually</u> create larger Galaxies of stars. The resulting relationship has been observed in <u>nearby</u> Galaxies <u>confirming</u> there is a correlation between how the SMBH formed and the eventual size of a Galaxy. **A SMBH may feed and grow a little larger while it is creating its Galaxy's Matter and it may lose a little mass due to Hawking's Radiation. Thus, the mass of a SMBH generally remains fairly constant or slightly grows while the combined mass of a Galaxy's stars significantly grows as more star making Matter is created and ejected from the polar regions of a Galaxy's central SMBH.** In the local universe, the mass of a Galaxy's stars to a Galaxy's SMBH's mass is said to be about 1,000 to 1. [122] **Search: Unexpectedly Massive Black Holes Dominate Small Galaxies in the Distant Universe.** *In the distant, early universe the ratio can reduce to 100 to 1, 10 to 1 or even as low as 1 to 1 which is described as an "unexpected discovery" from the James Webb Space Telescope.* Clearly if this trend continues back in time it will lead to very few or no stars around a SMBH (0 to 1). Thus, within a Natural Universe, which first creates a Galaxy's Super Massive Black Hole, this discovery is expected.

Scientists today are left to wonder why the two are related. [024] *"Why, they ask, is there this relationship"?* Since Galaxies have grown the total mass of their stars over time Galaxies of similar age will commonly share this relationship. **Because newly created protons will get to keep all of their inherited phase of virtual mass a SMBH is not required to grow massively larger while it is creating a far more massive Galaxy of stars.** The first stars are naturally provided with a crucial Dark matter Gravitational scaffold consisting of a central Super Massive Black Hole, a dark disk and a Dark matter halo which is required for quickly forming and shaping disk Galaxies. The progression is natural and there are no unexplainable loose ends.

With virtual protons one is not severely 'size' or time restricted by first having to form many hypothetical super stars; consequently, many of the first Black Holes can be constructed astonishingly quick and at an early time in the universe and can be 'Super Massive'. For example, Scientists say the Gravitational implosion which forms a 'Stellar' Black Hole may take only seconds to become a Black Hole. The incredibly rapid inwards collapse naturally unleashes a pulse of very powerful Gravitational Potential Energy which <u>accelerates vast quantities</u> of virtual protons remaining in the dark disk of the newly created Dark Galaxy to a high speed towards its Super Massive Black Hole (SMBH). This almost instantly begins the process which will <u>transform</u> COST units from their virtual proton state into the first regular protons and neutrons of real Matter. **Note how the Natural Model provides an explanation at the tiny level of Dark matter (virtual protons) exactly what it is that causes this Gravitational implosion, exactly why it is able to create <u>Super Massive</u> Black Holes and exactly why the passing of time stops inside of Black Holes**. This is possible for the reason that the model tells exactly what Dark matter is and exactly how it provides an actual source for <u>mass</u>, <u>Gravity</u> and the <u>passing of time</u>.

Any object squeezed into a sufficiently small volume will create a Black Hole. The object then continues to collapse under its own weight, crushing itself down to <u>zero size</u>. [023] However, according to <u>Einstein's theory</u>, the <u>object's 'mass' and 'Gravity' remain behind</u> as an extreme distortion of space and time. This distortion of Space-time is a Black Hole. But how can that same mass and Gravity remain after all Matter has crushed itself out of existence so as there is absolutely <u>nothing</u> left? No quarks survive to excite and be agitated at near light speed to create regular mass as within our own atoms but by some <u>unknown</u> means this very same mass and its Gravity mysteriously remains. [023] Simply saying Mass and Gravity remains behind is really not good enough and reflect that both the source of mass and the source of Gravity remains a mystery. When there appears to be <u>nothing</u> left what remains behind to provide mass and Gravity? <u>Only a Natural Model</u> can make sense of this and provide a clear, realistic and viable answer. The mass which remains behind is virtual. Virtual <u>mass</u> provides negative energy from a phase of a lower state of vacuum than the vacuum of dimensional Space which provides the <u>attraction part of Gravity and is the primary source of all mass</u>. This is how both mass and Gravity are left after all particles are crushed out of existence inside of a Black Hole.

Scientists have quite a few very different and elaborate theories on what happens inside of Black Holes. However, a Natural Model has no walls of fire or exotic gateways to parallel universes inside of Black Holes. A Black Hole isn't a hole that goes anywhere it is only a sphere of empty Space.

Because the escape speed inside of a Black Hole exceeds the speed of light the tiny Gravitational waves which were providing entangled dimensions to Space are now <u>not able to be produced</u>. Consequently, inside of the event horizon Space is now effectively <u>dimensionless</u>.

When a virtual proton becomes a regular proton its inherited phase of virtual mass provides **both** the strong attraction part of Gravity and the source of the strong force which <u>tightly binds</u> a proton's particles together while creating a proton's regular mass-energy. This sensibly reconciles Gravity with regular mass. A SMBH is made directly from virtual mass which the entangled dimensions from <u>all objects outside</u> of the Black Hole <u>are very much attracted to</u> which causes their entangled dimensions to be stretched and warped towards the Black Hole allowing all objects outside of the Black Hole to feel a Black Hole's Gravity. Inside of <u>all Black Holes</u> there is a likeness to the directly related strong force. At the center of a Black Hole all objects are <u>bound so extraordinarily tightly together</u> that they are now crushed out of existence, all of which <u>resembles</u> very closely how the strong force works. **Only now this <u>same force</u> is so strong and <u>relentless</u> that <u>instead of</u> just binding particles exceptionally tightly together, when creating a proton's regular mass, <u>the force now crushes all particles out of existence</u> leaving only virtual mass behind. Because virtual mass is a state of empty Space nothingness it cannot be crushed out of existence. One may now <u>rightly</u> say all Black Holes are created by the strong force.**

When creating a proton's regular mass a phase of virtual mass momentarily disappears <u>before</u> it can crush a proton's particles out of existence. The source of a Black Hole's mass, the source of an atom's regular mass and the source of a particle's mass is virtual mass provided by COST units (virtual protons). All mass should have a common and easily explainable <u>physical source or origin</u> which is what this model clearly provides.

Gravity is known to propagate at the speed of light but the escape speed of a Black Hole exceeds the speed of light so how can Gravity get <u>out</u> of a Black Hole? It cannot and for a Natural Model it does not need to. I find many difficult to understand answers to choose from when I Google; *how can Gravity get out of a Black Hole?* Because a Natural Model has two parts for Gravity it can rationally solve this seemingly unsolvable quandary. For objects outside a Black Hole Gravity works exactly the same as it does here on Earth where every atom <u>contributes</u> to Earth's mass and Gravity.

Gravity wise, <u>nothing</u> can come out of the Black Hole. However, acting like water falling over the <u>edge</u> of a waterfall, <u>everything</u> can get <u>into</u> a Black Hole. The attraction part of Gravity is provided by the Black Hole and the delivery part of Gravity is provided by objects <u>outside</u> of a Black Hole. Tiny Gravitational waves expanding from all objects <u>outside</u> of a Black Hole provide 'entangled' dimensions to Space which <u>provides the delivery part of Gravity</u>. The tiny Gravitational waves from all objects outside of a Black Hole can get <u>into</u> a Black Hole which allows their entangled dimensions of Space to become extremely stretched and warped as they naturally <u>flow into</u> a Black Hole. Thus, a Black Hole's Gravity, in a <u>seemingly back to front way</u>, is delivered to objects outside of a Black Hole. This is why nothing has to come out of the Black Hole for Gravity to act normally for objects outside of a Black Hole and can only be explained this way if Gravity has both an attraction part and a delivery part like this model clearly provides.

For objects outside of a Black Hole Gravity now operates much the way as described by General Relativity which allows a SMBH to remain centered within a Galaxy. Because a Black Hole can <u>only express the attraction part</u> of Gravity the model suggests two Black Holes are less likely to interact with each other and stellar Black Holes are less likely to combine to create intermediate Black Holes which may well solve a long standing <u>mystery</u> of why intermediate Black Holes are observed to be extraordinary rare. However, a Black Hole regularly acquires an <u>enormous accretion disk</u>, the volume of which dominates a Black Hole, effectively <u>providing</u> a Black Hole with a delivery part of Gravity which now feels another Black Hole's Gravity.

Because inside a Black Hole is now <u>effectively dimensionless</u>, a Black Hole can only be measured from the outside. If all the dimensions of Space were removed from outside of the Black Hole, all Space would be in a lower state similar to the Black Hole and the Black Hole would not exist.

[084] When one researches whether <u>singularities</u> of Black Holes are real or only artifacts of mathematical theories, one surprisingly finds that singularities are commonly the result of Mathematicians attempting to deal with the unmanageable consequences of 'infinity'. Because an equation which can handle infinity is yet to be developed, a 'singularity' is commonly an attempt to manage mathematics where the function of 'infinity' causes all current mathematical equations to break down. Infinity has no true mathematical denotation because infinity <u>is a concept and not a number</u> which can be used with any mathematical operator like the symbols -, +, x, or /. This is why the mechanisms operating inside of all Black Holes are said to be **beyond today's theories of physics and is why** General Relativity is said to break down at very small scales, which happens to be where the 'singularity' of Black Holes appears. Consequently, Scientists will often say that they do not really know exactly what goes on inside of Black Holes since all of their theories effectively break down. A Natural Model reveals today's theories effectively break down because they cannot cope within a state of Space which is effectively dimensionless. Within dimensionless Space the symbols -, +, x, or / are meaningless.

The solution to this mathematical uncertainty is to provide reasoning for an area where mathematical equations, rather than continue infinitely tiny, are designed to provide a calculable end result of 'zero' dimensions whenever the escape speed <u>exceeds the exact speed of light</u>. One can now calculate: At the same moment the escape speed exceeds the speed of light the passing of time stops, Space becomes dimensionless and a Black Hole is created. This now mirrors a mathematical singularity. Because the escape speed inside of a Black Hole exceeds the speed of light, the tiny, expanding Gravitational waves which provide <u>entangled</u> dimensions to Space cannot now escape from a COST. This provides an area inside of a Black Hole where, without <u>entangled</u> dimensions, the area is effectively dimensionless and without a passing of time. This is an area where every position is in the same exact dimensionally nowhere position which mirrors a singularity <u>without using</u> the erroneous and incorrect notion of infinity. Since the inside is effectively dimensionless this area can only be measured from the outside and clearly explains how a singularity may be both of a large area when measured from the outside and wrongly appear to be mathematically <u>infinitely</u> tiny to the point of being not measurable or incalculable from within. All of which mirrors the state of a dimensionless singularity. This is a similar to a photon of light where within a photon's own <u>state of Space</u> it travels nowhere in no time even if it has actually travelled for 100,000 years across an entire Galaxy within one's own spatial Dimensional Reality.

The Gravity process relies on vacuum fluctuations. As the virtual mass phase is filled by ground-state of dimensional Space, the lower state of vacuum of virtual mass is immediately reproduced by the next fluctuation. This is why vacuum 'fluctuations' are required so as the flow in of dimensional Space can be maintained. Travelling towards the outer event horizon from any position inside a Black hole is forbidden for the reason that to do so one would be required to exceed the speed of light. Special Relativity implies that time stops at event horizons of Black Holes and at the speed of light but time may appear normal here for a self-observer.

Inside of the event horizon there is now a calm of Space where all of the mechanisms responsible for the laws of Relativity have also slowed to a standstill. Without <u>entangled</u> dimensions of Space-time, this is a place where all Relativity theories and physics must also breakdown but where this model will later show why Quantum Theory may excel. Note how these 'singularities' have super, strong Gravity for the reason that COST units are continually producing their virtual mass phase which is providing the attraction part of Gravity and so are continually attracting the dimensions of Space and time from all objects outside of the Black Hole. Note again how well this model unravels the mystery of exactly what the singularity of a Black Hole is, how it is made and why the traditional laws and theories of the universe breakdown. Although technically not a true singularity to the point of being **incalculable,** infinitely small, Black Holes do mirror the same properties if measured from the inside. Inside, every position is in the same nowhere position. The only possible way to measure this type of singularity is from a Dimensional Reality immediately outside the singularity from where it may measure quite large. This kind of singularity now exists as a **calculable state** of Space which mirrors the far outer void called Anti-space.

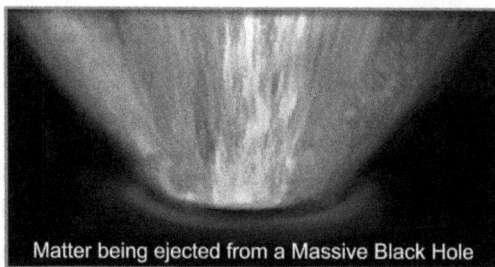

Matter being ejected from a Massive Black Hole

Dimensionless Space may at first seem absurd and irrational; this is for the reason that we are all forever, without exception, individually locked by entanglement to an inescapable realm of a Dimensional Reality. The Natural Model will show how the predictions of both Quantum Theory and Special Relativity expose the existence of a dimensionless state of Space which is simply Space where the dimensions of Space are totally ignored. The superb thing about vacuum fluctuations is as quickly as the lower state of vacuum is filled by dimensional Space the lower state of vacuum is immediately reproduced by the following fluctuation. [015] Quantum vacuum fluctuations have been observed in experiments fluctuating to a phase of a lower state of vacuum than the ground-state of empty Space; this is the Natural Model's negative energy <u>phase of virtual mass.</u> This is why the ground-state of dimensional Space which is positive energy continuously rushes into Black Holes. **One can now say Black Holes are made from virtual mass and are a negative state of energy which the positive energy of the expanding dimensions of Space are attracted to and so rush into.**

<u>Note;</u> stellar Black Holes are created from the <u>positive</u> mass-energy of true particles so are today <u>said to be positive</u>. However, inside of a Black Hole all of those same particles providing positive mass-energy are crushed out of existence so, of course, they do not continue to exist to provide positive mass-energy and <u>nothing stays behind</u> to provide mass and Gravity. Within a Natural Universe, the negative energy of virtual mass is a phase of pure nothingness which cannot be crushed out of existence and is <u>what stays behind</u> inside of a Black Hole to provide mass and Gravity. Consequently, within a Natural Universe all Black Holes will be a state of negative energy since they are clearly made from the negative energy state of virtual mass. The next chapter will show exactly how the negative energy of virtual mass creates the positive mass-energy of the particles of regular protons.

The very <u>first implosion</u> which created the first Super Massive Black Hole likely occurred within one of the very largest clouds of virtual protons and produced extraordinary large and powerful Gravitational waves and shock waves. Like the Cosmic Microwave Background radiation the ripple of the powerful Gravitational waves <u>may be detectable today</u>. The shock waves provided a domino effect, triggering the Gravitational implosions within numerous nearby clouds of virtual protons spontaneously creating numerous Super Massive Black Holes which powered up their regions with massive amounts of Gravitational Potential Energy. Although this provided a onetime peak to Galaxy creation a cloud of Dark matter may collapse and form a SMBH to make a new Galaxy <u>at any time</u> before or after this peak.

Our fledgling universe now has the conveyance for light, yet it has no light;

it already has the scaffolding and Gravity for atoms, yet it has no atoms. It has Dark Galaxies for stars to Gravitationally cling to and evolve into bright Galaxies, yet it has no stars. Most notably, our fledgling universe now has massive quantities of virtual protons accelerating towards newly created Super Massive Black Holes. Naked Super Massive Black Holes are now primed and ready to provide Galaxies with a powerful heart to form around and rapidly evolve from. Importantly, modeling clearly separates Super Massive Black Hole formation from their smaller Stellar Black Hole cousins.

Black Holes and the overall expansion of the dimensions of Space.

Within a Natural Universe light travels one-way into a Black Hole for the same reason that light radiates away in all directions. The reason for this is a Black Hole is essentially created from negative energy, the same of which is found within a far outer void of empty Space now called Anti-space.

[111, 017] **What drives light to speed away at 299,792,458 meters per second? Answer; the negative energy of the strong attraction part of Gravity.** (Chapter 13 explains why this is called Quantum Gravity). Naturally, this is why, from an event horizon, all light travels into a Black Hole and **is why the source of Gravity warps or bends light** in much the same way as the source of Gravity stretches and bends the tiny Gravitational waves which provide the dimensions of Space-time. Like a Black Hole the far outer void of Anti-Space is also in a lower, dimensionless state of empty Space which attracts photons so as they speed away in all directions at **299,792,458 meters per second.** The negative energy of the outer void of empty Space also attracts the dimensions of Space so as they expand in all directions. Consequently, photons (light) speed away from their source atoms in all directions at the same speed for the same reason that light travels one way pass the event horizon into a Black Hole. Because Quantum particles are **not trapped at stationary equilibrium** at the center of entangled dimensions allows photons of light to instantly speed away in all directions due to the strong attraction part of Outer Gravity and for the strong attraction part of Inner

Gravity to redirect their path when passing an object of mass. Light speeding away in all directions provides remarkable **evidence** of the source of Outer Gravity which is responsible for accelerating whole Galaxies away.

How can a void which is billions of light years away attract photons and tiny Gravitational waves? Answer: **At the <u>speed of light</u> dimensional distance is meaningless, since, the distance to the far outer void is <u>contracted</u> to a single point.** Einstein's theory of Special Relativity has told us photons at <u>light speed</u> completely <u>ignore</u> dimensional distance, consequently, to a photon the billions of light years of dimensional distance does not exist. For a photon or an <u>entangled</u>, tiny Gravitational wave the far outer void of Anti-Space is not distant <u>but instantly in the same place</u>. No other known process is capable of such a mammoth task of expanding our universe and a Natural Universe has it in place from the very beginning. **This is also why the very same tiny Gravitational waves which provide dimensions and are endowed with <u>positive</u> energy from the event which created them are <u>similarly attracted</u> to the <u>negative energy</u> state of the tiny phase of virtual mass at the center of Dark matter, protons and neutrons.** At chapter thirteen I will similarly show how a pair of Quantum <u>entangled</u> particles may **instantly react** to the orientation or actions of their entangled partner even if separated by light years within a Dimensional Reality. Many things which appear <u>totally</u> irrational or impossible within one's Dimensional Reality will become rational within what is effectively dimension<u>less</u> Space.

[026, 070] Today Physicists are attempting to come to grips with the reality that our home universe is apparently not put together the way Physicists have for long believed. In search of particles which agree with proposed theories, notably Supersymmetry, Physicists have ramped up the energies at the Large Hadron Collider to the high energies calculated to expose many of the most sought after and predicted particles. However, Physicists have failed to find particles which may be Dark matter, no leptoquarks, no sign of extra dimensions required for theories and <u>none of the most sought after</u>

<u>Supersymmetry particles.</u> Although Scientists have been predicting these things for decades the Large Hadron Collider failed to discover any sign of many of even the most cherished theories. The particles of Supersymmetry, among other things, in a most complex way was <u>said to solve the origins of mass</u>. Theorists are now facing the daunting task of confronting the not-predicted LHC results head on; Science is presently at crossroads; the LHC results will require a significant shift away from current ideas and theories for the results require long standing beliefs to be very reluctantly discarded as science currently scrambles for a new direction which they hope will lead to a new branch of physics. [027] However, decades of confounding experiments have Physicists considering a startling possibility; we may reside in an unnatural universe where natural does not make sense.

If the shock results from the LHC was not enough now the recently launched James Webb Space Telescope has almost immediately found massive, matured Galaxies existing at a time too early to be possible *"under current cosmological theory"*. The Galaxies from the dawn of the universe have *"blindsided theorists"* who are again scrambling to explain the unanticipated observations. Even the Big Bang theory is now said by many to be in real trouble. [091] The problem being the discovered Galaxies have not had enough time to form this massive within the short time available after the said Big Bang. The discoveries from the JWST include a Galaxy with a redshift of 20 placing it seemingly only 180 million years after the said Big Bang. A Galaxy of similar size of our Milky Way has also been observed with a redshift of 10 placing it less than 500 million years after the Big Bang. **Given its maturity it is like the Galaxy must have originated from a time <u>before</u> the proposed Big Bang.** *"Even if you took everything that was available to form stars and snapped your fingers instantaneously, you still wouldn't be able to get (a Galaxy) that big that early (after a Big Bang),"* says Michael Boylan-Kolchin, a cosmologist at the University of Texas. Others have responded with articles defending the Big Bang but the articles fail to address the Big Bang's Matter and Anti-matter asymmetry problem.

7 THE CREATION OF MASS.

Dark matter is totally invisible and is only <u>indirectly</u> exposed by its Gravity. Thus, whatever it is which provides Gravity <u>has to also be totally invisible.</u> Wherever one finds Gravity one finds mass. **So now to understand the source of mass we <u>require a totally invisible and not directly detectable kind of mass which also provides Gravity</u>.** *This totally invisible kind of mass has to provide as much as ten times more pressure inside of a proton than inside of a neutron star while at the same time madly agitating a proton's particles at <u>near the speed of light</u> while it is binding a proton's particles tightly together. The mysterious source of mass is also required to become weak to nothing when a proton's quarks are closest and extraordinary strong when a proton's quarks are farthermost apart. The origins of all mass and Gravity also needs be something which can <u>remain</u> after all <u>particles</u> are <u>crushed out of existence</u> inside of a Black Hole and, for a Natural Model, <u>has to be there at the beginning</u> to allow a virtual proton to <u>cling on</u> to three positive Matter quarks to create a regular positive proton. Finally, the source of mass and <u>Gravity</u> should ideally clearly explain why Galaxies are accelerating in the direction which the universe is expanding.*

[026, 070] With the recent failure of the Large Hadron Collider to expose the particles predicted by Supersymmetry Scientists today say that developing new physics in order to solve <u>the origin of mass</u>, is one of the biggest and most urgent challenges of today's astrophysics.

What is mass? Common answer is weight, right? No, not really correct, weight is really 'mass times Gravity'. Without Gravity you are weightless but you still retain your rest mass. **So, more Gravity more weight.** If you weigh 100kg on Earth you will weigh only 16.5kg on the Moon where there is less Gravity than on Earth. It is <u>obvious</u> that mass and Gravity are required to <u>share a common bond</u> but what physically provides this bond?

Einstein's Special Relativity refers to just two kinds of mass: Rest mass and relativistic mass. Relativistic mass is dependent on the **speed** of the observer and **includes** rest mass while rest mass is of course a particle's mass at rest or when compared to another particle with the same speed. Like Special Relativity my Natural Model will be referring mostly to just <u>two kinds of mass</u> which I will call regular mass and virtual mass. Regular mass is simply the rest mass and relativistic mass of all atoms and particles, however, virtual mass is a concept which is unique to my Natural Model.

Today, the source of the mass created by binding energy within the nucleus of atoms remains mysterious. It is said the energy from the particles of the nucleus of atoms being <u>highly excited and held tightly together</u> accounts for 99% of the mass of all atoms. The actual particles, the electrons, quarks and gluons, make up the other 1% of mass of an atom's total mass. This 1% of mass is said to be acquired from the particles drifting through the Higgs field, however, for reasons which will become clearly apparent, my Natural Model also uses binding energy to account for this 1% of mass.

The COST unit's vacuum fluctuations produce a Matter and Anti-matter pair of <u>virtual particles</u> which annihilate to provide a positive energy phase before fluctuating to a tiny phase of <u>virtual mass</u>. [016] Virtual mass is a tiny phase of deep vacuum empty Space and what is today considered ground state vacuum is a false vacuum because it contains the dimensions of Space-time in the form of a woven web of the <u>positive</u> energy of very tiny Gravitational waves. Virtual mass provides the attraction part of Gravity and interacts with particles to provide particles with regular mass. The phase of <u>virtual mass</u> provides a <u>common bond</u> between regular mass and

Gravity. Within a Natural Universe the mass of all particles, all atoms, Dark matter and all Black Holes is directly related to the vacuum fluctuations of the COST fluctuating to a tiny <u>negative energy</u> phase of virtual mass. [111] **Negative energy is a state of Space which has less energy than the ground state of empty dimensional Space.** Negative energy is powerful simply because it has the weight of all dimensional Space-time sitting above it.

Amazingly, the combined mass of all of an atom's particles themselves, accounts for only around 1% of an atoms mass. Actually, an atom's particles are said to be basically massless without the part played by the Higgs boson. <u>The Natural Model is not directly reliant on the Higgs</u>. If an atom's particles do not account for around 99% of an atom's mass than there is no getting around the fact that 99% of an atom's mass, or your own mass, is provided by other means which is mostly related to the mysterious 'binding energy' associated with the strong nuclear force. [003] Consequently, the majority of all regular mass which represents around 99% of your 'heavy' mass is actually a mysterious energy which <u>highly excites, agitates</u> and binds the particles inside of your protons and neutrons and tightly binds everything together to create the tight nucleus of atoms.

[011] *It is most important to understand that <u>not any</u> of* this mysterious 'binding energy' has *anything to do with the much talked about Higgs boson. As a result, 'one of the longest standing mysteries of physics, the exact origins of 99% of the mass of all known atoms-of-matter is still mysterious, even after the <u>much publicized</u> discovery of the Higgs'.* So one may ask how is this 'binding energy' created and exactly what creates it.

<u>The present common story</u>; the nucleus of atoms is made of neutrons and protons, each neutron and proton is very <u>loosely</u> said to have three quarks <u>bound tightly together by gluons</u>. The three quarks account for most of the 1% of the total mass of all of an atom's particles and the gluons themselves are essentially massless. The essentially massless gluons are said to be the carriers responsible for the strong nuclear force, which is said to 'give rise'

to the 'binding energy' which 'glues' and so binds the three quarks of protons and neutrons tightly together and maintain the structural integrity of the nucleus of atoms. A proton has now acquired energy from the gluons 'gluing' the quarks and the nucleus tightly together. It would now require energy to separate the tightly bound together quarks. Energy here equals mass, hence, mass-energy. [033] One may now at first believe all is good, we now know mass is 'somehow' created by simply gluing three quarks together but when one digs deep there is a very worrying short answer at the very bottom line which is; *"we do not understand how the gluons actually glue, consequently, nobody really understands the actual origins (source) of almost all of the mass within the universe"*. See [087] *there is a big problem that does not get much attention: gluons.*

[033] Although the theory says mass is created by the gluons gluing quarks tightly together there remains no answer to exactly how the gluons glue, consequently, the origins of 99% of all mass remains a mystery. The search for the origins of mass should also directly reconciled mass with the source of Gravity which is also a most urgent problem of modern particle physics which strangely gets little attention. What of Albert Einstein's famous $E=mc^2$, where the speed of light is squared? How can simply gluing three quarks together create so much energy that the sums require multiplying twice by the speed of light? Obviously, there is far more to it than the simple story of gluing three quarks together. I believe particle Physicists are very clever and obviously have amazing resources at their disposal but answers are more likely to come when one is pointed in the right direction.

[003; 028] **Today, there is much more to it than the common story of only gluing particles tightly together;** evidence from the Large Hadron Collider paints a far different image of a proton or a neutron than the images found online where three large Matter quarks dominates an image. It is now said there are actually far too many quarks and anti-quarks and emitted gluons to even count, all of which are somehow being excited and madly agitated about at near the speed of light. Their violent motion provides the particles with mass-energy, the mass of which relates <u>directly</u> to an acquired amount of Gravity. However, if all of the anti-quarks annihilate with a quark there are always just three extra quarks left behind which cannot annihilate since they have somehow lost their anti-quark partners. Because they <u>live to provide a proton's basic properties</u> these are the three most important quarks and are called valence quarks. Because these three leftover quarks are Matter quarks is why the universe is made from Matter. If there had been three leftover anti-quarks the universe would have been made from Anti-matter but everything would apparently appear the same as today. All of these quarks and anti-quarks as well as the gluons they emit are 'virtual' particles but they make up a proton which is a 'real' particle.

Protons have positive electric charge which causes them to have strong mutual electromagnetic <u>repulsion</u>. Consequently, Scientists postulated that a strong attractive force was somehow tightly binding, like a kind of glue, all within the nucleus of atoms and preventing the nucleus from flying apart.

Within the Natural Model gluons will play an important but <u>lesser role</u>. The COST, acting as a virtual proton, will provide the vacuum fluctuations for creating the sea quarks of regular protons, neutrons and so atoms: Firstly, for my model for creating mass I will match the invisible Cost (Dark matter) with an atom in this way; if one was to disassembled an atom by removing <u>each and every particle</u> from every proton and neutron which makes up an atom and measure each particle separately, the sum-total of all of the particles mass will only represent around 1% of an atoms mass. So now ask, as one has <u>removed every particle</u> and there is now <u>nothing left</u>, what was the '<u>invisible</u>' source or mechanism responsible for the other 99% of an atom's regular mass? **Answer; the <u>invisible</u> mechanism <u>directly inherited</u> from a virtual proton (Dark matter).** My Natural Model simply creates mass by exposing particles to a tiny, fluctuating sphere of pure empty Space which I have called virtual mass. Gluons may play an important role in interacting with quarks, however, a state of virtual mass will provide a phase of negative energy to do <u>all of the work</u> of creating regular mass.

Strangely, science has been unsuccessfully attempting to solve many of these mysteries; the composition of Dark matter, the source of Gravity, a proton's mass generation mechanism and how to reconcile mass and Gravity with different explanations. Now note the similarity of these mysteries. <u>First</u>, Dark matter is said to be an <u>invisible</u> particle which has Gravity and so should have mass. <u>Second</u>, 99% of an atom's mass is not the sum of an atom's particles; the actual, raw source of this 99% of an atom's mass and energy is produced by an <u>invisible</u>, unknown, mysterious mechanism said to be associated with gluons and the strong nuclear force. <u>Third</u>, the actual raw physical source of Gravity is also an unknown mystery and both Dark matter and atoms have Gravity. <u>Forth</u>, mass needs to be reconciled with Gravity but nobody knows how this is physically achieved. Fifth, the source of both mass and Gravity needs to be <u>the only thing</u> which cannot be crushed out of existence within a Black Hole.

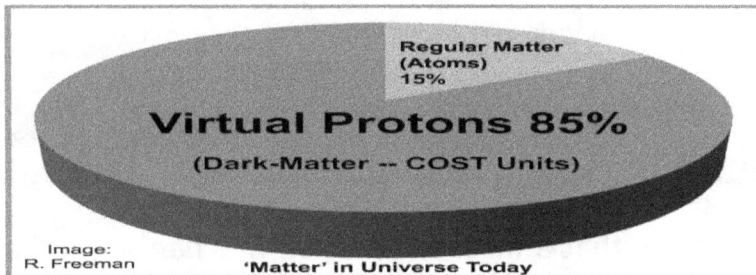

The COST being an invisible configuration of Space-time will provide the source of mass and energy from amplified Quantum vacuum fluctuations. The fluctuations of the COST provide a positive energy phase of virtual particles and a negative energy phase of virtual mass. The negative energy phase of virtual mass, by attracting the positive dimensions of Space, provides the source of the attraction part of Gravity. The same negative energy phase of virtual mass will provide near light speed agitation of a proton's particles and bind the particles tightly together which provides the particles with energy which equals regular, positive mass. Virtual mass is a tiny phase of <u>pristine Space</u> which provides <u>negative energy</u> and sits <u>below the ground state</u> of dimensional Space.

From empty Space, particle-less Configurations of Space and Time set the wheels in motion to build an entire universe from mass, energy and Gravity. **The COST, in its Dark matter state, can now be called a <u>virtual proton</u>, however, to be transformed into a regular proton many virtual protons will need to first progress a destined but rather complex path.**

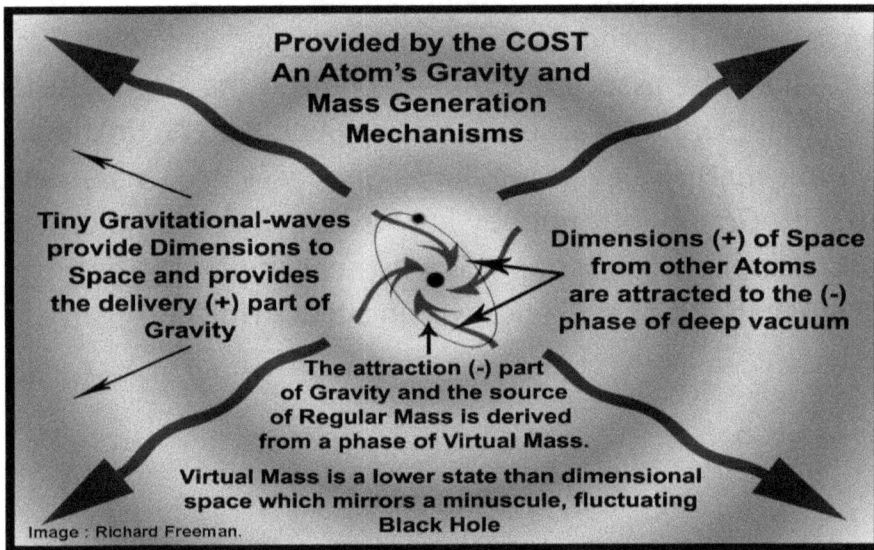

It really is imperative that a mass generation mechanism and a mechanism for Gravity go together like two peas in a pod. With the Natural Model, virtual protons gain energy when their Quantum vacuum fluctuations are squeezed by the warped Space-time of their own combined Gravity. [015] *Note: This <u>same</u> squeezing affect on Quantum vacuum fluctuations has been observed in experiments.* Energy is increased by pair production of Matter and Anti-matter pairs of virtual particles which annihilate causing the energy phase to become louder with energy. The energy phase is followed by an <u>opposite reaction</u> phase of virtual mass which by attracting the dimensions of Space provides the attraction part of Gravity. Thus, as a

Quantum vacuum fluctuation becomes louder with energy, the vacuum phase increases its volume below the ground state of dimensional Space creating a larger abnormality in dimensional Space which increases the attraction of the dimensions of Space which in turn delivers more Gravity.

Now ask; how can empty Space provide all mass? Answer; the empty Space of a sub-ground state, negative energy phase of virtual mass which effectively mimics a rapid fluctuating, in a state of appearing and collapsing, minuscule Black Hole at the very hearts of the protons and neutrons of atoms, which delivers strong, binding, agitating negative energy at the shortest of range at the nucleus of atoms and provides the attraction part of Gravity. [111] Negative energy is simply a state of Space which has less energy than the ground state of dimensional Space. The phase of virtual mass is technically not a fully formed Black Hole but is best described this way because it will provide the tiniest seed which makes all Black Holes. [021] **Theories allow a Planck scale *virtual* Black Hole such as this to appear spontaneously but only very briefly.** It may not be very wrong to say; *"we are all powered by tiny Black Holes"*.

Consider a stellar Black Hole which size is measured by its mass: Because all Matter has been crushed out of existence there are no gluons and no quarks for gluons to bind tightly together to create mass and there are no particles to drift through the Higgs field to gather mass. A Black Hole is basically a sphere of empty Space. The big unanswered question is how can a sphere of empty Space be a sphere of enormous mass and why does Space-time rush into a Black Hole? Answer is no mystery; all Black Holes are simply made from the fluctuations of virtual mass which is a tiny phase of pure empty Space. [023] **The fact that inside a Black Hole all particles are crushed out of existence leaving only a sphere of pure empty Space which is also a sphere of pure mass and Gravity is all the evidence one needs to know that empty Space is indeed the source of both mass and Gravity.**

If the Earth was crushed into neutron star material its diameter would be reduced to just 305 meters of <u>super dense material</u> of which one teaspoon full would weigh 4 billion tons. However, if the Earth was crushed down to a Black Hole all of the Earth's Matter would be crushed out of existence and the Earth would be reduced to an 18 mm sphere of <u>empty Space</u>. The Earth would now be a sphere of pure mass. This is why all mass is acquired from the empty Space of the fluctuations of virtual mass. [111] The negative energy of virtual mass is powerful simply because it is a tiny phase of pure nothingness residing below the **weight of all dimensional Space-time**. Dimensional Space-time is not nothingness but, like Einstein said, is like a fabric which curvature of can be stretched and warped to deliver Gravity.

Atoms are 99.99999% <u>normal</u> empty Space. A neutron star is created from the collapsing core of a large exploding star. In a neutron star the Gravity is so great that nearly all of that empty Space within atoms has now vanished and protons have adsorbed an electron to become neutrons which together with existing neutrons are forced tightly together to become a star made from neutrons. In a neutron star Matter is exposed to Gravitational Pressure which is so strong that Matter is at the <u>last survivable step before becoming a Black Hole</u> where all Matter, including these surviving neutrons, is crushed out of existence. [002] One teaspoon of super-dense neutron star material weighs around 4 billion tons which is like more than 26,500 ocean liners as large as the luxurious Queen Mary 2 all in one teaspoon. However, Scientists have discovered that the pressure <u>inside of a proton amazingly peaks ten times greater</u> than the incredible pressure inside of a neutron star. What could possibly be so powerfully strong to create even more pressure inside of a tiny proton than inside of a neutron star which is just one step away from becoming a Black Hole? My Natural Model provides probably the only possible and obvious answer; one will effectively find a minuscule 'fluctuating', in a state of rapidly collapsing and reappearing, Black Hole at the center of protons which is directly responsible for the pressure inside of a proton <u>peaking ten times greater</u> than inside of a neutron star. It is <u>very unlikely</u> that any other <u>known invisible source</u> is able to provide such tremendous pressure within such a tiny area responsible for providing a proton with far more pressure than inside of a neutron star.

So what is it that creates pure mass? Answer; pure nothingness; the empty Space of a tiny phase of **virtual mass** which <u>needs to mirror</u> a minuscule, rapidly collapsing and reappearing Black Hole which will provide the seed for all Black Holes. [021] **Theories allow a Planck scale *virtual* Black Hole such as this to appear spontaneously but only very briefly.** As a phase within a proton, this now allows the pressure inside of a proton to peak far higher than <u>at the heart of a neutron star</u>, however, <u>before binding energy crushes particles out of existence</u> the minuscule virtual Black Hole collapses and momentarily releases particles only to reappear when quarks are farthermost apart and again drives particles tightly together creating more resistance pressure, from quarks fighting back, than inside of a neutron star. The total agitation cycle or <u>fluctuation</u> naturally violently, at near the speed of light, excites the particles where the positive energy of the agitation cycle is <u>counterbalanced by the negative energy</u> phase of virtual mass which created it. This counterbalancing is <u>essential</u> for conforming to the conservation of energy law which is required for a sum-total zero mass-energy universe to exist. The created 'positive energy' <u>of the very excited particles</u> can now be expressed as regular mass or simply as mass-energy.

For decades I have read that gluons somehow simply bind or glue a proton's particles tightly together and in doing so created mass. [003] The current view from the LHC is the near <u>light speed</u> violent, pandemonium-like agitation of a proton's particles provides particles with energy which equals mass but I have yet to read how gluons now manage to create this near <u>the speed of light</u> violent agitation of particles or how gluons naturally counterbalance the created mass-energy to conform to the conservation of energy law. [029] The bottom line is the full story of where almost all known mass comes from remains one of the longest standing mysteries of physics.

Now ask what could possibly provide the near <u>the speed of light</u> violent agitation of particles? The near light speed is important since Einstein has shown the energy here equals mass by the speed of light squared ($E=mc^2$). It is unlikely there is any other known and fitting source capable of providing near light speed agitation and the described pandemonium of particles other than a minuscule fluctuating, in a state of rapidly collapsing and reappearing, Black Hole. This is exactly what Black Holes do; they give particles near light speed.

There is no shell or bubble around a proton to contain its particles; all of this tremendous pressure, binding energy and near light speed of particles needs to be provided by something which resides at the center. It simply makes rational, common sense that this phenomenon which creates regular mass, should also be the source of the attraction part of Gravity and the source of all Black Holes. Note; binding energy is **different** to Einstein's Gravity which will have both an attraction part and a delivery part.

My Natural Model provides a very fitting resource which works on all types of particles within the nucleus. The binding energy of the strong nuclear force which accounts for all regular mass is no more than the COST rapidly fluctuating from 'expanding energy' to a lower state of deep vacuum. The strong negative energy phase of a deep vacuum provides a phase which mimics a minuscule, rapidly fluctuating Black Hole at the very hearts of protons. This provides all protons and neutrons and so atoms with a tiny, fluctuating seed Black Hole which constrains particles tightly together and if an immense amount of atoms come together the same collective binding will cause atoms to crush themselves out of existence and create a regular Black Hole. With this model all Black Holes, regardless of their size, are made from virtual mass which is a tiny phase of negative energy which is why ground-state positive dimensional Space flows into all Black Holes and in so doing so provides and delivers massive Gravity.

[030] Reference links to two images. At the first address there's an amazing image of a Hydrogen atom and the next address there's an extraordinary

image of a Black Hole. Both images look astonishingly similar to this image which I have created from a donut, the Black Hole image is outlined by the glowing gas that surrounds its event horizon and the Hydrogen atom image has the haze of its lone electron creating a field surrounding a dark nucleus.

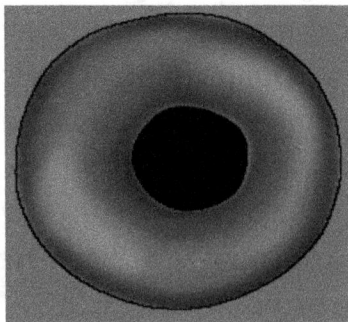

[029] One can be almost completely certain that the strong force which binds a proton's particles tightly together is being created by a rapidly fluctuating phase of a negative energy state of the deep vacuum of virtual mass simply because it mysteriously works back-to-front to all other forces. Scientists have discovered that the strong force of binding energy strangely acts very differently to all other forces such as electric, magnetic or Gravitational, all of which are <u>stronger when objects are closer</u> and weaker when objects are farther apart. However, the way a proton's quarks are bought together to form a proton is fundamentally different from all other known interactions or forces. Rather than acting like a regular force, the binding force strangely becomes weak to nothing when quarks are closest and mysteriously becomes extraordinary strong when quarks are farthermost apart.

The binding force then strangely disappears outside of the nucleus. Scientists today remain baffled how a force can act in this seemingly back-to-front way. A strong attractive 'force' would naturally be expected to be strongest, like a magnet, when particles are closest, instead, when quarks are closest the force becomes weak and puzzlingly momentarily disappears. Strangely, the binding force reappears to be inescapably strong as quarks again move to be farthermost apart. **Consequently, the <u>unsolvable</u> and**

baffling part has long been; what could cause a strong force to become weak and disappear when quarks are closest only to reappear again, in its very strongest state, when quarks have moved farthermost apart?

Virtual mass is a phase of negative energy of a deep vacuum which <u>because of the task is has to perform</u> needs to have the <u>characteristics</u> of a rapidly collapsing and reappearing Quantum Black Hole. [021] **Theories allow a Planck scale _virtual_ Black Hole such as this to appear spontaneously but only very briefly.** The quarks move apart when the phase of virtual mass disappears. When farthermost apart the negative energy phase of virtual mass is <u>reinstated in full</u> and now it's Black Hole like grip <u>prevents</u> any quarks escaping from a proton, to escape, a quark now needs to momentarily exceed the speed of light which is not possible. This is why the strong force is inescapably strong when quarks are most apart.

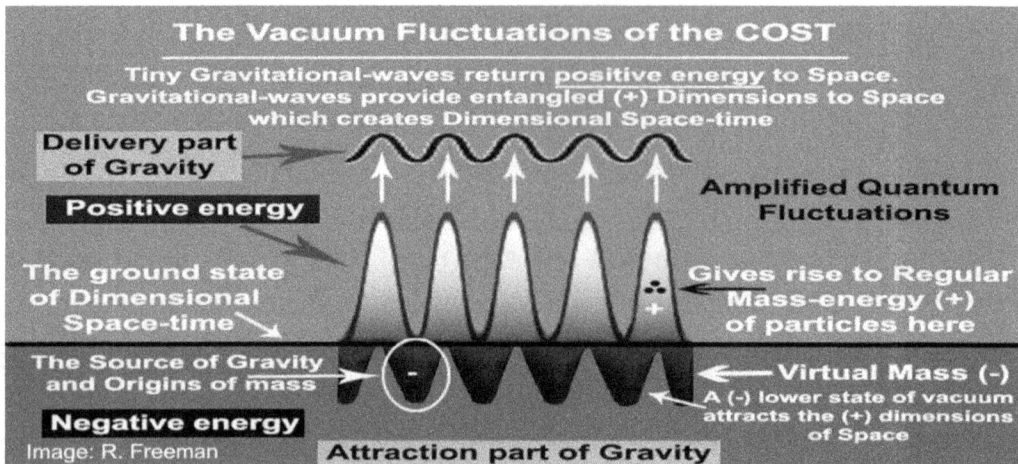

The Vacuum Fluctuations of the COST

Tiny Gravitational-waves return positive energy to Space. Gravitational-waves provide entangled (+) Dimensions to Space which creates Dimensional Space-time

Delivery part of Gravity
Positive energy

Amplified Quantum Fluctuations

The ground state of Dimensional Space-time

Gives rise to Regular Mass-energy (+) of particles here

The Source of Gravity and Origins of mass
Negative energy

Virtual Mass (-)
A (-) lower state of vacuum attracts the (+) dimensions of Space

Image: R. Freeman
Attraction part of Gravity

Now a flow of positive <u>dimensional</u> Space begins rapidly spiraling in from the outside as the quarks chase the rapidly disappearing negative energy of virtual mass to the center and in doing so hold a proton intact. The quarks now become tightly close together, almost <u>inescapably as one</u> but the quarks are <u>aggressively fighting back</u> from being <u>crushed out of existence</u> which is when the central positive pressure peaks 10 times higher than within a neutron star. **A proton is provided with spin from the <u>dimensions</u>**

of Space rapidly spiraling in from the outside which solves the proton's spin crisis. Now the negative energy phase of virtual mass <u>becomes exhausted</u> which is when the positive quarks are tightly bound closest together. This is why the strong force mysteriously disappears <u>when quarks are closest together</u>. Momentarily without the negative energy phase of virtual mass to tightly retain them the positive quarks, <u>with great might</u>, <u>repel</u> and rebound with positive pressure from each other but before the positive quarks can escape from a proton a new negative energy phase of virtual mass reappears in full and the negative pressure immediately intensely drags the quarks back to the center. All of which naturally creates internal pandemonium by rapidly agitating particles at near light speeds. [003] One complete <u>fluctuation</u> cycle of this near light speed <u>motion-energy</u> of a proton's quarks and their emitted or created gluons represents a proton's regular mass-energy. This is how the <u>negative</u> energy of virtual mass physically creates <u>positive</u> regular mass. [002] Scientists are today aware of both the inward and outward pressure phases within a proton but admit this is a *"fundamental aspect of the proton that we know very little about"* and *"where all the (mass) energy comes from is a big complicated mess"*.

The phase of virtual mass effectively provides a minuscule 'fluctuating', in a state of rapidly collapsing and reappearing, Quantum size Black Hole at the center of protons and this is what Black Holes do, they give particles near light speeds. **Note**: With my Natural Model it would have been impossible for binding energy to work like a regular force even if I had wished it to.

[011] This extraordinary video **actually reveals empty Space is the source of most of our mass**. In the video **'Your Mass is Not from the Higgs Boson'**, Prof. Derek Leinweber speaks about the findings of their research at the University of Adelaide. The research focuses on a super computer simulation of the theory of Quantum Chromodynamics which describes the action of the strong force within a proton. At time 3.57 one will find an image of quarks sitting on 'lumps' in the gluon field where the quarks are

positioned around a central area of empty Space nothingness. It is this actual empty Space, <u>vacuum nothingness</u>, which the research **clearly identifies** as being the source of the strong force which agitates and binds quarks to <u>create a proton's mass</u>. **The conclusion of the super computer simulation is that it is** *"extraordinary, because what we think of ordinary empty Space that turns out to be the thing that gives us all, <u>most of our mass</u>"* (99%) Another video; 'CSSM and CoEPP: The Standard Model and Beyond' relates to empty Space vacuum containing 'areas of positive charged density and <u>negative charged density</u>'.

With my Natural Model this **'extraordinary' empty Space** nothingness <u>exposed</u> by this computer simulation is <u>derived</u> from the same [015] **'astonishing phenomenon'** <u>discovered</u> by Prof. Alfred Leitenstorfer from the University of Konstanz in Germany. [016] This is also the same <u>lower state of deep vacuum</u> which Scientists have <u>discovered</u> by a calculation from the potential energy of the Higgs field and the masses of the Higgs and top quark which has been described as one of the **'biggest unsolved mysteries in physics'**. That is, the extraordinary, astonishing phenomenon which is one of the biggest unsolved mysteries in physics is actually a phase of virtual mass provided by my COST. <u>Note</u>; virtual mass would be expected to be a dimensionless state of Space which is why it is a lower state of Space.

But there is more! I believe the most compelling evidence, or <u>proof</u> if you now prefer, that a tiny phase of nothingness <u>truly provides both</u> mass and Gravity is the discovery that Galaxies are accelerating in the direction the universe is expanding. When a phase of pure empty Space provides both mass and Gravity and the universe is expanding into a seemingly endless void of pure empty Space the universe <u>will have</u> a massive resource for Gravitational Acceleration which acts in the direction which the universe is expanding. That is, due to Gravitational Attraction, Galaxies are simply accelerating as they free fall in the direction which the universe is expanding and there is absolutely no need for the mystifying and <u>totally,</u>

<u>hypothetical</u> Dark energy to appear from a never to be revealed source.

Today the exact movement of the particles inside of protons is not a done deal but it is generally agreed that particles are excited and agitated about at near light speeds which provides the particles with energy which equals regular mass. The working mechanism, because it is supplied by a COST, has to be the same inside both a proton and a neutron. To change a proton into a neutron one up quark changes into, or is replaced by, a down quark.

Within protons, the fluctuating negative energy of virtual mass now equals a similar amount of regular positive mass-energy of a proton. To prevent <u>each fluctuation cycle</u> of this positive energy **disastrously accumulating and very possibly destroying a proton** the mechanism stabilizes by creating tiny positive Gravitational waves which <u>carries energy away</u> from each <u>energy fluctuation cycle</u> and provides the system with entangled dimensions to Space which are naturally <u>attracted</u> to the tiny <u>negative energy phase</u> of virtual mass of other protons and so atoms. The negative energy phase of virtual mass provides the <u>strong</u> attraction part of Gravity and the tiny positive Gravitational waves provide the <u>variable delivery</u> part of Gravity. Virtual mass provides Gravity but the regular mass created by wildly exciting the quarks of a proton <u>has no Gravity</u>. However, regular mass cannot come into being without the phase of virtual mass, thus, wherever one finds regular mass one will also find Gravity. **Negative energy is simply a state of having less energy than the vacuum of dimensional Space** [see 111].

[033] Even though gluons are theorize to convey the strong force by gluing the quarks together it remains a <u>complete mystery</u> how gluons actually glue. If gluons did somehow bring the quarks of protons tightly together it is <u>very improbable</u> that they would somehow turn this glue off and <u>release quarks</u> when they are closest together and somehow <u>wait to only reinstate</u> their fullest influence when quarks have had the opportunity to move farthest apart. It would be very difficult to explain how gluons turn themselves <u>on and off</u> like this. However, this is exactly how a tiny phase of virtual mass

works. The negative energy phase of virtual mass comes and goes with fluctuations. The strong force is primarily driven by a tiny negative energy phase of virtual mass which is only a state of pure empty Space. Virtual mass <u>stays behind</u> to provide mass and Gravity when all particles, including quarks and gluons, are <u>crushed out of existence inside</u> of a Black Hole.

Although today's theory of a proton's mass, Quantum Chromodynamics (QCD), is <u>mathematically extraordinarily complicated</u>, Scientists say they have made some progress with it. The QCD describes how a proton's mass is created by a strong interaction between quarks and gluons which create <u>flux tubes</u> of empty Space. Scientists have attempted to learn from the QCD equation by calculating <u>approximate</u> solutions using super computers but the QCD's complexity is yet to give up its final definitive answer. If one adds the <u>uncertainty principle</u> to the QCD's complexity, like some have implied, the final answer may never be certain but may reflect a proton's energy level when measured. Scientists say *"in experiments at both the Relativistic Heavy Ion Collider and the Large Hadron Collider they see patterns that they think QCD should explain, but are unable to because of the sheer complexity of the theory"*. Although the QCD theory generally works <u>differently</u> to my greatly more simple phase of virtual mass both are expressing regions of empty Space as being *"the thing that gives us all, <u>most of our mass</u>"*.

Ideally, if not mandatory, a theory for the <u>source</u> of the strong force should provide a clear and understandable correlation with Gravity, however, the QCD theory itself fails to reconcile its created amount of mass with Gravity. The QCD theory fails to explain a proton's spin or how both mass and Gravity are provided in Black Holes where these very same quarks and gluons have been <u>crushed out of existence.</u> The theory of QCD cannot account for the mass and Gravity of <u>invisible</u> Dark matter which is clearly responsible for at least 85% of the universe's mass and Gravity. The QCD cannot explain why Galaxies are accelerating or why Matter prevailed over Anti-matter. The QCD theory cannot directly explain how Quantum Gravity works. **The <u>Natural Model</u> has a very clear and understandable path to superbly correlating all of these things, <u>including Quantum Gravity</u> directly to the '<u>source</u>' of the strong force. I show at page 347 how Quantum Gravity works for the <u>exact same reason</u> the strong force works.**

A Natural Model is all about looking at the big, fully correlated picture and clearly reconciles all mass created by the strong force with <u>both</u> Einstein's Gravity and Quantum Gravity. Mother Nature does not first sit down and design over complex things; she simply stands aside and lets objects interact with each other allowing a universe to evolve naturally by its own means. Simple Quantum vacuum fluctuations evolved into virtual protons which provided the blue print to evolve into regular protons which evolved into atoms which evolved into everything else, including ourselves.

So what role can gluons play? There appears to be no free quarks so it is apparent that <u>valence quarks</u> need to interact with partners and do this by emitting and adsorbing gluons through a configuration of flux tubes. The gluon field creates flux tubes between <u>valence quarks</u> which may <u>distort and direct</u> the phase of virtual mass and allow quarks to interact to exist.

Teams of particle Physicists have searched for many decades attempting to discover the means deployed by the strong force to create nearly all mass. [117] see: ***Quark Math Still Conflicts With Experiments -- QCD simply fails to produce a meaningful answer.*** Some refer to negative potential energy while others talk of areas of negative charged density. [041; 003] Some attempt to understand quark activity with experiments using the world's most powerful particle colliders where quark movement is described as near light speed pandemonium while others have pursued theories and equations using powerful super computers which results show three quarks being moved more orderly apart by flux tubes. [032] However, theories like the QCD, in its original state, have become <u>so incredibly complex</u> that it is said today's <u>most powerful supercomputers</u> would take many decades or possibly lifetimes of continuous computation to fully unravel and provide definitive answers. I don't have a super computer or a collider machine only a tiny phase of pure empty Space which I have called virtual mass. A million dollar prize remains awaiting for anyone who discovers the final answer.

[015] One really needs to go back to the research by Prof. Alfred Leitenstorfer and his team at the University of Konstanz which clearly show Quantum Vacuum Fluctuations simply phasing to a vacuum lower than the ground state of empty Space. It is easy to understand the progression is natural; simple Quantum Vacuum Fluctuations which naturally create pairs of virtual particles evolved into virtual protons which now also naturally created pairs of virtual particles. Virtual protons now evolved into regular protons which now also naturally create pairs of virtual particles.

One may ask; would not a fluctuating, minuscule Black Hole at the hearts of nucleons turn all objects into Black Holes? No, compared to the <u>fluctuating</u>, minuscule Black Hole atoms are so big that this is like finding a black poppy seed at the center of a tennis court. Only if all of the poppy seeds were to come together would they create a regular Black Hole; and is exactly how this model created the first Super Massive Black Holes. This is why stellar Black Holes are made the same way. When creating a Stellar Black Hole all Matter is crushed out of existence leaving only COST units fluctuating to a phase of virtual mass. Now strong Gravity prevents the production of their tiny, Gravitational waves allowing COST units to come very close together.

The tiny phase of **virtual mass** does not extend pass the nucleus of atoms which <u>**solves**</u> why the strongest part of binding force does not extend pass the nucleus but acts strongest on all particles within the nucleus. [031] The COST by providing an incoming whirlpool flow of ground-state dimensional Space is responsible for a proton's spin which <u>**solves**</u> a mystery of why the proton's spin has be found to be more than the spin of the quarks. The rapidly fluctuating mechanism of the COST is responsible for all of the rapid movement of particles. This activity is controlled and regulated or timed by the very rapidly fluctuating COST which mechanism is based on Quantum

<u>vacuum</u> fluctuations. The COST in its raw Dark matter state, before it acquired three additional Matter quarks, was indeed a virtual proton.

[032] Scientists have used several methods attempting to measure the exact radius of a proton. Due to the standard model of particle physics Scientists expected the size of a proton to be a <u>fixed</u> measurement. Experimental results have mysteriously varied greatly, however, Scientists have <u>remained determined</u> to discover what this <u>fixed</u> measurement is.

Naturally the phasing to the negative energy phase of virtual mass affects the <u>size</u> and mass of a proton. Not only the proton but the quarks themselves may vary in size. When quarks and gluons are momentarily <u>tightly bound</u> together a proton has a <u>smaller radius</u> and its regular mass momentarily peaks the instant before virtual mass momentarily disappears. When quarks are momentarily <u>farthermost apart</u> a proton has a <u>larger radius</u> and its regular mass is least but now a fresh phase of virtual mass immediately peaks and begins bringing the quarks back together and in doing so creates regular mass. Thus, the size of a proton is not fixed as theorized but is variable, consequently, radius, <u>as experimental results have shown, depends on when or how it is measured</u>. The reason why quarks themselves would be expected to expand is because of the ramifications of how this model solves wave and particle duality. Particles are not hard, tiny, permanent marbles. **When exposed to a tiny phase of strong, <u>negative</u> energy of a deep vacuum a particle's wave-like form is wrapped into a very dense <u>spherical field</u> to become a particle state which now has regular mass from <u>being tightly constrained.</u>** If a particle is momentarily less constrained its tight particle state loosens up causing a particle to expand. Left unconstrained an unobserved particle quickly becomes a wave of potentials. To be a particle requires binding energy.

The bottom line; for a model for the origins of regular mass to be successful the model <u>has to clearly show</u> how it also physically creates an appropriate amount of Gravity relating to an alike amount of mass. The model must

<u>also clearly show</u> why and how Gravity physically slows the passing of time.

So what creates the regular mass of all known Matter? Answer; a tiny phase of empty Space at the hearts of nucleons. Sure, it sounds a little crazy that a tiny phase of invisible nothingness provides all mass but please think about it. Place a piece of paper on the nozzle of a vacuum cleaner, the paper now requires a force to remove it, it is as if the paper is now heavier. Similarly, quarks become heavy with mass by just being where they are.

[034] I can hear one say, what of the Higgs particle, does it not provide mass? From what I read the Higgs boson is far from being a 'done deal' for it is presently said there are many unsolved problems relating to the 'found' Higgs boson and Physicists say *"they understand little about the 'present everywhere' Higgs field"*. A successful finalized model for mass should work in a clear and understandable way and not be surrounded by many presently unsolvable loose ends which prevent it from working the way it has to work. Like the Matter and Anti-matter asymmetry problem of the Big Bang I struggle to understand why scientific theories are loudly proclaimed to be factual and accurate when they obviously contain disturbing flaws.

Scientists know that the Standard Model of Particle Physics <u>forecast that all of its particles are massless</u> which is in disagreement with what we observe around us. Due to the <u>double slit experiment</u> one can confidently deduce that <u>all particles are waves</u> which have been constrained as particles. The Higgs boson has <u>nothing to do</u> with solving wave and particle duality. Unfortunately, without a rational, commonsense model for wave and particle duality, <u>particle</u> Physicists may be looking at only half the picture. Quantum Theory says a particle <u>has no properties</u>, <u>including the property of mass, until it is observed or measured</u>. If the famed Higgs field was the final source of a particle's mass all particles would surely have the <u>property</u> of mass at all times and observing or measuring would have nothing to do with it, Quantum Theory and experiments say this is not so. It is obvious that observing or measuring has a profound physical effect on a particle.

Just the act of contracting a tiny Quantum 'wave' so as it is <u>bound up tightly</u> as a tiny dimensional particle should surely require binding <u>energy</u> which will provide regular <u>mass</u> to a particle. It is simply <u>common sense</u> that to actually physically bind a wave into a particle requires some form of a compression or binding together mechanism similar to the strong force. <u>If one accepts</u> that it obviously <u>requires energy to tightly bind</u> or squeeze a wave of potentials into a tightly bound together particle state than particles have obtained the property of regular mass-energy from their wave state being physically contracted to a particle state. See <u>chapter 13.</u>

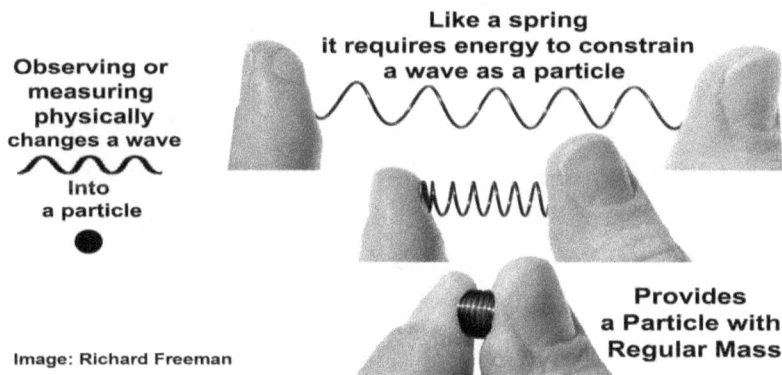

Observing or measuring physically changes a wave ～～～ Into a particle ●

Like a spring it requires energy to constrain a wave as a particle

Provides a Particle with Regular Mass

Image: Richard Freeman

The contraction to a particle would be expected to <u>directly cause</u> both spin and regular mass to be naturally built into particles <u>at the time they are formed from waves</u>. Actually, the fact that particles have spin is evidence that this modeling is correct. The Higgs boson cannot provide particles with spin and I doubt the Higgs has anything to do with changing a wave into a tightly bound together particle. A Natural Universe has no use for thick sticky syrup or sticky fly paper, regular mass is provided to a particle in a **rational and understandable manner** for the same reason as compressing a wave like spring between your fingers where the compressed spring retains the energy required to compress it. The bigger the spring and the tighter it is compressed the more energy it can retain. Bigger particles such as quarks would likely retain more energy than much smaller gluons which they do.

Energy here has an equal amount of mass which means <u>this model</u> **cannot**

avoid providing a particle with mass even if it leaves us in a 'sticky' position with the Higgs. This also provides a clear understanding as to why the Standard Model of Particle Physics appears to predict that all particles should be massless. Unobserved and left to their own devices particles are waves without properties until they are tightly restrained as particles by the act of observing or measuring **which interacts with a Dimensional Reality**.

One could argue very strongly that any modeling for mass which fails to directly and clearly reconcile its mass with Gravity will eventually be a failure. It is just as important, in order to understand the very fundamental nature of particles, to finally solve the cause of wave and particle duality. Without a rational model for wave and particle duality Scientists searching for the primary source of regular mass appear to be ignoring a most fundamental property of particles. I cannot express enough the importance of solving wave and particle duality. Without solving wave and particle duality one cannot explain why we can observe all of the wonderful things around us. Mother Nature has a way of turning waves of potential energy into the particles with the properties which make up everything which we can all observe. Solving wave and particle duality is obviously fundamental if one is to understand the very nature of the very same particles which make up you and me. Without a rational method which allows the simple act of observing or measuring to physically transform a wave into a particle all Matter would exist as invisible waves and all Matter as we know it could not exist. Today, wave and particle duality remains a mystery of physics.

While particle Physicists without a doubt undertake brilliant research it seems they are attempting to solve everything as particles and are ignoring wave-particle duality, there are not many 'wave' Physicists around. Truth is, particle Physicists have little choice because once one attempts to observe or measure a wave it immediately contracts to a particle. The double slit experiment leaves little doubt that all particles reside as fuzzy waves unless they are being observed or measured. In searching for clues of the working of particles Scientists smash protons into each other in massive colliding machines with the use of extremely strong magnetic **fields which are purposely designed <u>to hold particles together</u>.** Protons smash apart producing an array of sub-atomic particles which are said to have very short life spans. Particle Physicists then observe and measure the results. It is plausible that once out of the environment of where restrained waves were held as particles, the particles very quickly disappear as waves of potential, thus, agreeing with the confirmed results of the double slit experiment.

One may now wish to ask; what restrains the waves as particles so as atoms can build objects made from atoms? I am very confident that my car does not disappear into waves when I'm not looking at it. Inside of a proton the powerful phase of virtual mass would be expected to do exactly this as part of the way it powerfully binds together all within the nucleus of atoms. This naturally provides the particles with their particle mass. The orbiting electrons being far from the nucleus of atoms are free to act as waves.

Nevertheless, a 1 % of an atom's mass is theorized to arise from a complex and said <u>too difficult to rationally describe</u> interaction of an <u>atom's particles</u> with the Higgs field. The Higgs boson has been a much celebrated particle, however, with such modeling there remains <u>many unsolved issues</u> and too many to reflect on here. Particle Physicists say particles attain mass by being slowed by the Higgs field which is akin to thick sticky syrup or sticky fly paper. Others say forget the metaphors as it is **<u>not possible to fittingly or rationally describe</u>** the theorized interaction with the Higgs field other than to say 'the theory' tells us the 'interaction' provides mass.

027; 070 <u>Scientists remain mystified</u> as to why the 'recently discovered' particle, designated as being the Higgs boson, was <u>observed</u> (LHC) to be so light. The Standard Model of Particle Physics represents all known particles. <u>Scientists say</u> interactions between the Higgs and the particles of the Standard Model of Particle Physics would make the Higgs very much heavier than the 'found' Higgs particle. The found Higgs boson has a mass of 126 giga-electron-volts. However, interactions with the other known particles should add about 10,000,000,000,000,000,000 giga-electron-volts to its mass. This is a little like searching for a lost solid gold bowling ball and finding a polystyrene ball and being mystified why your prized golden bowling ball is now so lightweight. For decades Scientists had referred to a cherished theory, called Supersymmetry, which was said to <u>solve</u> this.

The Supersymmetry theories that are said to explain properties relating to the discovered light Higgs boson predict a family of new particles with masses similar to those of the W and Z bosons. These are the predicted particles which the Large Hadron Collider was built to discover. Supersymmetry says <u>every</u> particle had a Supersymmetry partner. All of these 'extra' <u>theorized</u> particles would have <u>canceled out the contributions to the Higgs 'mass' allowing a light Higgs boson to exist</u>. To finally prove their well respected, solve-all Supersymmetry theory beyond doubt Scientists built the largest, most sophisticated machine ever constructed, the Large Hadron Collider (LHC). [086] At the time, Scientists said <u>if their theory was correct,</u> Supersymmetry particles should appear in collisions at the LHC. By <u>fixing</u> the mass of the <u>Higgs boson</u> this would have solved a major problem with the Standard Model of Particle Physics. Physicists had <u>little doubt</u>, if they really, truly existed, the LHC was <u>specifically designed</u> to have the capability to thoroughly explore the power levels required to expose the much sought after Supersymmetry particles.

[026, 070] The experiments have now been carried out and Physicists have now looked precisely where these particles need to reside but in the collision debris, Physicists have found no particles that could be Dark matter, no siblings or cousins of the Higgs boson, no sign of extra dimensions, no leptoquarks and <u>no trace of any of the many desperately sought after Supersymmetry particles.</u> Some Physicists have described this as their worst nightmare for the unexpected results have killed off many of their long standing beliefs, predictions and cherished theories. This may well be the greatest failure in modern particle physics. However, by pointing one in a new direction, success will often come from failure.

[027] Theories say that when the Higgs interacts with other elementary particles the normal mass of the boson should be subjected to large fluctuations which cause its mass to grow bigger than any value which Scientists have observed. To solve this, Supersymmetry says that the Supersymmetry partners of every elementary particle also interact with the Higgs in such a way which almost exactly cancels out the fluctuations of normal partners. Consequently, many of the seemingly insurmountable problems relating to the found light Higgs providing particles with mass was said to be nothing a swarm of Supersymmetry particles at the LHC couldn't cure. **Unfortunately for particle Physicists <u>no trace</u> of any of the 'many' Supersymmetry particles was discovered during the LHC experiments which were specifically designed to expose them.** Particle Physicists now say it will take additional work to establish whether or not the discovered particle is the predicted Higgs boson. [088] In an effort to solve this problem relating to the Higgs some are turning to new theories which have been described as both clever and far-fetched. [070;] *A Deepening Crisis Forces Physicists to Rethink Structure of Nature's Laws.* [027] *We may reside in an unnatural universe where natural does not make sense.* Consequently, there remain very fundamental, unsolved problems with the 'found' Higgs boson.

The bottom line is the <u>observed results</u> do not add up for the Higgs. [036] The fact that the particle said to be the Higgs has been observed to be **<u>unaccountably light</u>** tells us there is <u>required to be something additional at play</u> or to put it in the words of a most respected particle Physicist *"there is an additional intrinsic contribution whose source is not known or easily discovered. As for why its mass is what it is we do not know."* A Natural Model's *"additional intrinsic contribution"* is the virtual mass phase of a deep vacuum which provides negative energy which does all of the work of creating regular mass. Particle Physicists spend much of their time debating how 'particles' allow all to operate. My virtual mass is not a particle it is a just a tiny state of complete empty Space which has now been discovered in experiments but unfortunately nobody understands why it should exist.

Today's theories relating to mass are exceedingly complex and all have serious, unsolved loose ends, many of which have remained for many decades as the biggest, unsolved mysteries in physics. To solve these mysteries, one may need to step sideways from particles and explore the nature of Space-time. To find something one needs to look in the right place, unfortunately, Scientists today actually have no idea what the fabric of dimensional Space-time is or how it was made.

I truly respect particle Physicists, I have provided numerous links to their dedicated and brilliant research but I sometimes think that they are inclined to have too many independent theories attempting to solve what is expressed to be independent mysteries. Over the years many of these proposed theories, many of which I have also previously taken on board, have failed. I now see the unsolved mysteries differently. The origins of mass, the source of Gravity, wave-particle duality, Space, dimensions and the passing of time, are obviously all freely and closely interacting with each other. Thus, all related theories should also freely and closely interact and relate directly to every other related theory; there should be no loose ends which is surely the way Mother Nature operates a Natural Universe.

A Natural Universe has no unworkable theories which do not agree with actual observed results of experiments, no sticky fly paper, metaphors or new exotic particles, just a tiny phase of pure, vacuum empty Space. This allows atoms to have far more mass-energy than the sum of their particles which allows a universe to be built from very little energy. Amazingly, an atom is now in essence a bunch of near massless particles with a few electrons buzzing around and all mass is provided free by particles just being where they are. The all important delightful electrons buzzing around creating an electron cloud make atoms and objects appear solid.

Our universe is basically no more than one big free lunch! When the source of an atom's mass, energy and Gravity is a tiny phase of pure empty Space and 99.99% of a whole atom is also loosely said to be empty Space, one can begin to understand that a universe full of Matter can come into being from nothingness. No Big Bang of immense energy is required for we are all made from mostly empty Space! Sounds crazy but in a way is true.

[026, 070] The failure of the LHC to find Supersymmetry particles has left Scientists with no other long standing theories to solve these mysteries related to the origins of mass. Interestingly, I have read without the modeling of Supersymmetry, another popular theory called String Theory, which was for many years the talk of the town, cannot describe the observed Universe. However, many particle Physicists remain fixated on discovering new exotic particles and unfathomable dimensions which they hope will lead to new exotic theories to explain the origins of mass.

In an attempt to understand how Matter exists without the particles the LHC failed to expose, Scientists are now proposing to build a much bigger colliding machine to be called the Future Circular Collider. Scientists say their proposed 100 kilometers circumference Collider will also search for new physics and evidence beyond the Standard Model to account for Dark matter and the domination of Matter over Anti-matter (after the Big Bang).

But there is some light; the team of Scientists led by Prof. Dr. Alfred Leitenstorfer from the University of Konstanz in Germany may be tantalizingly close to figuring this out. Compared to CERN (LHC), which at times had 6000 scientists plus administrative staff members, and hosted about 12,000 users, Prof. Alfred Leitenstorfer's team can be said to be the underdogs. [015] **However, they may make the most significant discoveries simply because they are looking in a different place, they are looking into pure nothingness.** I myself was astounded to read what they had discovered, for in my mind they have clearly discovered my Natural Model's phase of virtual mass which I had originally predicted over thirty years ago.

8 THE CREATION OF THE FIRST ATOMS.

Astronomers have discovered Galaxies reside within vast regions of Dark matter. Consequently, one should first ask; **why do Galaxies reside within regions richest with Dark matter and why is almost all regular Matter confined to Galaxies?** One should then more seriously ask why Super Massive Black Holes are found <u>only at the hearts of Galaxies</u>. One should ask these things because there really has to be a very good reason why Super Massive Black Holes are found at the hearts of even the very first small, adolescent Galaxies. The obvious answer; Matter was created within **regions richest with Dark matter** because Dark matter provided a massive resource for creating the Super Massive Black Holes which provided the power houses to transform Dark matter into regular Matter. When one accepts this very obvious and simple reasoning Matter creation and Galaxy formation becomes understandable. Today, the amazingly powerful regions at the poles of Super Massive Black Holes and the truly extraordinary large amounts of <u>star forming Matter they eject</u> are indeed shrouded in mystery.

I have used actual experimental evidence to explained how Gravity, powered by nothing more than <u>empty Space</u>, slowly arose from squeezing numerous Quantum vacuum fluctuations and, in doing so, created <u>virtual protons</u> (Dark-matter) armed with vacuum fluctuations. I have explained how virtual protons created Super Massive Black Holes. Now it is time to explain how virtual protons, by acquiring three additional Matter quarks, are transformed into the regular protons which make hydrogen atoms.

[015] As already expressed virtual protons are based on Quantum vacuum fluctuations. Related research by Alfred Leitenstorfer from the University of Konstanz in Germany has revealed *'when the vacuum was squeezed'* fluctuations fluctuated to a level below the background noise level, which is *'lower than the ground-state (the vacuum) of empty Space'*. The Scientists also observed *the fluctuations became way louder,* (an increase of energy).

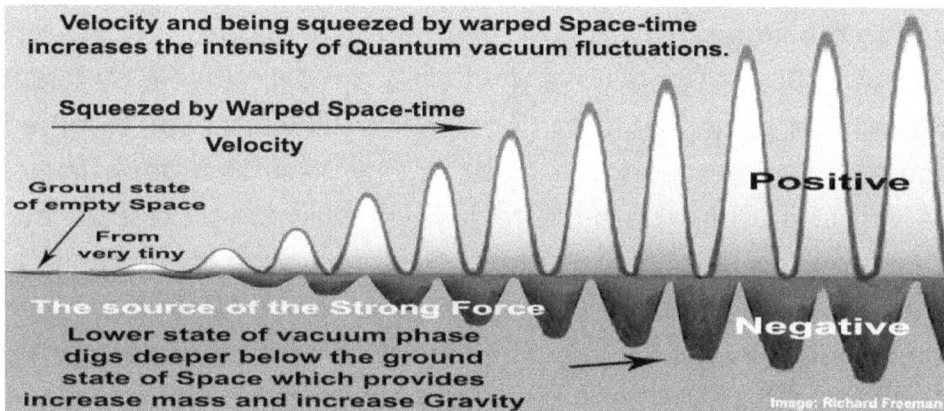

The two most important observed properties which we are most interested in are fluctuations have the ability to become very significantly amplified which <u>means they became many times larger</u> and fluctuated to momentarily provide a <u>phase of a lower state of the vacuum</u> which directly aligns to my phase of virtual mass. The energy phase would be expected to be larger because it is a phase of expanding energy. The lower state phase, although equal in being opposite, is more compact. Virtual protons will again increase their energy by being squeezed by Gravity's warped Space-time which will provide the squeezing of Space observed by Prof. Dr. Alfred Leitenstorfer and his team. When exposed to strong spiraling magnetic fields Matter particles will be separated from anti-particles. This all remains closely aligned with my first book which I published in 2014, only now the results from several different fields of more recent research allow the very same modeling to be reinforced by the actual results of scientific experiments and observations.

Our fledgling universe has now progressed to where it has the conveyance for light, yet it has no light; it already has Gravity for atoms, yet it has no atoms and it has Dark Galaxies with central Super Massive Black Holes with outer halos of Dark matter for Galaxies of stars to evolve within, yet it has no stars. Only now, by first having these vital components in place, can our young, growing universe begin to produce its first real particles of Matter.

All life, Planets and Stars began and developed from the smallest and advanced to the largest, this is how Mother Nature constructs everything in the universe. A Big Bang is an event which is not explainable by scientific or natural laws and somehow instantly supplies all Matter for all time in less than one astonishing second. In contrast, my Natural Model first created invisible virtual protons (Dark matter) which will now be transformed into regular protons to make regular atoms. Beginning with the insignificant and advancing to the significant is truly Mother Nature's way. One cannot instantly have a fully grown oak tree complete with all of its acorns for all time. As wonderful as Mother Nature is, she does not have a magic wand.

We will obviously require a massive amount of energy to transform virtual protons into regular protons to create regular Matter. The most radical and inexplicable method would be to suggest that a colossal, dense form of an unknown-to-man form of energy just instantaneously appeared everywhere from absolutely nowhere as hypothesized at the 'Big Bang'. However, we obviously need to do a lot better than that if we wish anyone to believe in the commonsense approach of a naturally evolving universe. I am sure no one would believe me if, for this model, I now simply wrote, *"From truly nowhere and for no known or actual reason, an unknown-to-man form of all-powerful energy just bizarrely appeared and within a few, astonishing seconds had transformed virtual protons (Dark matter) into a sea of regular Matter, thus creating our whole universe"*. This could very easily be said but would never, ever be explainable.

My Natural Model offers a far less radical approach where the actual source of energy sensibly matches the fundamental requirements of the known characteristics of the Matter which will diversify our vast universe today. It is truly only common sense to obtain energy from a source and with a method which will mirror the same known characteristics and behavior of Matter. One is then able to understand why it is natural for Matter particles to behave and have the characteristics and properties they are observed to have. Energy, mass and Gravity are surely the most fundamental properties of all atoms, which is why it is truly most fitting and compelling that our first atoms appropriately come into being from a resource directly related to the same fundamental properties of all atoms, a resource so common that it is prevalent and available everywhere throughout the universe which is, energy enabled by Gravity.

Today many of science's theories, which for decades had been enthusiastically expressed with bountiful amounts of confidence, are disappointingly incomplete or broken and even in a state of complete failure. In a relentless search to understand everything, the resource of many thousands of Scientists, many billions of dollars and massive machinery has produced results which have been both exciting and disappointing. My aim has been to use this knowledge by including the success and more importantly understand the reasons for the failure.

Many of the failures have greatly reinforced the concepts of my Natural Model. For instance, if the theorized WIMPs were found by sophisticated detectors than my virtual protons may not be Dark matter, if the theorized Supersymmetry particles were found within sophisticated colliding machines the COST may not play its role in the creation of regular mass, if the theorized slowing of the expansion of the universe was found by mighty telescopes than my modeling which required Outer Gravity was wrong and if the hypothetical Graviton was found my model for Gravity would again be wrong. If the present theory of how Quasars are powered my model for creating Matter would likely not be correct. Scientists have also failed to directly detect Dark energy which if they did would again mean that my model for Outer Gravity was wrong. If the theorized hierarchical merging of colliding tiny adolescent Galaxies to grow larger Galaxies was discovered by the James Webb Space Telescope than my Fountain Formation model of Galaxies quickly and neatly growing themselves was wrong. If my Natural Model was correct all of these theories or models were required to fail. Over the decades it has taken me to compile this model, all of these enthusiastically supported ideas have failed or have remained a mystery.

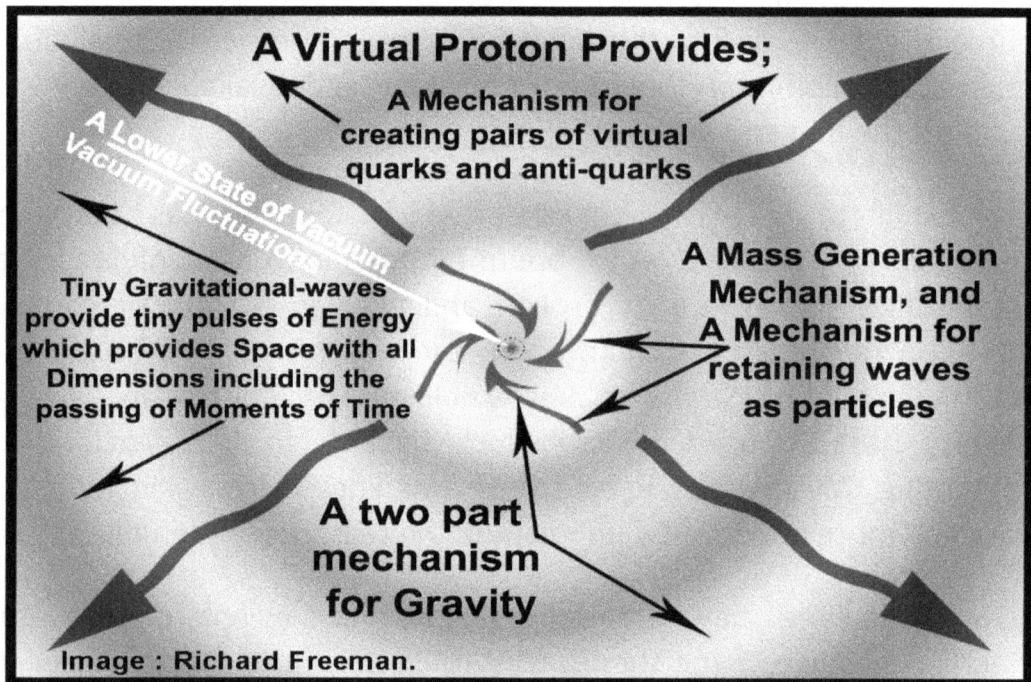

A Virtual Proton Provides;

A Mechanism for creating pairs of virtual quarks and anti-quarks

A Lower State of Vacuum Vacuum Fluctuations

Tiny Gravitational-waves provide tiny pulses of Energy which provides Space with all Dimensions including the passing of Moments of Time

A Mass Generation Mechanism, and A Mechanism for retaining waves as particles

A two part mechanism for Gravity

Image : Richard Freeman.

I have respected the predictions of the most successful theories and used the actual results of experiments and observations to piece this model together. There is no doubt one will require a method to provide a virtual proton with far more substance and mass than invisible Dark matter. However, a virtual proton will provide **regular protons** with the essential vacuum fluctuations for creating virtual pairs of sea quark and anti-quarks which come and go with annihilation. **Virtual protons will also provide regular protons with their essential Gravity and mass generation engines.**

The formation of Super Massive Black Holes has unleashed a super, massive pulse of Gravitational Potential Energy to everything within <u>their regions</u> and is most probably the greatest Gravitational event which will ever occur within our universe. Billions of these types of events occurring throughout our adolescent universe unleash Gravitational Potential Energy on a scale like the universe may never see again. This may trigger a powerful beginning that initiates the creation of regular Matter at numerous places where Galaxies will form throughout the universe.

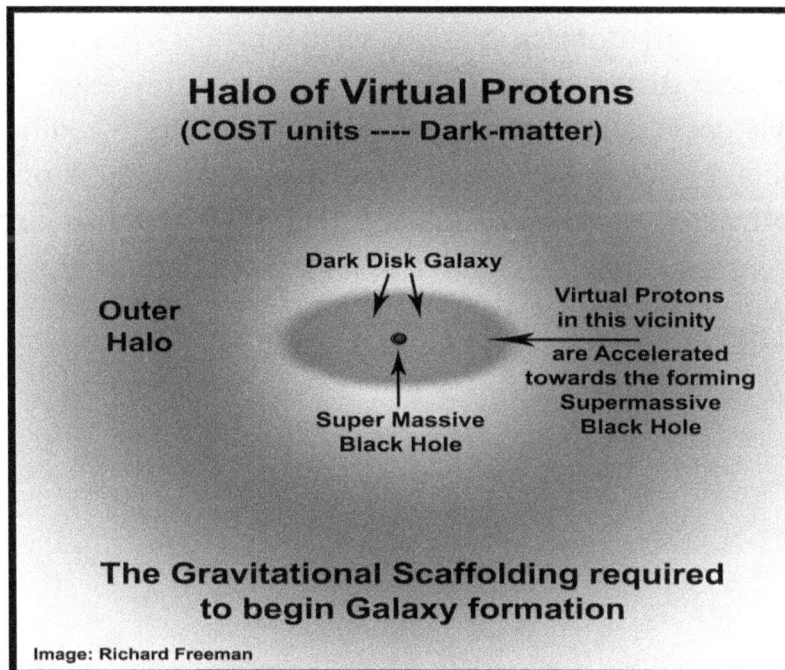

Halo of Virtual Protons
(COST units ---- Dark-matter)

Dark Disk Galaxy

Outer Halo

Virtual Protons in this vicinity are Accelerated towards the forming Supermassive Black Hole

Super Massive Black Hole

The Gravitational Scaffolding required to begin Galaxy formation

Image: Richard Freeman

The Gravitational collapse at centers of large clouds of virtual protons (Dark matter) first created Super Massive Black Holes surrounded by outer halos of Dark matter. This first created disk shaped Dark Galaxies made from Dark matter which provided the Gravitational scaffold essential for beginning the formation of regular bright disk shaped Galaxies of stars. The significance of the direct relationship of providing 'both' the first atoms as well as the Gravitational scaffolding essential for new Galaxies at the exact time for Galaxies to develop is strong evidence of how Galaxies were first created.

The actual creation of Super Massive Black Holes provides the <u>initial pulse</u> of acceleration of virtual protons, acceleration of which the newly created Super Massive Black Holes provide the massive Gravity to sustain. Virtual protons fluctuate from a phase of particle and anti-particle pairs of virtual particles to a phase of virtual mass. **The virtual particles will provide the source of the particles of regular protons and the phase of virtual mass will provide a source of all regular mass. Virtual protons also provide a two part mechanism for Gravity.** This is why virtual protons provide the perfect, precise scaffolding and mechanisms essential for regular protons and neutrons to form atoms and is why virtual protons (Dark matter) had to be created before true atoms. The road map ahead is clear and straight forward; a virtual proton only needs to acquire three additional Matter quarks to become a regular proton and a regular proton only needs to acquire an attracted electron to make a hydrogen atom. Hydrogen atoms make stars where a repeated chain of fusion creates all other elements.

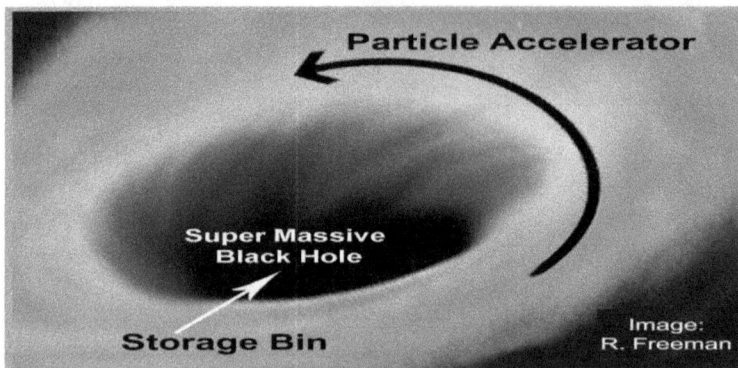

Super Massive Black Holes will provide essential storage bins to lock away created anti-particles to prevent anti-particles from annihilating our precious Matter particles. Depriving virtual Matter particles of their virtual anti-particle counterparts allows virtual Matter particles to become <u>fully exposed</u> to the <u>complete</u> virtual mass cyclic phase of the COST which will allow virtual particles to act more like regular Matter particles.

Matter and Anti-matter generally spiral <u>away</u> from each other inside of spiraling magnetic fields. [081] Super Massive Black Holes power Quasars which provide powerful, magnetic fields which will be essential for separating anti-particles from Matter particles and directing anti-particles into a Super Massive Black Hole. Scientists say the regions at the poles of Super Massive Black Holes are extremely complex and require more research and observation before Scientists unravels their full implications. In my first book, **'The Big Stretch, A Universe Without The Big Bang'**, this <u>same</u> modeling was based on self-reasoning of existing knowledge. Many new discoveries and observational evidence from new telescopes and Space craft have now closely aligned with much of my previous modeling.

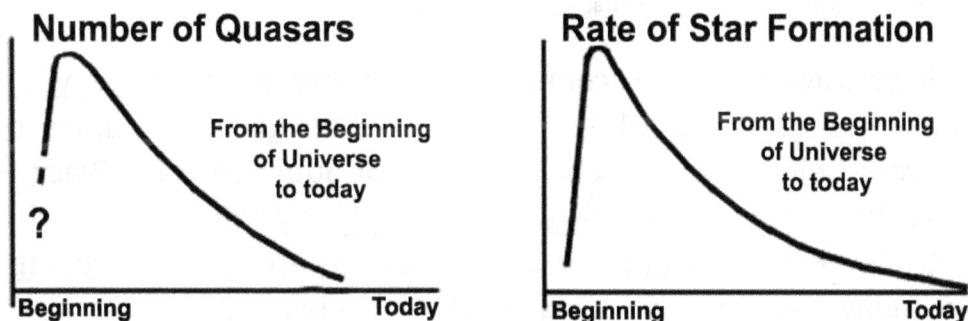

Number of Quasars

From the Beginning
of Universe
to today

?

Beginning   Today

Rate of Star Formation

From the Beginning
of Universe
to today

Beginning   Today

When a Quasar's jet is pointing directly towards you a Quasar is called a Blazar. Both Quasars and Blazars are found in Radio Galaxies. Developed Quasars are observed to be most prolific near the beginning of the universe and reside in Galaxies with an active Galactic nucleus which is exactly what is required for a universe which creates its own Matter in these regions.

[081] Powered by Super Massive Black Holes Scientists remain puzzled how it was possible for these systems to exist when there is not the required time from a Big Bang for them to grow given <u>current theories and understanding</u>. A Natural Universe forms Super Massive Black Holes very quickly from a Gravitational implosion of central regions of large clouds of virtual protons. For example, Scientists say the Gravitational implosion of Matter which forms a 'Stellar' Black Hole happens at near the speed of light and may take only seconds to become a Black Hole.

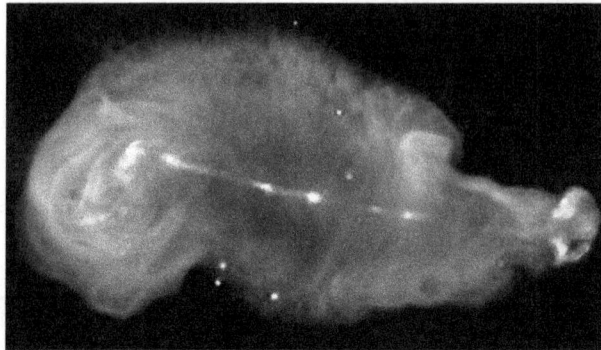

[039] The impressive radio telescopes of today allow one to <u>actually observe</u> back near the beginning of the universe when gigantic plumes of hot Matter were being ejected from the poles of Super Massive Black Holes. One can actually observe <u>exactly where and how Matter is being created</u> to form a Galaxy. Even though these amazing events are <u>clearly observable</u>, <u>no science today</u> has been directed to the possibility that all Matter was created this way and so these regions remain shrouded in mystery.

If legendary sleuth Sherlock Holmes, who was famous for his astute logical way of thinking and his forensic science skills, was attempting to locate the source of Matter and had a telescope, rather than a magnifying glass, he may have said that it is *"Elementary my dear Watson, all Matter is, in effect, entirely confined to Galaxies, so that is where we must apply our astute observation and deduction, in our quest to locate the very source of all regular Matter".* Indeed, this evidence is clear and indisputable.

Just a few parts of a second into the life of a Big Bang universe, energy is required to transform into mandatory, <u>exactly equal</u>, parts of Matter and Anti-matter. Meaning everything which Matter has, Anti-matter mirrors. What happened to swing the balance away from Anti-matter remains as one of the greatest unsolved mysteries in physics.

008 <u>We shouldn't be here; the physics says so</u>! Over many decades I have read these same words in many different articles. The laws of physics and many decades of undeniable evidence from <u>many</u> experiments <u>absolutely confirms</u> that moments after the Big Bang all Anti-matter and Matter would have indisputably, completely annihilated each other leaving zero Matter to construct a universe. Meaning the universe cannot have begun with a Big Bang. <u>All Physicists know this is true.</u> This is not a small issue, nor is it an issue which has ever gone away. I truly cannot understand why these stringent laws of physics which are today backed up by many decades of associated experimental evidence are conveniently not adhered to and are seemingly arrogantly ignored by the Big Bang theory. Many <u>very precise</u> experiments attempting to prove this is not true have all proved it is true; all Matter <u>would have been</u> annihilated by Anti-matter if there had been a Big Bang. We live within a universe where everything is made of Matter. The whereabouts of the 'missing' Anti-matter which did not annihilate all 'remaining' Matter moments after the Big Bang is a most important question <u>which, despite the enormous resources of modern science, nobody has yet been able to provide a definitive answer</u>. Despite the overwhelming experimental evidence Scientists continue to believe that all the Matter in the universe is made from a percentage of Matter which bizarrely came to exist without its <u>mandatory</u> Anti-Matter counterpart.

Scientists today remain stubbornly searching for ways to solve the Matter and Anti-matter asymmetry problem. In actual fact, <u>there is only one known sure way</u> to solve the asymmetry problem. At the moment Matter transforms from energy, Matter must immediately be separated from Anti-matter. Anti-particles must now be forcefully directed into a secure storage bin from where anti-particles can never escape. [037] This is exactly why, when Scientists at CERN create anti-hydrogen atoms they use magnetic fields to guide the anti-hydrogen atoms into a storage bin in the form of an electric, <u>magnetic</u> and or '<u>vacuum</u>' trap **which is purposely designed <u>to</u> hold particles together.** This prevents their very precious Anti-matter from colliding with ordinary Matter and being immediately annihilated. [120 see video] *Anti-matter is said to cost $2700 USD trillion per gram to make.* Like the Scientists at CERN, to create a universe full of regular Matter, modeling will require magnetic fields to forcibly spiral anti-particles away from Matter particles and guide the anti-particles into a storage bin which will act like a vacuum trap. All Scientists know this is true but such things are impossible for a Big Bang to supply. Our Super Massive Black Holes have been made from a deep vacuum phase of virtual mass which will provide a massive vacuum trap which is obviously an ideal place to store anti-particles.

If our universe is to obey the laws of physics and current experimental evidence it must produce anti-particles which must immediately be dealt with or anti-particles will destroy all Matter particles. It is of no use side stepping the problem or beating about the bush, if a Natural Universe is to conform to the stringent laws of physics it must have an initial separator which is able to instantly separate Matter particles from anti-particles and a storage bin to eternally lock away anti-particles. Without an initial separator and storage bin all Matter will undeniably be annihilated and you or I can never be. [009] There is no other known way; surely this stringent law of physics must be obeyed and the results of associated experiments should be respected. Strangely, it seems that no one has actually contemplated how it may have been possible to separate Matter particles from anti-particles at the time Matter was created. Protons and neutrons are hadrons. All hadrons are made of quarks held together by the strong force. All quarks are first created in quark and anti-quark pairs which usually annihilate each other. A neutron can become a proton and a proton can become a neutron. As one can change into the other and Scientists mostly relate research to protons we will also now relate mostly to protons.

Almost all Matter in the universe today is confined to Galaxies. Each Galaxy should have had a mirror amount of Anti-matter. There is truly only one obvious place within a Galaxy where all of this Anti-matter could hide and never be a threat to normal Matter. This is at a Galaxy's center within its Super Massive Black Hole. Yes I know the Super Massive Black Holes appear to be way too small to be containing a mirror amount of all of a Galaxy's Matter but there is a way. The solution is only possible if Matter is created by simply transforming virtual protons into regular protons. At this time Hawking's Radiation may actually cause the Black Hole's mass to slightly reduce. [028] An asymmetry proton has three valence Matter quarks, numerous gluons and numerous sea quarks in the form of pairs of quarks and anti-quarks. By cleverly transforming a natural, symmetry virtual proton into an unnatural, asymmetry proton the only Anti-matter we will need to account for is the three lost virtual anti-quarks which were the symmetry partners to the three valence quarks of a newly created proton (or neutron). A proton is at least 99 times more massive than its three 'lost' anti-quarks. The gluons don't need to be accounted for because there are no anti-gluons, rather it is said they act like their own anti-particle. With a Natural Model almost all of a Super Massive Black Hole's mass is provided by phases of virtual mass. Created close to an event horizon, a newly created and ejected proton retains its tiny phase of **virtual mass** inherited from its virtual proton state. **Thus, newly created protons get to keep all of their mass and a SMBH is not required to grow beyond its observed size.**

Amazingly, at least 99% of Anti-matter does not need to be created. 99% of a proton's regular mass-energy will come from the phase of virtual mass simply jiggling newly acquired valence quarks about at near the speed of light and unlike a Big Bang there is no need to create anti-protons and anti-neutrons. However, one will need to account for the CMB radiation created during annihilation of Matter and Anti-matter.

One must finally solve the Matter and Anti-matter asymmetry problem which has confounded Scientists ever since a Big Bang was first proposed back in 1927. This is why Super Massive Black Holes had to be created before the production of regular Matter. [024; 025] The evidence clearly supports this. Every massive Galaxy has a Super Massive Black Hole at its center. In general, the more massive the Galaxy, the more massive it's Super Massive Black Hole. Now ask; what caused this direct relationship? The relationship and evidence is very obvious, the bigger the Super Massive Black Hole the more Matter the system was able to produce. Like finding mangos under a mango tree the reasoning is simple; all Matter has to be created here because this is where all Matter is generally observed to be.

The enormous Gravity of our Sun provides the means for the Sun to produce its tremendous energy by way of nuclear fusion. The Sun's Gravity is plainly puny compared to a Super Massive Black Hole which may be well over 40,000,000,000 times greater. This is the truly gigantic **Gravitational** resource which source itself is pure empty Space and will provide the means for virtual protons to be transformed into regular Matter protons.

[005] The Quasar J0313-1806 is powered by a Super Massive Black Hole existing only few hundred million years after the said Big Bang, however, Scientists say Black holes could not have grown Super Massive by way of their proposed methods in only a few hundred million years from the Big Bang. Powered by Super Massive Black Holes Quasars are without doubt one of the most powerful objects ever observed in our universe today. Quasars can be 27 trillion times brighter than our Sun and may exceed a thousand times more than the energy output of 200 billion stars. Quasars emit this colossal amount of energy from a relatively small area which necessitates far more efficiency than even the nuclear fusion which powers our Sun. Scientists admit there remains much they still do not understand about these incredible power houses which were most prevalent near the beginning of the universe. Their ability to display unimaginable energy and residing where and when they are required clearly makes the mechanism which powers Quasars the most likely contender as a source of a Galaxy's Matter. [116] **The evidence is today clearly observable; the cores at the poles of Super Massive Black Holes generate massive amounts of energy and**

release colossal amounts of the <u>exact kind of Matter</u> which creates stars and this was clearly peaking near the beginning of the universe.

Scientists have developed a complex theory for Quasars which involves Matter forming an accretion disk where Matter is exposed to incredible heat and angular momentum causing intense Gravitational stress and friction. The method has been said to release up to <u>10%</u> of Matter's energy before the Matter is fed into the Super Massive Black Hole. This provided a <u>maximum theoretically allowed</u> temperature from a Quasar emission of <u>100 billion Kelvin</u>. Such a process is also said <u>not to be sustainable</u> for the period of time Quasars were most prolific, that is from 9 billion to 13 billion years ago. It has been said that Quasars would simply consume their entire Galaxy, long before this period elapses, which is why Quasars are said to have lifetimes of 100 to 1000 million years, even though Quasars are actually observed to be most prolific during a period of billions of years.

Thus, the maximum theoretically possible temperature from a Quasar emission is <u>100 billion Kelvin</u>. However, the creation of Matter will ideally require a far greater temperature which like the Big Bang allows Matter to evolve during cooling and allows the Cosmic Microwave Background radiation to cool over 13.8 billion years to now be a very cold 2.725 Kelvin. This will allow Matter to evolve from extremely hot plasma where free protons are created and cool to temperatures which allows protons to grab an attracted electron and form hydrogen atoms and finally molecular hydrogen both of which are ideal for making stars. Readily forming naturally from hydrogen atoms, molecular hydrogen is a molecule composed of two hydrogen atoms held together by a covalent bond.

The photons of the Cosmic Microwave Background radiation tells us that during the creation of Matter there was a lot of Matter and Anti-matter annihilation creating the photons of the CMB. Obviously, if my Natural Model is correct it will also have a lot of annihilation creating the same photons. To account for all of this annihilation energy the temperature in these regions needs to be increased from the theorized <u>100 billion Kelvin</u> to ideally match the <u>10 trillion Kelvin</u> temperature which is said to relate to a <u>similar time</u> when Anti-protons were annihilating with protons within the first second after the said Big Bang. **This is immensely hotter than any temperature which had ever been recorded anywhere in the universe.**

[005] Truly amazing, an international team of astronomers have now used the Russian radio space telescope 'Radio Astron' and combined it with telescopes in Germany, US, Mexico and Puerto Rico to construct a virtual telescope 170,000 kilometers across. This allowed researchers to make a truly mind-boggling observation; they discovered temperatures <u>inside the jets</u> of Quasar 3C 273 were of <u>10 trillion Kelvin which is an astounding 100 times hotter than current theories for Quasars allow</u> and is a trillion times hotter than the surface of the sun and easily the hottest temperature to now be recorded in the universe. Scientists are puzzled for such results plainly violate the theoretical upper temperature limit of Quasars. There is obviously a very mysterious activity happening within Quasars which is powerfully more intense than anyone had previously believed possible. Researchers say, although exciting, this means that established theories relating to how a Quasar gives off light are clearly incorrect. Scientists are now scrambling to model new theories capable of providing Quasars with unfathomably more powerful engines to account for the observed output.

100,000,000,000 K. The theorized upper temperature limit of a Quasar. 10,000,000,000,000 K.[005] The measured so <u>factual</u> temperature of a Quasar. 10,000,000,000,000 K. [038] The theorized temperature at 0.0001 of a second after the Big Bang when Anti-protons were <u>annihilating</u> with protons.

One needs to seriously ask; what fuel could actually increase the theorized energy output and temperature of a Quasar by 100 times and be sustained over the period which Quasars are actually observed to be most prolific? The best and probably the only plausible answer is, like at the said Big Bang, Matter and Anti-matter <u>annihilation</u> which is the most powerful fuel known to exist and releases all 100% of Matter's energy. However, this naturally means both Matter and Anti-matter are being created here which, of course, is why the temperature here is <u>exactly the same</u> as a similar phase of the theorized Big Bang and obviously provides the strongest evidence that this is where all Matter is truly created. This is also exactly where this Matter <u>needs to be</u> to form a regular Galaxy which to actually form a regular Galaxy requires the support of a Gravitational scaffold consisting of a central Super Massive Black Hole and an outer halo of Dark matter which adds even stronger evidence and superb correlation that this is true.

This kind of annihilation and temperatures are said to have only occurred at the Big Bang during the creation of Matter. The Natural Model provides powerful Matter creation cores at the poles of Super Massive Black Holes where powerful magnetic fields are used to separate Matter particles from anti-particles. However, overcrowding will obviously cause numerous Matter and anti-particle collisions resulting in the release of a colossal amount of annihilation energy. Super Massive Black Holes provide the massive Gravity to power the system which is in turn motorized by annihilation energy. Due to the existence of the Cosmic Microwave Background radiation Scientists have theorized only one particle of Matter per one billion survived being annihilated by anti-particles at the Big Bang. This suggests the process which separated Matter particles from anti-particles was very inefficient at creating Matter but very efficient at <u>releasing colossal amounts of energy</u>. Given the colossal, almost unfathomable energy required to power these systems and to eventually blast new Matter far from the massive Gravity near event horizons such inefficiency, due to annihilation, may very well be exactly what is required.

<u>Ideally</u>, one would expect a Natural Universe to create Matter at a similar extremely high temperature so as the evolution of Matter occurs <u>during cooling</u> in a similar way as from a Big Bang. Because the Big Bang was born from an unfathomable, infinitely dense and hot singularity like point the Big

Bang was born even <u>hotter</u> but very quickly <u>cooled</u> to the temperature of the annihilation of Matter and Anti-matter period. Whereas within a Matter creation core first powered by Gravitational Attraction and Pressure the temperature would be expected to begin cooler before <u>increasing</u> to the annihilation of Matter and Anti-matter period. [038] The temperature just 0.0001 part of a second, that's one ten thousandth of a second, after the Big Bang is <u>calculated</u> to have been an astonishing <u>10 trillion Kelvin</u> which occurred during the annihilation of Matter and Anti-matter. Consequently, our Matter creation cores should ideally reach the same 10 trillion Kelvin.

[005] Thus, it is a very <u>unlikely coincidence</u> that this happens to be <u>the exact same</u> temperature which has now been discovered at the core of Quasar 3C 273 which is surely an indication to where and how all Matter was truly created. [043 video] This will allow the evolution of Matter and the release of the Cosmic Microwave Background radiation to occur as Matter is exposed to a similar range of cooling temperatures and events as with the proposed Big Bang. The Cosmic Microwave Background radiation with its near perfect blackbody spectrum is thought to have originated from when Matter was in an extremely hot dense state of plasma that existed billions of years ago. Now 13.8 billion later this radiation has cooled to a very cold 2.725 Kelvin.

[005] Although this discovery at Quasar 3C 273 defies current theories, there is now little doubt that at the active poles of Super Massive Black Holes there exists cores of Matter in a **very similar state of hot plasma which is thought to have <u>only existed</u> at the proposed Big Bang.** [119, 116] <u>Observable evidence</u> clearly shows this same hot, star forming Matter and radiation being ejected in massive quantities from these same regions of the very earliest observed Galaxies from near the beginning of the universe. This allows Super Massive Black Holes to seed and grow their own Galaxies with their own self-created Matter. **The evidence is obvious and observable.**

Creating Matter in the regions where regular Quasars will form will be like trillions of Little Bangs rather than one massive Big Bang, however, there are <u>absolutely vital</u> things which the Matter from the Big Bang <u>did not have</u>; an actual <u>source</u> for this colossal energy and Matter, a sound reason for Matter to clearly prevail over Anti-matter and a provided Gravitational scaffold for newly created Matter to create normal, regularized Galaxies.

The first Matter will be explosively released directly from Matter creation cores. If many of these events occurred almost simultaneously and close together than <u>this is the actual Big Bang</u>. Many of these events may have occurred at much the same time because the shock waves from the Gravitational implosion which created the first Super Massive Black Holes likely triggered the Gravitational implosions of numerous Super Massive Black Holes. This would have flooded these regions with <u>Gravitational Potential Energy</u> which triggered a massive pulse of Matter production.

The Big Bang had no possible source for its immense energy which is why nobody knows what kind of energy created the Matter of the Big Bang. Matter in the form of protons will obviously require a mechanism to <u>actually create their own mass and Gravity.</u> The protons forming after the proposed Big Bang are not clearly provided with such a mechanism, consequently, the source of mass and Gravity, like the source of the Big Bang, is not clear and remains shrouded in mystery.

Evidence clearly implies that the universe was made from virtual particles.

[014] The energy of Quantum vacuum fluctuations has been explained by the <u>uncertainty principle</u> first put forward in 1927 by German Physicist Werner Heisenberg. It says that at any observed tiny point in empty ground-state Space there are temporary changes in energy. Occasionally this energy is changed into mass, in the form of particle and anti-particle pairs. Usually these pairs of virtual particles combine and then mutually annihilate. However, due <u>to external energy forcing them apart</u> they can sometimes avoid annihilation and become actual, real particles. The physics and experimental evidence support this.

[040] *A team of Physicists, following months of super computer modeling, say their research <u>decisively confirm that it is true</u> that all Matter, the same stuff that you and I are made from, <u>is actually no more than fluctuations within the Quantum vacuum</u>. Stephan Dürr's team used months of time on a parallel computer network at Jülich, which can handle 200 trillion arithmetical calculations per second. Their method of super computer modeling permits the complexities of <u>the strong force</u> to be simulated. They say their results will mean that <u>all reality is virtual</u>.*

Quarks and gluons are expressed as being virtual particles. Virtual particles appear out of empty Space during vacuum fluctuations. <u>A proton is a real, regular particle which is made from an amalgamation of these very same virtual particles</u>. If all Matter and mass was created by the 'fluctuations within the Quantum vacuum' method as expressed by my Natural Model than the super computer simulation is in agreement. [041] *Further research; Phiala Shanahan an Assistant Professor of Physics at MIT has via a more recent public lecture webcast stated that <u>ongoing</u> super computer modeling confirms; "protons and neutrons are made from quarks and gluons – that's it. There is no deeper structure to be found. Quarks and gluons are the fundamental building blocks of the universe". Shanahan also revealed the pressure inside of a proton peaks higher than inside of a neutron star.*

<u>It is true, quarks and gluons are first created as virtual particles.</u> The COST (Dark matter) is nothing more than a Configuration of Space-time and is based on Quantum vacuum fluctuations which produce particle and anti-particle pairs of virtual particles. It turns out that <u>we can call these virtual particles quarks</u>. One can find a remarkable video online which says this. [042] The video is called 'Empty Space is NOT empty'.

Protons, like neutrons, have <u>three</u> all important primary Matter valence <u>quarks</u> plus a sea of <u>virtual</u> quark and virtual anti-quark pairs which appear and disappear during vacuum fluctuations. Because they have lost their anti-quark partners, the three valence quarks now act as real particles and

live to provide a proton with its basic properties. The three <u>primary</u> valence Matter quarks are the three extra Matter quarks that are left, at the very basic level, if the sea of virtual quark and virtual anti-quark pairs annihilate without being replaced. In protons the three primary valence quarks are Matter quarks and in anti-protons they are anti-quarks. This is why the universe is made from Matter and not Anti-matter. **Without a doubt, each of the three <u>primary</u> valence Matter quarks <u>must have been born with an anti-quark partner</u> so the big question one really needs to ask; how did the three Matter quarks become separated from their anti-quark birth partners? The answer will tell how a universe of Matter was created and <u>finally solve</u> the Matter and Anti-matter asymmetry problem which today remains terminal for a Big Bang theory.**

Take away the chief three valence Matter quarks which are providing a proton with its <u>basic properties</u> and a proton now <u>without its basic properties</u> acts much like Dark matter which has <u>no known basic properties</u> besides Gravity. One is now left with the rapidly fluctuating nature of the COST producing virtual pairs of quarks and anti-quarks. This is <u>further evidence</u> that the COST in its Dark matter state is a virtual proton which transforms into the regular protons and neutrons to make atoms. All that is required to transform a virtual proton into a regular proton is to provide a way to give a virtual proton <u>three additional Matter quarks</u> which because they have lost their anti-quark partners act as regular particles which are able to provide a proton with its basic properties.

Quarks and gluons are the fundamental building blocks of the universe. During Quantum vacuum fluctuations quarks and anti-quarks materialize out of empty Space. Quarks emit and adsorb gluons. No anti-Gluons are emitted by quarks; rather it is like a gluon is its own anti-particle. Gluons, it seems, materialize out of Space as a bound interaction between quarks. So, we need only to create three stable quarks and their gluons will appear.

We know from theories and experimental evidence that the energy phase of Quantum vacuum fluctuations produce Matter and Anti-matter pairs of virtual particles which materialize out of the nothingness of Space and then annihilate in a tiny pulse of energy. We now know we can call these pairs of virtual particles quarks and anti-quarks. We know that as energy is increased (speed) and as a working part of a proton, Quantum vacuum fluctuations are responsible for increasing the numbers of pairs of sea quarks and anti-quarks and associated gluons simply materializing out of the nothingness of Space-time. Quantum vacuum fluctuations provide the vital, operational mechanism at the heart of all protons and neutrons located in the nucleus of atoms. All is evidence that virtual protons transform into the regular protons and neutrons to make atoms.

For the first tiny moments after the said Big Bang, the universe is said to have existed as a very hot, dense soup of <u>quarks and gluons</u>. Quarks and gluons are the building blocks of protons and neutrons. Protons, neutrons and electrons are the building blocks of all atoms in the universe.

[043] We know the photons of the Cosmic Microwave Background radiation were produced within a confined and hot, dense plasma like state which provided a near perfect black body spectrum. We also know that the very early creation of Super Massive Black Holes would have supplied a colossal amount of Gravitational Attraction. The aim is to use this knowledge to produce the following roadmap to the creation of all regular Matter.

The empty Space from which a virtual proton will create quarks and gluons is not really nothingness. This is now an area which is full of very powerful <u>Gravitational fields</u> near the event horizons of Super Massive Black Holes as well as the positive energy of the <u>dimensions</u> of Space-time. This is the energy which the vacuum fluctuations of the COST has access to in order to be transformed from a virtual proton to a regular proton. Because quarks interact with each other by <u>emitting</u> and absorbing gluons we will focus on the three <u>extra</u> valence quarks which will create the first protons and neutrons and provide protons and neutrons with their basic properties.

All protons and neutrons have three primary Matter valence quarks <u>plus</u> their Quantum vacuum fluctuations <u>bring into being</u>, out of empty Space, a sea of quarks, anti-quarks and gluons. Protons are much like neutrons except protons are positively charged and neutrons are neutral. A neutron can change into a proton and a proton can change into a neutron.

The Matter creation systems at the poles of Super Massive Black Holes likely first created protons. Neutrons were created from a high-energy collision of a proton and an electron or neutrino. However, a <u>free</u> neutron is unstable and will quickly decay into a <u>proton</u>, an <u>electron</u> and an anti-neutrino. If protons and neutrons did not undergo fusion to form heavier atomic nuclei than neutrons, after about 15 minutes, default back into protons again. We need not worry about gluons because they are emitted by quarks so we can assume that an interaction between quarks naturally created gluons. As one can easy change into the other and Scientists mostly relate this time to protons I will also now relate mostly to protons.

Scientists have remained fixated and so have focused <u>exclusively</u> on a Big Bang; so no one has even attempted to explain how Matter may have been created at the poles of Super Massive Black Holes. Consequently, despite bountiful evidence that this really occurred, there is currently <u>no research</u> to refer to relating to exactly how this occurred. It had to occur in a way which <u>accounts for</u>; the CMB radiation, Matter prevailing over Anti-matter, the configuration of different Galaxies and in a way Super Massive Black Holes would not grow more massive than they are observed to be.

Near the very beginning of the Big Bang the universe was said to be in a very hot, dense state which somehow emerged from a singularity like point. In this state, Matter and Anti-matter is said to have been <u>attracted</u> to each other. With a Natural Model, Matter and Anti-matter are also first in a similar very hot, dense state where positive quarks were also attracted to negative anti-quarks while quarks repel each other and anti-quarks repel each other.

This may change when regular protons are formed. Three virtual quarks created by a virtual proton will become <u>valence quarks</u> and acquire newly assigned fractional electric charge values or color charge. According to the theory of quantum chromodynamics (QCD) quarks have a property called color charge. There are three types of color charge which are blue, green, and red. Each quark is complemented by an anti-color; anti-blue, anti-green, and anti-red. All quarks carry a color, while every anti-quark carries an anti-color. The attraction and repulsion between quarks charged with different combinations of the three colors is called strong interaction. The interaction is said to be carried out by particles known as gluons.

For a <u>Big Bang</u>, let's <u>just pretend</u> the laws of physics are not adhered to and all anti-quarks mysteriously disappeared. The moment after all anti-quarks inexplicably disappeared the remaining quarks are speeding apart at the speed of light plus. Now we have a very <u>rapidly expanding</u> quark and gluon soup. As quarks are now speeding apart, how do you get positive quarks which also **repel each other**, to neatly come together in <u>lots of three</u> and for gluons to 'glue' the three positive quarks together to form a proton? All of which is difficult when <u>nobody even knows</u> how the gluon glue 'sticks' in the first place [033]. Thus it is said with seemingly very little reasoning, within the first second, the remaining quarks simply *"aggregated to produce protons"*. <u>Exactly</u> how <u>hadrons</u> such as protons and neutrons first formed, from a quark, anti-quark and gluon plasma, moments after the Big Bang appears to be another of those 'biggest unsolved mysteries in astrophysics'.

If we did find three quarks together how would it be possible to provide them with the <u>vacuum fluctuations</u> to create both the strong force and temporary quark and anti-quark pairs of sea quarks to create and power a proton? Do quarks somehow turn on this mechanism by themselves? Turns out I have not been able to find a precise, commonsense answer other than when Matter cooled it just happened. **In contrast, the path of creating a regular proton from a virtual proton is relatively straight forward and <u>fully</u>**

explained within a Natural Universe. Dark matter sensibly acts as a virtual proton which directly provides the essential mechanism, <u>inherited</u> from a common Quantum <u>vacuum fluctuation</u>, to now <u>power the throbbing heart</u> of a regular Proton. By acquiring three extra Matter quarks a virtual proton, which <u>already creates</u> quark and anti-quark pairs, can now be transformed into a regular proton. The <u>vacuum fluctuations</u> powering a regular proton now provide <u>evidence</u> of how a proton was first created.

Protons <u>do not</u> have an outer, bubble-like shell providing a sphere-like container with quarks and gluons inside. A proton or a neutron is basically only a collection of three valence quarks and emitted gluons plus a sea of sea quarks and anti-quarks. <u>However</u>, this collection of particles and virtual particles <u>does require</u> an invisible sphere-like configuration of Space-time to take the place of a container and provide the three primary quarks with a sea of virtual quarks. The COST provides protons and neutrons with an important configuration of Space-time which provide vacuum fluctuations to create numerous temporary quark and anti-quark pairs of sea quarks and a virtual mass phase of the strong, negative energy of a deep vacuum which <u>keeps the whole act together</u> without a container. All is why, in order to create a universe of regular Matter, <u>the universe had to have a massive supply of virtual protons</u> acting as Dark matter. As this model has no use for the mystifying Dark energy the content of our universe began as 100% Dark matter and today has a rounded 16% regular Matter and 84% Dark matter.

Because a quark <u>stubbornly refuses</u> to be separated from partners, it is very unlikely that independent free quarks existed so as the COST could have grabbed exactly three <u>free</u> Matter quarks to form a proton. The model will use Gravitational Attraction and Pressure to increase the COST to an energy level where it is producing at least <u>three</u> virtual quark and anti-quark <u>pairs</u> which is important because quark and anti-quark <u>pairs</u> are natural, normal and above all is <u>symmetry</u>. The first problem is at this early time Matter quarks are positive and anti-quarks are negative which means they are naturally attracted to each other and quickly come together and annihilate. The process will need to <u>stall</u> this annihilation just long enough to provide a tiny window of opportunity for three anti-quarks to be ripped away by powerful spiraling magnetic fields so as the three remaining quarks cannot be annihilated. The three remaining quarks will naturally become a proton's valence quarks which <u>will live to provide</u> a proton with its basic properties.

The COST retains its <u>vacuum fluctuations</u> to produce extra pairs of quarks and anti-quarks but will now have three <u>additional</u> primary valence Matter quarks which are stable enough to provide a proton's properties. Now with three additional Matter quarks the COST is transformed from a virtual proton into a regular proton. This needs to happen where the asymmetry of Matter over Anti-matter can be solved and where photons created from annihilation activity can be embedded with a near perfect black body.

[044] The aim is to combine the previous pages with current research relating to existing Quasars, particularly to the Quasar known as Q0957+561 which Astronomers at the Harvard-Smithsonian Center for Astrophysics have monitored and studied for 20 years. A timeline relating to the temperature of events calculated for a Big Bang will naturally provide the order of events. The quarks inside of protons will tell us exactly <u>when</u> anti-particles were separated from Matter particles and why this caused virtual protons to become hadrons, that is, protons and neutrons.

<u>Rather than being a problem</u> the near perfect blackbody spectrum of the Cosmic Microwave Background radiation <u>tells us much about the Matter creation system</u>. It tells us that the Matter creation core was initially a <u>closed confined system</u> and it had a very hot plasma core. It tells us that there was a lot of particle and anti-particle annihilation which was creating photons. [043] In order to obtain this near perfect blackbody spectrum the photons of the CMB radiation are thought to have been <u>trapped within a hot, dense plasma state</u> until the universe expanded and cooled to the point which allows the first atoms to form. With the formation of hydrogen atoms the plasma became a gas and the universe cleared sufficiently which allowed the photons of the CMB radiation to travel throughout unobstructed. For a Natural Universe this infers that Matter was at first trapped in a very similar hot, dense, confined, plasma state while in very close association with a Super Massive Black Hole which remains at the heart of a Galaxy which forms from the Matter created. [005] The core of Quasars observed today do indicate a similar hot, dense, confined plasma core, consequently, we know there exists Matter here in a similar state which Matter is said to have existed moments after the said Big Bang. However, existing Quasars do not show a near perfect blackbody spectrum.

To provide a near perfect blackbody spectrum like that of the CMB radiation the cores of developing primeval Quasars would have remained tightly closed and possibly cloaked by massive Gravity. The first batch of created Matter would have been initially trapped within a hot, dense plasma state and in close association with its power source. Powerful spiraling magnetic fields drove Matter to the surface of the core until a buildup of energy broke this detainment and caused an inflationary event by explosively releasing colossal amounts of energy, photons and newly created Matter. This would have naturally provided newly created Matter with an extremely powerful inflation phase. Amazingly, there is evidence that a similar explosive event may have occurred relatively recently at the center of our own Milky Way Galaxy. [046] I have managed to uncover evidence that the magnetic fields in these very regions can actually act in a precise way which allowed this to happen. Hence, the goal here is to accommodate the blackbody (thermal) spectrum of the CMB radiation in a way which incorporates data directly derived from observational research.

A Super Massive Black Hole (SMBH) remains surrounded by virtual protons which are migrating from a Dark Galaxy's dense dark disk. Virtual protons now begin creating an invisible accretion disk around the SMBH. Driven by the initial Gravitational collapse of the SMBH virtual protons now obtain further energy from extreme speed within an accumulating accretion disk. Virtual protons may now <u>begin to glow</u> from the intensity of their annihilating virtual particles producing <u>glowing gluons</u>. From an <u>intensely glowing</u> and <u>overflowing</u> accretion disk, high energy <u>virtual protons</u> begin to funnel <u>towards the poles</u> of a Super massive Black Hole (SMBH).

A SMBH has now created at each pole a storm system of Space-time harboring a Matter creation core of a developing primeval Quasar. At rest virtual protons produce a Matter and Anti-matter pair of virtual sea quarks which annihilate in a pulse of energy. Powered by <u>vacuum fluctuations,</u> virtual protons gain speed and are <u>squeezed</u> by the warped Space-time of strong Gravity and developing magnetic fields causing virtual protons to correspondingly gain energy, <u>like a regular proton</u>, by producing extra pairs of sea quarks before the pairs annihilate. [015] Vacuum fluctuations becoming 'louder' when 'squeezed' in a vacuum has been observed in experiments.

Near the poles of Super Massive Black Holes the <u>vacuum fluctuation</u>s of virtual protons are being aggressively squeezed by warped dimensional Space-time and tightly funneled around both poles. Virtual protons are now producing at least three <u>temporary</u> pairs of sea quarks. In close proximity to the event horizon the vacuum fluctuations of virtual protons naturally produce additional <u>pairs</u> of virtual quarks and anti-quarks which <u>annihilate</u> each other and <u>produce gluons</u> in a similar manner to which they will when they become a regular proton or a neutron. However, a virtual proton producing additional 'pairs' of sea quarks <u>is not yet a proton or a neutron</u>. Annihilation of all virtual quarks and anti-quarks continues but a virtual proton continues to <u>reproduce</u> extra pairs of sea quarks. To become a regular proton a virtual proton needs to acquire three <u>additional Matter valence quarks</u> which are simply <u>three</u> sea quarks which have <u>lost</u> their anti-quark partners. Quarks which lose their anti-quark partners <u>cannot annihilate</u> with partners so they now live to be fully 'exposed' to the deep vacuum phase of virtual mass which <u>restrains them</u> as primary, particle-like quarks which now <u>survive to provide</u> a regular proton with its properties.

Because anti-particles are opposite charged they <u>spiral away in opposite directions to Matter particles</u> inside of magnetic fields. Consequently, the powerful magnetic fields found within these regions will provide the ideal tool to force Matter and Anti-matter apart and drive Matter and Anti-matter in opposite directions. Because a virtual proton is a tiny configuration of Space-time it is as if the Matter and Anti-matter <u>pairs</u> of sea quarks <u>are simply materializing out of empty Space</u> and time. Quarks <u>annihilate and produce</u> gluons; thus, we now have quarks, anti-quarks and gluons. In order to separate a quantity of anti-quarks from quarks, one will first <u>require a way to create magnetic fields;</u> however, at first there are only quarks and gluons with which to create the required magnetic fields.

[045] A team of Scientists led by Yigit Dallilar at the University of Florida have studied similar central region of an accretion disk at the pole of a Black Hole. *They say "Black Holes themselves don't generate magnetic fields. This means that the accretion disk corona magnetic fields are 'somehow' generated by the 'Space' around the Black Hole but the process is not well understood."* Note that this research implies magnetic fields are *'<u>somehow generated by the 'Space'</u>'* in the regions required for this model. One needs to ask; how can magnetic fields simply appear to originate from Space?

Scientists have yet to understand how these types of magnetic fields were first created. One may well ask; what <u>invisible unknown mechanism</u> within Space could be responsible for first creating these magnetic fields? Gluons interact with other gluons and quarks. The gluon field is said to act very similar to the electromagnetic field. Within a forming, high energy, concentrated, crowded core numerous gluons created from <u>annihilating</u> quarks may <u>strongly contribute</u> to the creation of magnetic fields by a method relating to flux tubes. While there requires further research, one can find several astrophysics papers online detailing how intense electromagnetic fields are created in quark-gluon plasma by <u>ultra-</u>

relativistic related processes. Astrophysics deals with the science of Space by applying the laws of physics and chemistry to explain the birth, life and death of Galaxies, stars, planets, nebulae and Quasar like objects in the universe. While I am not clever enough to understand many of these very extensive and complex papers, an important clue is a particle is called ultra-relativistic when its speed is very close to the speed of light. Near the event horizons and at the poles of a Super Massive Black Holes is an ideal place to find particles travelling at these extreme speeds and this happens to be where we need our magnetic fields to be focused. This is where many of the astrophysics, research papers are indeed directed.

An overfull swirling accretion disc, made from invisible, ultra-relativistic virtual protons (Dark matter), is now overflowing and squeezing its contents into a confined, spinning storm-like center at the poles of Super Massive Black Holes. When a virtual proton's pairs of virtual quarks annihilate they are immediately being replaced by the following fluctuation. Exposed to ultra-relativistic speeds of an accretion disk and directed to the poles of a Super Massive Black Hole the fluctuation energy of virtual protons continues to increase. Virtual protons increase their energy by increasing their pair production of quarks and anti-quarks, self annihilation of which is now producing numerous gluons. Gluons have been observed to glow at ultra-relativistic speeds which cause a swirling accretion disk to glow as it overflows, funnels and squeezes virtual protons into developing Matter creation cores at each pole. The temperature rises within the cores which have now developed spiraling magnetic fields. **The Matter creation cores remain tightly constrained and possibly cloaked by powerful magnetic fields and the massive Gravity near an event horizon.**

This first creates a quark, anti-quark and gluon annihilating soup which resembles the recipe of the first few moments after the said Big Bang. Annihilation and high energy collisions added leptons which include electrons and neutrinos as well as photons to this soup. Virtual protons continue to <u>overflow</u> from an accretion disk into central cores at both poles of a Super Massive Black Hole. Annihilation energy and friction heats this core into a quark and gluon plasma. Temperature within this plasma core grows to at least 10 trillion Kelvin which increases the intensity of developing spiraling magnetic fields which are now focused around an intensifying dense and elongated core.

If one was to study the building blocks of Matter attempting to discover how Matter was created, one only needs to ask; why do protons have three additional quarks when <u>quarks are always created in quark and anti-quark pairs</u>? Answer; this is clearly telling the energy level of virtual protons when they were exposed to powerful magnetic fields and is clearly telling exactly how protons and neutrons were first created.

Within each rotating plasma core, created from ultra-relativistic virtual protons, the essence of creation now occurs. [015] **The research by Alfred Leitenstorfer and his team at the University of Konstanz is now again <u>most valuable</u>.** Their experiments revealed as they *'squeezed the vacuum'* Quantum vacuum fluctuations became louder and a phase *'lower than the ground state of empty Space'* <u>**greatly increases its volume**</u> which for this model is a negative density state below the positive state of Space-time. **By creating the <u>strong force</u> this same negative energy phase will now play a key part in separating quarks from anti-quarks and finally allow a universe created from Matter to exist in a way which <u>obeys the laws of physics</u>.**

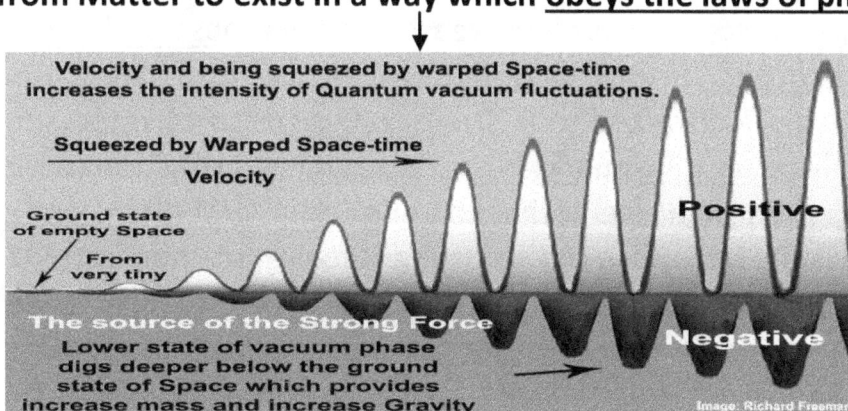

Velocity and being squeezed by warped Space-time increases the intensity of Quantum vacuum fluctuations.

Squeezed by Warped Space-time
Velocity

Ground state of empty Space
From very tiny

Positive

The source of the Strong Force
Lower state of vacuum phase digs deeper below the ground state of Space which provides increase mass and increase Gravity

Negative

Image: Richard Freeman

Now exposed to increased motion and aggressively squeezed by the warped Space-time near a Super Massive Black Hole the energy phase of virtual protons becomes louder by increasing the pairs of quarks and anti-quarks and the strong force becomes stronger when the negative energy phase of deep vacuum increases its volume of negative energy. The negative energy increases and peaks when the positive phase is least. When the phase of negative energy <u>peaks,</u> there is momentarily no positive state to cancel the effectiveness of the negative state. This momentarily gives <u>unimpeded attraction</u> of positive quarks to the negative charged energy of a deep vacuum phase of virtual mass.

Quarks can be changed from positive or negative but at this stage <u>Matter quarks are positive</u> because Matter is positive and anti-quarks are negative because Anti-matter is negative. A virtual proton continues to produce virtual quark (Matter) and anti-quark (Anti-matter) <u>pairs,</u> which is <u>symmetry.</u> Within the newly created elongated cores the magnetic fields are attempting to spiral Matter particles and anti-particles <u>away from each other</u>, however, mutual attraction continues to cause annihilation.

In an <u>ideal situation</u> the majority of virtual protons are now creating just <u>three</u> quark and anti-quark pairs. The <u>evidence</u> for this is the abundance of proton and neutrons all of which have just three primary valence quarks.

[015] As the <u>negative</u> charged density phase of virtual mass is <u>squeezed by warped Space-time</u> its volume increases and like <u>observed in experiments becomes stronger</u>. Virtual mass is now attempting to cling on to <u>positive</u> quarks which are attracted to it. The <u>negative</u> phase of virtual mass is now providing three <u>positive</u> quarks with the strongest attraction to the heart of a virtual proton and attempting to <u>repel</u> their anti-quark partners. **This attraction is the beginning of the strong force** which momentarily stalls the annihilation process by <u>momentarily</u> clinging on to the three <u>positive Matter quarks</u> just long enough for the powerful, spiraling magnetic fields to rip the three anti-quarks from a virtual proton. The three anti-quarks

spiral away and are absorbed by the Super Massive Black Hole. The virtual proton's negative energy phase of virtual mass now tightly restrains the three positive quarks which now provide a newly created regular proton with properties which changes a virtual proton from a neutral state to the positive state of a regular proton. Spiraling magnetic fields now speeds the newly created protons to the surface of the closed Matter creation core.

The retained three positive Matter quarks, having lost their anti-quark partners, cannot now be annihilated. The three positive Matter quarks are now, for the first time, fully exposed to the fluctuating negative energy phase of virtual mass which violently agitates the three quarks about at near light speed providing them with newly found motion energy which provides the newly created proton with a massive increase of positive mass-energy. These are the three Matter valence quarks which are left within a regular proton if every available anti-quark annihilates a quark.

Now harboring three additional Matter quarks a newly born positive proton continues to create from vacuum fluctuations virtual pairs of Matter and Anti-matter sea-quarks as it did when it was a virtual proton and as it did when it was once a simple Quantum vacuum fluctuation. If an anti sea-quark comes in contact with a valence quark both may immediately annihilate and the annihilated anti-quark's partner quark replaces the annihilated valence quark. Quarks can become up or down quarks and this change occurs when a proton becomes a neutron or a neutron becomes a proton. Conforming to their new asymmetry environment two of the three primary valence quarks become up quarks and one becomes a down quark. The quarks emit and absorb numerous gluons, thus a regular positive proton is born. Because a proton now has three additional Matter quarks is why this is now an asymmetry proton and is why the universe is made from Matter and not Anti-Matter.

Symmetry is good, sound arithmetic and above all it is natural. With the symmetry of pair production the books of Mother Nature are balanced by design since all Matter can be annihilated with all Anti-matter returning symmetry to Space. The only thing wrong with symmetry is that it cannot provide a universe made from Matter. However asymmetry is not good or natural so the books do not balance. An asymmetry proton is now no more than a freak of nature which naturally occurred many times over. Unable to unite with their partners and annihilate, quarks now reacted with each other in ways which Scientist have assigned different terms and mixed color prefixes. A proton has two up quarks and one down quark and a neutron has two down quarks and one up quark. When a proton changes into a neutron, one up quark changes into a down quark. Proton and neutrons are called nucleons. A virtual proton transforms into a nucleon which may take the state of a proton or a neutron and <u>each can be changed into the other</u>.

Why did virtual protons retain three Matter quarks and not anti-quarks? If virtual protons had retained three anti-quarks the universe would have been made from Anti-matter. With a Natural Model there is no mysterious one in a billion or more chance of Matter prevailing over Anti-matter as with a Big Bang. Matter was always going to clearly prevail. So why did virtual protons so clearly favor Matter quarks, so much so that no natural occurring Anti-matter atoms have ever been found? The reason for retaining three Matter quarks is the phase of virtual mass of a virtual proton is not only a <u>negative pressure</u> state but is also of a <u>negative charged density state.</u> Like charges <u>repel</u> each other and unlike charges <u>attract</u>. Thus, two negative charges repel one another; two positive charges repel one another while a positive charge and negative charges attract.

When I first modeled this I was searching for a reason to why virtual protons held on to three quarks and not anti-quarks. I had no idea that my previous modeling for how the strong force created binding mass played a part. I questioned what the chances would be that everything would be in the right order for I cannot change things to suit a result. However, the modeling allowed me to realize that the strong force, by clinging tightly to three quarks, played a major role in creating regular Matter.

[109] In a highly excited environment the strong, <u>negative energy</u> phase of a virtual proton is also of a <u>negative charged density</u> which naturally <u>attracted</u> the three <u>positive</u> Matter virtual quarks to the center of a virtual proton where they are <u>momentarily</u> held tightly by the strong, negative pressure of <u>deep vacuum</u> provided by virtual mass. This is the very **beginning of the strong force** which, by attracting and momentarily strongly clinging on to three positive quarks at the center of a virtual proton, obviously played a <u>vital part</u> in allowing their anti-quark partners to be ripped away from a virtual proton.

Separation.

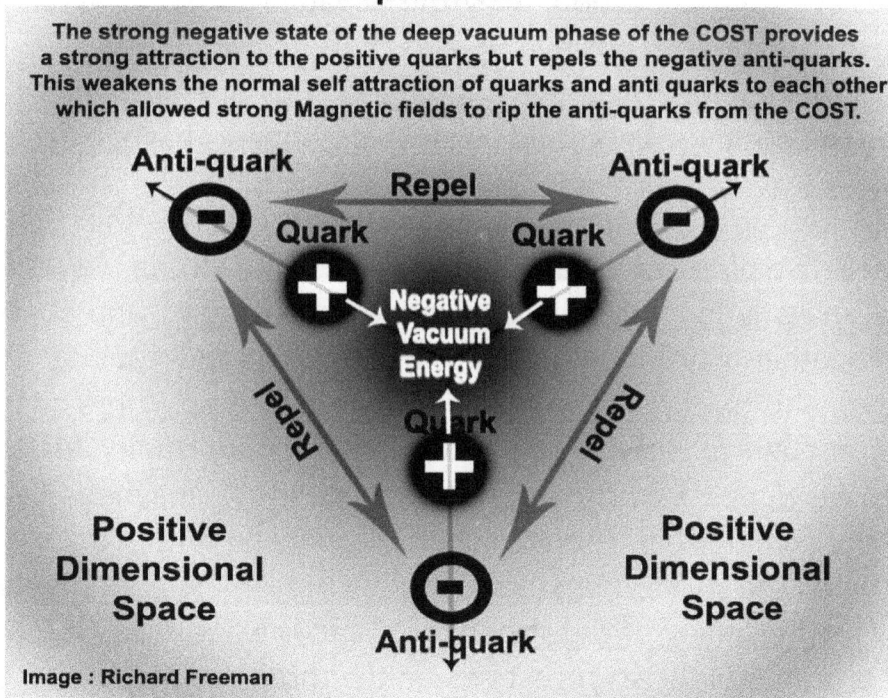

The strong negative state of the deep vacuum phase of the COST provides a strong attraction to the positive quarks but repels the negative anti-quarks. This weakens the normal self attraction of quarks and anti quarks to each other which allowed strong Magnetic fields to rip the anti-quarks from the COST.

Image : Richard Freeman

Separation Detailed: Three <u>positive</u> quarks are momentarily held tightly at the center by a strong <u>negative energy</u> phase of virtual mass. The three negative virtual anti-quarks are now on the outside and their <u>negative</u> state is attempting to <u>move the anti-quarks apart</u> which keeps them <u>evenly</u>

separated. This forms a three arm 'Y' shape with the three Matter quarks momentarily held tightly at the center by the newly created strong force and the three anti-quarks at the end of each arm. The strong, <u>negative energy</u> phase of virtual mass of a virtual proton momentarily engulfs and retains the <u>positive quarks</u> while at the same time the same negative state of virtual mass <u>repels</u> the <u>negative anti-quarks</u>. This momentarily <u>weakens</u> the attraction of the <u>negative</u> anti-quarks to the positive quarks which are engulfed within a tiny phase of <u>negative energy</u> from a deeper vacuum than the ground state of dimensional Space-time. The <u>negative</u> anti-quarks may also be attracted to the surrounding <u>positive</u> dimensional Space.

The anti-quarks are now farther from the negative phase of virtual mass feeling less <u>vacuum attraction</u>. Now the negative phase of virtual mass is strongly attracting the positive <u>quarks</u> but its same <u>negative state</u> is attempting to repel the negative <u>anti-quarks</u>. This briefly weakens the self attraction the positive quarks and negative anti-quarks have to each other.

Now the three <u>positive</u> quarks are being strongly held within the central, strong, negative phase of virtual mass and the same strong, negative phase of virtual mass <u>is momentarily preventing the negative</u> anti-quarks from coming together with the three positive quarks and causing all to be <u>annihilated.</u> This <u>briefly suspends annihilation and keeps three quarks alive</u> just long enough to provide a tiny window of opportunity for the anti-quarks to be ripped from a virtual proton by strong magnetic fields. The opportunity comes when the attraction of quark to anti-quark is weakest which is when the strong, <u>negative phase of virtual mass is strongest</u>. This is naturally <u>the beginning of the strong force</u> and is when the COST, acting as a virtual proton, has the strongest grip on the three positive Matter quarks.

If the anti-quarks <u>were not</u> ripped away at this precise instant the central negative phase of virtual mass momentarily disappears which releases the positive Matter quarks to the attraction of the negative anti-quarks causing quarks and anti-quarks to now come together and be annihilated by each other. However, now another three pairs reappear and the process again attempts to rip the three anti-quarks from a virtual proton. The process may continue many times until it is successful. Within the Matter creation core <u>annihilation</u> may have occurred several <u>millions</u> of times for every successfully created proton. **<u>Note</u>: This process <u>likely also occurs</u> within the <u>powerful magnetic fields</u> wrapping the accretion disk, from where newly created protons may be directly released or fed to the poles of a SMBH. However, attention of this activity has been directed to the poles.**

When exposed to <u>strong, spiraling magnetic fields</u> the negative energy phase of virtual mass which creates the strong force provided the positive quarks with a far stronger bond to a virtual proton than it provided the farther away and negative anti-quarks which were naturally attracted to the surrounding positive dimensional Space. This is why a virtual proton held on to the three Matter quarks and why the magnetic fields were able to easily rip the three anti-quarks away. This is also why the universe is made from Matter and not Anti-Matter and how <u>the beginning of the strong force within a proton</u>, by <u>tightly</u> clinging on to three positive quarks, played a vital part in allowing the universe to be made from Matter and not Anti-matter. **The strong force actually <u>dictated</u> that the universe was made from Matter.** But for the strong force all Matter would been annihilated before it was separated and you and I would have never existed. This superb correlation provides further evidence that the previous modeling of how the strong force creates regular mass is surely correct.

Without their anti-quark partners, the three <u>retained</u> regular Matter quarks now survive by means of interacting with each other by emitting and absorbing gluons. Protons retain their overall positive state by splitting the charge of <u>two</u> up quarks each of +2/3 and one down quark +1 which equal +1 or positive. For neutrons one up quark +2/3 and two down quarks each -1/3 equals 0 or neutral. Both positive and neutral remain attracted to the negative state of virtual mass. The negative state of virtual mass made it near impossible for it to cling onto three negative anti-quarks; it was far more likely for it to make positive protons and neutral neutrons. This is because both positive and neutral are attracted to negative. When Scientists create anti-hydrogen atoms they use an electric, magnetic and or 'vacuum' trap **which is purposely designed to hold particles together.**

Strong, spiraling magnetic fields at the poles of Super Massive Black Hole provided a powerful mechanism capable of separating vast amounts of Matter and Anti-matter and immediately providing a storage bin for Anti-matter. Even if anti-protons were created the spiraling magnetic fields may have allowed regular protons to prevail over anti-protons.

The strong, spiraling magnetic fields rip and spiral away negative anti-quarks into the Super Massive Black Hole which provides a storage bin from where Anti-matter can never escape and destroy all Matter. Many of the spiraling magnetic fields will eventually grow long jets which will carry newly created protons and electrons far from the Matter creation cores.

An electron's mass is approximately 1836 times smaller than that of the proton. High energy collisions within the Matter creation core created electrons which are naturally attracted to the newly created regular, positive protons. Free protons and the attracted free electrons are now driven by magnetic fields in a direction towards the surface of the elongated but closed core. Matter remains too hot to form atoms.

Because the Matter creation core has become elongated the spiraling magnetic fields have closed over the surface of the core. By creating (from fluctuations) additional pairs of sea quarks the COST, although now a regular proton, continues to fluctuate and create quark and anti-quark pairs of sea-quarks as it did when it was a virtual proton (Dark matter).

It is important to Note: The ripped away anti-quarks and the retained quarks were both born with exactly the same state of very little mass-energy. However, the ripped away anti-quarks are now residing without their COST and without a near interaction with a <u>complete</u> phase of virtual <u>mass</u> the virtual anti-quarks act as <u>massless waves</u> which allows anti-quarks to be directed into the Black Hole but contribute very little to its mass <u>or</u> as theorized with <u>Stephen Hawking's Radiation</u> the Black Hole may now possibly <u>lose</u> a tiny amount of its mass.

<u>In sharp contrast;</u> <u>because</u> the three Matter quarks now cannot be annihilated by their departed virtual, anti-quark partners they <u>now live to enjoy a strong, near interaction</u> with a <u>complete</u> phase of virtual mass which by mimicking a tiny appearing and collapsing Black Hole <u>constrains</u> them as <u>regular particles</u> and <u>violently agitates</u> the same three quarks about at near the speed of light providing the three regular quarks with newly found motion energy which provide the newly created regular protons with a comparatively <u>massive amount</u> of mass-energy <u>given</u> to them by a <u>tiny phase of empty Space</u>. This mass is increased by an inherited quark and anti-quark pairs of sea-quarks. The sharp contrast of the mass of the now discarded three anti-quarks and the <u>massive increase</u> of the mass-energy of the newly created regular proton is the <u>key reason</u> why every object we observe has so much positive mass-energy, most of which is <u>given free</u> by a tiny phase of vacuum nothingness. **Since newly created protons get to <u>keep all of their inherited phase of virtual mass</u> a Super Massive Black Hole is not required to grow beyond its observed size.**

This is how Mother Nature found a very clever way to create tiny sphere like packages, crammed full of mass and energy, created from near light speed motion energy, $E=mc^2$, for building a universe endowed with massive amounts of mass-energy. (See: wave and particle duality chapter 13).

Regular protons now naturally have <u>very significantly higher mass-energy</u> than virtual protons which allowed Super Massive Black Holes to create Galaxies of stars with <u>significantly more mass</u> than their central SMBH which had been created directly from virtual protons. In a virtual proton quarks do not survive long enough to enjoy a complete phase of virtual mass, thus, virtual protons (Dark matter) have very little mass-energy compared to the positive mass-energy now given to a regular proton.

Why 'three' quarks? It so happens to survive a quark requires a partner or partners to interact with and will stubbornly resist all attempts to isolate it and so it is said a quark can never escape its partners. Consequently, we cannot rip anti-quarks one at a time from a COST to create a proton.

Two quarks were also difficult. With the energy of just two quark and anti-quark pairs the phase of virtual mass was naturally weaker, consequently, the strong force was not fully developed and the strong, negative energy attraction of the two positive quarks to the center was weaker. The two negative anti-quarks are now on the outside and their negative state is attempting to move them apart which keep them evenly apart.

Because the two anti-quarks repel each other but are attracted to their positive quarks partners the two anti-quarks move to where they are directly opposite each other. The positive quarks repel each other and the weaker strong force was in all probability not strong enough to provide a sufficient and extended grip to hold the positive quarks in place. The result is a very unstable particle which is useless for building a universe.

With a 'Y' configuration of three quarks and anti-quark pairs the positive energy phase is louder which caused a reaction which gave a stronger, negative energy phase to provide the strong force with adequate strength to momentarily retain the three positive quarks. This proved to be the ideal and so most common configuration where the three quarks were firmly held at the center of a virtual proton and the three outer anti-quarks were in the most vulnerable position to be ripped away by the magnetic fields. Any created anti-protons may also be most unstable owing to anti-quarks being naturally negative leaving only vacuum attraction to the strong, negative charged energy of virtual mass. Man made anti-atoms are held in a near-perfect vacuum with electric and magnetic fields which may prolong their life. No natural forming anti-hydrogen atoms have been observed.

Migrating in from the dark disk of a Dark Galaxy a massive supply of virtual protons continues to stream into the accretion disk which is feeding a Matter creation core at each pole of a Super Massive Black Hole.

Caused from a central interaction with a surrounding accretion disk, the spiraling magnetic fields have wound up tightly coming together and effectively forming a noose over the elongated Matter creation core preventing the release of Matter which is why it is a closed system. Due to the massive Gravity near the event horizon and the nature of the powerful magnetic fields any escaping photons are mostly absorbed by the Super Massive Black Hole. This closed system of dense plasma contributed to photons of the CMB radiation having a near perfect blackbody spectrum. Now exposed to spiraling magnetic fields, within an elongated closed core, protons and attracted electrons begin to build a growing bulge above the event horizon. Within the growing hot, highly energized photons react with newly formed nuclei to form numerous pairs of electrons and their Anti-Matter equivalent positrons. As the energy grows the spiraling magnetic fields become more intense. Like twisting close a plastic bag, the spinning magnetic fields continue to wind up and tightly close the plasma core.

Within the central layers of the dense core Matter-particles and anti-particles, having initially been separated by magnetic fields, are now propelled in opposite directions and many now collide and instantly annihilate releasing all 100% of their energy which is responsible for maintaining the temperature at an astounding 10 trillion Kelvin. This corresponds to the 10 trillion Kelvin annihilation phase which occurred within the first second after the said Big Bang. [120 video] Just one teaspoon of Anti-matter annihilating with Matter releases as much energy as 10 nuclear bombs. Much of this annihilation may be caused by anti-quarks, which have initially lost their quark partners, now colliding with a newly formed proton and destroying it by annihilating its three primary quarks. All of this intense annihilation creates numerous photons for every surviving proton.

[125] Scientists have actually discovered a colossal Anti-Matter fountain <u>at the center</u> of our Milky Way Galaxy which is spewing out <u>15 billion tons per second</u> of Anti-Matter in the form of positrons (anti-electrons). Scientists say they have <u>eliminated all</u> the *"usual suspects"* as none could have produced such a colossal amount of Anti-Matter. Thus, they say *"no one has a clue"* of the source of this Anti-matter. **Is it only a coincidence that this truly colossal fountain is spewing Anti-Matter from the same region where a Natural Universe creates its Matter and <u>Anti-Matter</u>? I doubt it.**

How were the photons of the Cosmic Microwave Background radiation provided with a near perfect black body spectrum? [043] Near the beginning of the proposed Big Bang the photons of the CMB radiation were confined within exceptionally dense plasma which provided a near perfect black body spectrum. Within a Natural Universe Matter is created at the poles of Super Massive Black Holes within confined, closed and exceptionally dense, plasma cores where, <u>like within the said Big Bang,</u> the tightly confined plasma provided the system with a most uniform temperature. [046] The photons which will become the CMB were scattering off the crowded, hot, charged particles within plasma and could not travel far without colliding with another particle or were absorbed by the Super Massive Black Hole.

The Matter creation core is primarily powered by the massive Gravity of a Super Massive Black Hole which enables the energy from annihilation to power the core of the system which is <u>actually being fueled by virtual protons (Dark matter).</u> This is a similar sequence to how the Sun is also primarily powered by **Gravity** which enables the energy from nuclear fusion to power the core of the sun which is actually being fueled by the <u>regular protons</u> from hydrogen atoms. The opposing tugs of Matter and anti-particles causes the <u>growing</u> bulge of newly created Matter to rise farther which extends the magnetic fields farther above the event horizon while anti-particles may extend the magnetic fields down to the event horizon.

[046] Conforming to this modeling Astronomer Rudy Schild of the Harvard-Smithsonian Center for Astrophysics and his colleagues have found evidence that these types of magnetic fields actually extend and penetrate right through the surface of the 'collapsed central object' which they studied. This study is best explained by inherited magnetic fields contained <u>within</u> the central compact object. Obviously due to the ramifications of an event horizon the astronomers <u>have been reluctant to call this a Black Hole</u>. However, the object which they studied has a mass of 3 to 4 billion Suns which for all reasoning needs to be a massive Black Hole. This implication infers the magnetic fields are fittingly penetrating right through the surface of the massive object which, if a Black Hole, implies seemingly through the event horizon. To accommodate this abnormality our Matter creation cores are composed of a tornado storm-like structure which may possibly create a well in the event horizon at each pole of a Super Massive Black Hole.

As more Matter is forced by magnetic fields towards the surface of the elongated core the energy and pressure continues to grow a swelling <u>bulge</u> above the event horizon. This is the same tremendous energy which provides a regular Quasar with similar energy to a trillion, trillion atomic bombs exploding every second but for now has now been tightly contained.

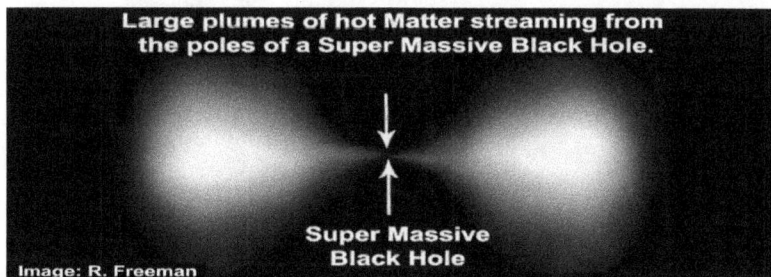

Large plumes of hot Matter streaming from the poles of a Super Massive Black Hole.

Super Massive Black Hole

Image: R. Freeman

[046] Unable to contain the enormous growing energy any longer <u>the spinning spooled up magnetic fields break</u> and in a fraction of a second <u>explosively release</u> colossal amounts of energy, Matter particles and radiation into

Space. This corresponds to the inflationary phase of the Big Bang when Matter was said to be escaping from a <u>singularity like point</u>. The high speed from being explosively released is required to avoid newly created Matter being captured near the massive Gravity of the Black Hole's <u>event horizon</u>. Because of the powerful confinement of the core it had remained semi-cloaked and now the released Matter possibly appears out of darkness. Observation evidence suggests Matter was first released this way in a succession of explosive like releases. If many of these events occurred almost simultaneously and close together than <u>this is the actual Big Bang</u>.

A percentage of the released Matter now finds its way back to the accretion disk and into the Matter creation core which complicated the system and enhanced the spiraling magnetic fields which now <u>grow by simply following</u> the hot Matter. This provided <u>two phases</u> for the release of Matter; **first an explosively release <u>directly from central regions</u>**, like an inflationary phase, and a second phase when Matter is <u>carried</u> great distances by long spiraling magnetic fields. The <u>two phases</u> are required to explain how different types of Galaxies are created and **why the protons of the CMB radiation created slightly cooler and slightly warmer regions.**

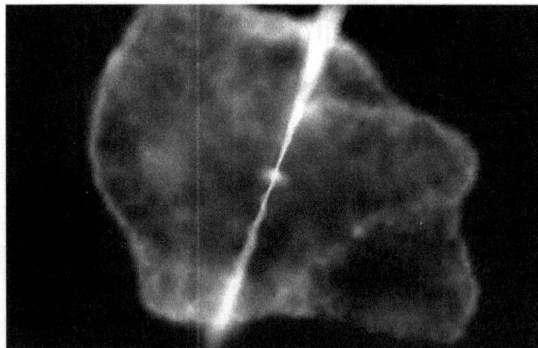

*This illustration depicts a gas halo surrounding a quasar in the early Universe. The quasar, in orange, has two powerful jets and a supermassive black hole at its center, which is surrounded by a dusty disk. The halo of glowing hydrogen is represented in blue. **Image credit: M. Kornmesser / ESO.** [047] Credit text ESO. The astronomers found that 12 quasars were surrounded by enormous gas reservoirs: halos of cool, dense hydrogen gas extending 100,000 light years from the central black holes and with billions of times the mass of the Sun. The team, from Germany, the US, Italy and Chile, also found that these gas halos were tightly bound to the galaxies, providing the food source to sustain the growth of supermassive black holes and vigorous star formation.*

Amazingly, evidence of both phases of the release of hot Matter appears to have been observed in the early universe by Astronomers using ESO's Very Large Telescope (VLT). The previous illustration, I believe, provides wonderful evidence that Matter was created by way of the Natural Model. The older first phase has formed halos of <u>glowing</u> hydrogen and the younger second phase has developed long spiraling magnetic fields carrying Matter afar. The reported discovery of the 12 Quasars which are located over 12.5 billion light-years from Earth is said to represent a major challenge to Astronomer's understanding of Super Massive Black Holes as well as formation and evolution of Galaxies. However, the discovery agrees very well to my Natural Model. Visually, one may suggest, instead of feeding the Massive Black Hole the hydrogen gas appears to have originated from the central Super Massive Black Hole which position is highlighted by the bright Quasar at the center. The gas has unbalanced leading edges inferring several explosive releases of Matter.

Powered by a Super Massive Black Hole of at least 12 billion solar masses, Quasar J0100+2802 resides near the beginning of the universe and has an energy output of about 420 trillion times the rate of the Sun. This is just a hint of the colossal amount of energy which may be explosively released during the first phase of Matter release. Such an super powerful event <u>may easily create elements</u> such as <u>helium</u> and other nuclei normally only associated with Big Bang nucleosynthesis, stellar nucleosynthesis and <u>explosive</u> supernova nucleosynthesis which may well explain the elements and metals found in the universe's <u>oldest observed stars</u>.

The magnetic fields may have been opened for a <u>relatively short time</u> and closed again until the pressure of newly created Matter repeatedly broke the magnetic fields. The process may be repeated many times over resulting in Matter being released in the <u>observed dense pulses</u> which is ideal for creating stars. The width of the core would be expected to measure more than <u>the distance across our whole solar system</u>; large enough to seed a Galaxy with enormous quantities of newly created Matter. [046] Astronomers at the Harvard-Smithsonian Center for Astrophysics have found evidence that similar magnetic fields by interacting with the accretion disk can act precisely this way and explosively release enormous amounts of energy. The Astronomers gained this information by continually monitoring Quasar Q0957+561 for 20 years. [039] The most detailed images of these regions from the Low-Frequency Array radio telescope clearly shows hot Matter being released in explosive like pulsations or blobs. The observed dense pulses or blobs are most important; a Big Bang cannot supply Matter in this way which clearly aided star formation. The density of Matter within these 'blobs' allowed their Gravity to attract more Matter which aided the early creation of stars.

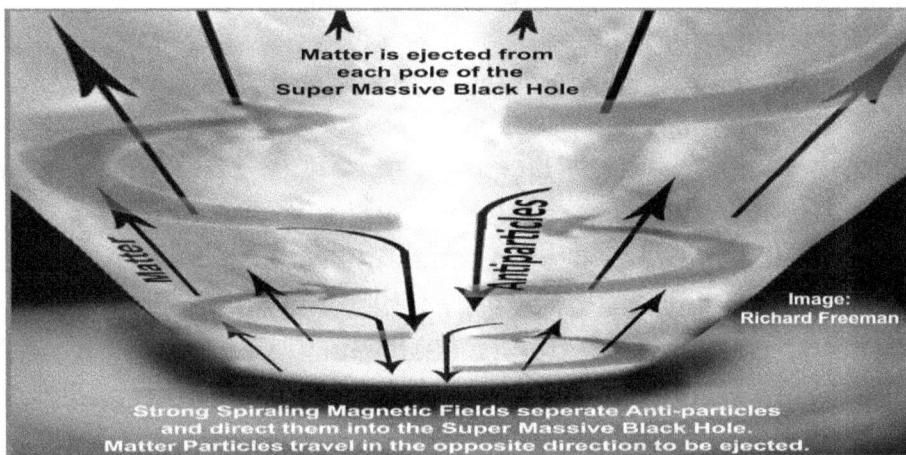

Matter is ejected from each pole of the Super Massive Black Hole

Matter

Antiparticles

Image: Richard Freeman

Strong Spiraling Magnetic Fields seperate Anti-particles and direct them into the Super Massive Black Hole. Matter Particles travel in the opposite direction to be ejected.

From a hot Big Bang the temperature takes 380,000 years to slowly cool, throughout all Space, to the 3,000 Kelvin where atoms can form. Within a Natural Universe, Matter is blasted away from hot Matter creation cores and released directly into cold empty Space which allowed Matter to very quickly cool. At about 1 billion Kelvin the temperature and density quickly became that which allowed protons to combine with neutrons to form small amounts of deuterium and helium nuclei. The temperature continued to very quickly cool to the 3,000 Kelvin where atomic nuclei are able to grab attracted electrons to form atoms.

[038; 079] Like after the said Big Bang, at 3,000 Kelvin the temperature of Matter had now cooled sufficiently for protons to grab an attracted electron to form hydrogen atoms, many of which 'pair up' to form molecular hydrogen. **This transitions the plasma to a gas and is when the photons of the Cosmic Microwave Background radiation are set free.** The photons of the CMB radiation became slightly cooler from losing energy escaping from the strong Gravity near the Matter creation cores but remained slightly warmer when released from far longer spiraling magnetic fields farther away from the stronger Gravity found nearer to the Matter creation cores and the forming Galaxy. **This caused the CMR photons to fill Space with ever so slight temperature variations which provided the CMB image it's not uniform appearance.** The hydrogen gas cools and form the first stars of a developing Galaxy which is already provided with its essential Gravitational scaffolding comprising of a parent central, Super Massive Black Hole, a disk of Dark Matter and an outer halo of Dark matter.

Is there any direct evidence when Matter creation cores developed regular Quasars they continued to create and eject new Matter? Established Quasars shine like light houses from very near the beginning of the universe and are, without a doubt, one of the most powerful objects ever observed in our universe. Quasars are powered by the massive Gravity of Super Massive Black Holes. [116 see Quasar J2054-0005] By intensely *"spewing out molecular gas, the raw material needed to form new stars"* this Quasar is truly displaying a Quasar's ability to seed its own Galaxy with newly created Matter but, for some very strange reason, very few are taking any notice.

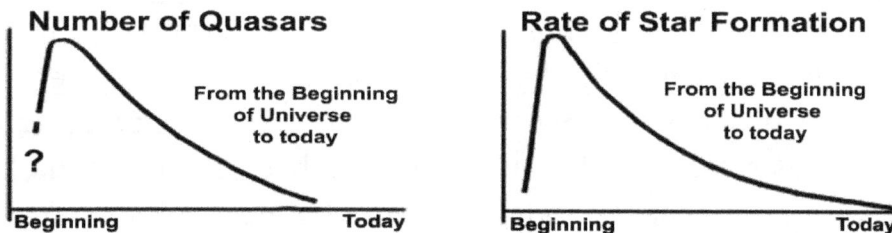

Above image: As one can clearly see from near the beginning of the universe to present day, there has always been a <u>direct correlation</u> between the number of Quasars and the availability of star forming Matter. This observation appears to have gone <u>unnoticed</u> as I had to obtain data from different sources to draw this image.

The data infers that whenever a Galaxy's Super Massive Black Hole has recently had a Quasar the Galaxy has more available Matter to form stars. As the number of Quasars declined so did the corresponding rate of star formation. So much so that an international team of astronomers have revealed that the rate of formation of new stars in the Universe is now only 3.33% of its peak and that this decline will continue. This research <u>did not</u> correlate the decline of star formation with the decline of Quasars. The <u>evidence</u> is clear to see; Quasars continued to supply their Galaxies with new star forming Matter. The evidence clearly suggests **both Galaxy and star formation <u>peaked</u>** within the early universe which means Matter production also likely peaked. This peak production of Matter is when the universe most resembled a Big Bang. The <u>hot</u> Matter from Quasars at first hindered star formation but quickly <u>cools</u> to be suitable for star formation.

A young Galaxy's dense, dark disk supplies dense loads of Dark-matter (virtual protons) to power a young Galaxy's Matter creation cores. As a Galaxy's dense, dark disk becomes depleted it becomes increasingly more difficult to fire up its Matter creation cores. [114 See: A dark matter disk in our Galaxy]. *An international team of scientists say their supercomputer simulation has exposed the presence of the disk of Dark matter within our own Milky Way Galaxy but the rotating dark disk is (today) only about <u>half</u> of the density of the Dark matter of the outer, non-rotating halo.* <u>Now ask;</u> why the dark disk appears to have <u>lost half its density</u>? Answer; originally <u>denser</u> than the halo, the dark disk has lost density due to its Dark matter being used to create a Galaxy's regular Matter during which much of the dark disk is <u>replaced</u> with a disk shaped, bright Galaxy of stars. Dark matter may also help star formation by stabilizing a Galaxy's reserves of hydrogen gas. **The**

end result is stars in Galaxies like our Milky Way are now dying from old age faster than they are being replaced with young new stars.

Galaxy Henize 2-10 is only 30 million light-years from Earth. About 3,000 light-years across it resembles what Scientists think were some of the first Galaxies to form in the early universe and is <u>forming stars very rapidly</u>.

Galaxy Henize 2-10. *CREDITS: SCIENCE: NASA, ESA, Zachary Schutte (XGI), Amy Reines (XGI). IMAGE PROCESSING: Alyssa Pagan (STScI).* **The enlarged window shows jets of material spewing outward from areas close to a Super Massive Black Hole.**

Because it is only 30 million light-years from Earth, Galaxy Henize 2-10 provides evidence that the process of first creating Super Massive Black Holes which then seeded their own Galaxies with Matter may have continued well passed its peak at near the beginning of the universe.

The longer, tighter, spiraling, magnetic fields of regular Quasars created jets containing many new regular protons which have transformed from virtual protons. The jets have been said to average <u>1.5 light years in diameter</u> and can grow to extend for hundreds of <u>thousands of light years in length</u>. Given the enormous size of these jets and the near light speed which they carry Matter they can easily account for colossal amounts of Matter which allowed the first small Galaxies to simply continue to grow themselves into very large Galaxies. In order to conform to this modeling the jets are required to contain many newly created protons and electrons which will make hydrogen atoms to make stars. But is there any evidence that this is really true? I am truly blessed to live in an age where I need only to ask a search engine to find the answer. [006] NASA's Swift Satellite was launched November 20, 2004. According to the Swift team, the jets are <u>made of protons and electrons</u>, which Scientists say <u>solves a mystery</u> (of what the jets are made from) as old as the discovery of jets themselves in the 1970s.

[048] Termed as being bizarre because they lack any significant amount of Dark matter, Galaxies NGC 1052-DF2 and NGC 1052-DF4 have, for their size, hundreds to thousands of times less stars. Scientists say the Galaxies challenges the standard ideas of how we think Galaxies work. The Galaxies provide further evidence that virtual protons (Dark matter) are truly transformed into the regular protons which will make hydrogen atoms to supply the fuel to build stars. Consequently, if a Galaxy lacks a significant amount of Dark matter (virtual protons) it will be naturally limited in its ability to create an ongoing supply of regular protons to create hydrogen atoms and molecular gas to make stars.

First matter appears in plumes from poles of a Naked-Super-Massive-Black-Hole.
Image: Author.

I had created the above image well over ten years earlier when I was writing my first book. This is the type of image I expected to find if Matter was first created from the poles of Super Massive Black holes. However, the amazing deep field images of early Galaxies from the Hubble Space Telescope disappointedly showed almost no signs of large plumes of hot Matter and radiation streaming out from the centers of first Galaxies.

Years later while holidaying at North Stradbroke Island my wife, Beverley, called me to hurry in and look at the evening news channel; I was ecstatic with excitement for there were the images which I had been searching for. It turns out to see these hot plumes of Matter from early Galaxies you need a very large low-frequency <u>radio</u> telescope. Many years searching for such images of early Galaxies on the internet and here they are on the evening news channel, I said to my wife Beverley, *"We are witnessing the creation of first Matter, just amazing"*. The breathtaking images are detailed ultra-high definition and created from data collected by the Low-Frequency Array <u>radio</u> telescope which is an array of 70000 small antennae spread across nine European counties. **The images clearly reveal massive plumes of hot Matter and radiation streaming from the Super Massive Black Holes of numerous early Galaxies which is exactly what this model requires.**

[039] Dr Neal Jackson, from the University of Manchester says the images reveal early Galaxies are *"dynamic sun and planet making factories, powered by (Super Massive) Black Holes"* and *"It's become very clear that, in order to understand Galaxy evolution, we need to understand the Black Hole right at the very centre, because it appears to have a fairly fundamental influence on how galaxies evolve"*. Amazingly, in one image one can even see what Scientist describe as *'periods of relative inactivity by the Black Hole when it spits out less material'* which conforms to my modeling of the Matter creation core being a closed system until a buildup of pressure from newly created Matter broke it's confinement and was explosively released as a wave like dense pulse. The wave like clumps are very important because as they cool the clump provide the density required where Gravity can drive the formation of stars. A Big Bang cannot provide Matter with such a logical and clearly <u>observable</u> procedure which <u>actually</u>

produces Matter in dense clumps which are required to form the first stars.

Scientists suggest Matter which they believe was created at the theorized Big Bang is being attracted towards a Super Massive Black Hole and by some understood process heated up and expelled within jets and enormous plumes which are so large that they may visually totally overwhelm the size of an already created small Galaxy of stars. However, **there is no observable Matter moving towards any of the SMBHs which can account for this massive output of Matter**. An accretion disk is tiny in relation to the amount of visible material being ejected. It is not like a Galaxy is almost concealed by massive clouds of hydrogen gas from a Big Bang which is now being attracted towards its central SMBH. Rather, it truly conforms to Matter actually being created at the poles of Super Massive Black Holes. This activity also clearly peaks at a time when the universe was in its infancy forming early Galaxies. The evidence is clearly observable.

[082] CERN says there is six times more invisible Dark matter than regular visible Matter. The facts are all in line and are undeniable. From all observable evidence hot Matter is simply coming from polar regions of Super Massive Black Holes and within the early universe this evidence is truly everywhere; but you need the frequency of a radio telescope to see it.

The observations match the images which I had created many years previous for my Natural Model which I had originally called 'The Big Stretch'. Scientists know there is far more invisible Dark matter associated with Galaxies than regular Matter which allows a juvenile Galaxy of stars to be freely visible while massive quantities of invisible virtual protons are fed to Matter creation cores and transformed into positive protons and then expelled from the polar regions of a Super Massive Black Hole (SMBH). The positive protons cool and attract an electron which creates a hydrogen atom. Hydrogen atoms and molecular hydrogen form stars. Scientists have discovered enormous clouds of hydrogen which align with the jets from Quasars. [116] Astronomers have even discovered a Quasar, J2054-0005, powered by a SMBH and existing in the early universe which they say is ***"spewing out molecular gas, the raw material needed to form new stars"***.

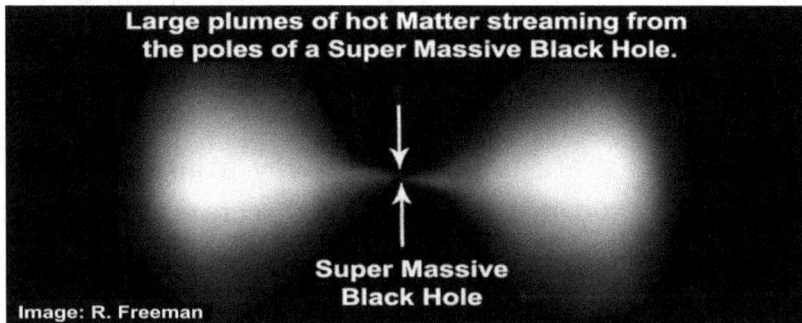

Large plumes of hot Matter streaming from the poles of a Super Massive Black Hole.

Super Massive Black Hole

Image: R. Freeman

Adolescent Galaxies can be expected to first retain a resemblance to the double lobe or elongated appearance of how Matter is at first introduced. However, its central SMBH, the Dark matter disk and outer halo provides a Gravitational scaffold to quickly regularize a young Galaxy's appearance. The number of observable Quasars appears to fall off at the very beginning of the universe. This would be when Matter creation systems were building their tightly constrained cores of Matter and the direct evidence of these tightly constrained cores of plasma is the near perfect black body spectrum of the Cosmic Microwave Background radiation. The first Matter was released, in explosive pulses from very close to the event horizon of Super Massive Black Holes. This allows a small young Galaxy to come into being before it creates a Quasar and not all Galaxies necessary created a long Quasar. Is there any evidence that similar explosive ejection of Matter occurring in more recent times? Such an explosive blast of Matter would surely not go unnoticed. Evidence of such an event could be a colossal, explosive release of Gamma rays and newly created hydrogen atoms.

[050] Amazingly, NASA's Fermi Gamma-ray Space Telescope has unveiled a completely unexpected discovery. The Fermi Bubbles span an incredible 50,000 light-years and appears to be the remnant of an explosive eruption from the Super Massive Black Hole at the center of our home Milky Way Galaxy. The explosive event likely occurred just a few million years ago and

released a similar amount of energy to a <u>hundred thousand exploding stars</u>. NASA says the structure's shape and emissions suggest it was formed as a result of a large and relatively rapid energy release, **the source of which remains a mystery**. The plumes extend 25,000 light years from both sides of our home Galaxy's central Super Massive Black hole.

With a Natural Model there is no reason why a large amount of virtual protons (Dark matter) did not drift into the accretion disk of the central Super Massive Black Hole and created a massive blast of newly created Matter. Newly created Matter is at first restrained in plasma cores at polar regions of a Super Massive Black Hole. Without a developed Quasar the pressure of the restrained Matter builds up until it is explosively released as one event which is why these plumes are well defined and sharp edged as they are observed to be. The restraining magnetic fields may remain open or close again. In the early universe these events continued because there was an enormous ongoing supply of virtual protons (Dark matter).

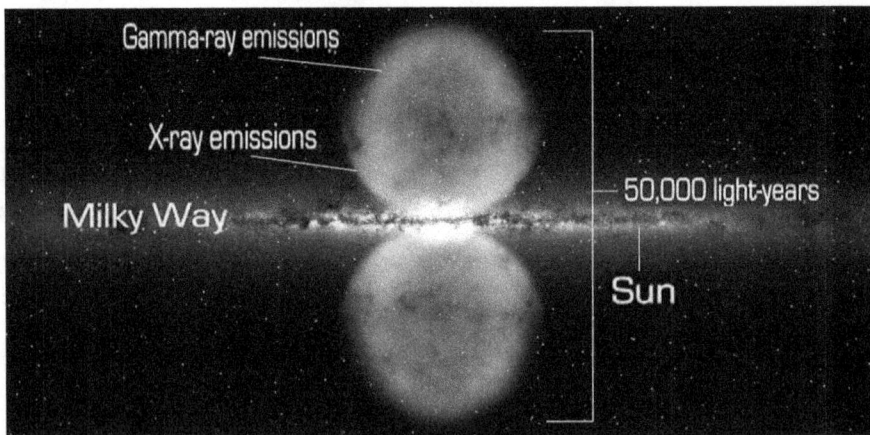

Credit: NASA Goddard Flight Center.

050 Above image *Huge <u>bubbles of gas</u> flowing from the center of Milky Way but still no idea what caused them.* A team of astronomers used the Wisconsin <u>H-Alpha</u> Mapper telescope to study the mysterious plumes and discovered they contained Hydrogen-Alpha atoms. These are Hydrogen atoms where their electron is in a higher energy orbit which could possibly be a state

from when a proton first captures a high speed electron to form a Hydrogen atom. Astronomers say it will require future observational and theoretical work to understand what is causing this phenomenon. (It seems every time I come across an event which agrees to my modeling the event is always a mystery to Scientists who are fixated on Matter being created from an inexplicable Big Bang). **Although Quasar jets likely carried Matter a great distance, plumes like these may be the preferred method for releasing colossal amounts of newly created Matter to where it can both form stars and fall to a Galaxy's Galactic plane.**

Within a Natural Model, a virtual particle does not become a regular particle just because it has lost its virtual anti-particle. Having lost its anti-partner only means it is not annihilated. To become a regular particle with certain properties the uncertain, virtual state of a 'wave' of potentials has to be exposed to the negative pressure of a deep vacuum phase of a COST. [014] Werner Heisenberg's 'uncertainty principle' explains the virtual particle phenomenon. Particle Physicists regularly fiercely debate the properties of the virtual realm and many discard the notion that virtual particles may have somehow transitioned into the regular particles from which we are all made from. However, Physicists know that the protons and neutrons we are all made from consist of many virtual particles. It seems everyone agrees that protons are real particles which are made from virtual particles called quarks and gluons.

Computer simulations of the magnetic fields in these regions clearly show Matter and Anti-matter particles traveling in opposite directions. A Quasar's powerful magnetic fields forcibly eject massive amounts of Matter away from Super Massive Black Holes which means anti-particles generally travel the opposite direction and into the Super Massive Black Holes. Consequently, at these exact sites existing at near the beginning of the universe, we know there was exactly the right mechanism to forcibly separate massive amounts of anti-particles from Matter particles. A Super

Massive Black Hole (SMBH) then provided a storage bin for massive amounts of anti-particles from where anti-particles could never escape and destroy all created Matter. If one wishes to <u>obey the stringent laws of physics</u>, than this is the <u>only place</u> in the early universe where Matter could have been created and it was created at exactly the right place to form a Galaxy which will obviously harbor a SMBH at its center. **Note: <u>ejected protons retain</u> their entire tiny phase of virtual mass <u>inherited</u> from their virtual proton state. Thus, newly created protons get to <u>keep all of their mass</u> and a SMBH is not required to grow beyond its observed size.**

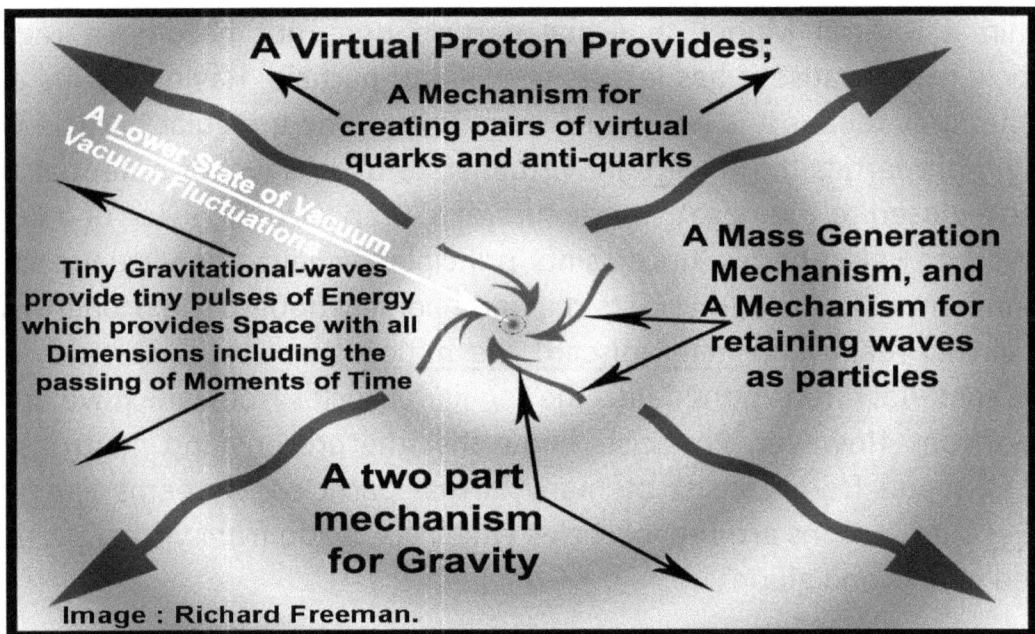

A Virtual Proton Provides;

A Mechanism for creating pairs of virtual quarks and anti-quarks

A Lower State of Vacuum Vacuum Fluctuations

Tiny Gravitational-waves provide tiny pulses of Energy which provides Space with all Dimensions including the passing of Moments of Time

A Mass Generation Mechanism, and A Mechanism for retaining waves as particles

A two part mechanism for Gravity

Image : Richard Freeman.

<u>The laws of physics absolutely confirm</u> that if there had been a <u>Big Bang,</u> Anti-matter would have annihilated all Matter and you and I could not have ever existed. [009] Many experiments have <u>confirmed this is true</u> and now Scientists at CERN's Baryon Antibaryon Symmetry Experiment <u>have recently verified</u> *'with unparalleled precision'* that *'Matter and Anti-matter are perfect mirror images'* of each other, with only their charges reversed. <u>Note</u> that Scientists use the words *'<u>perfect</u> mirror images'*. The word 'perfect' here has a <u>big meaning</u> for it means there is <u>absolutely no wriggle room</u> for

Matter from a Big Bang to have prevailed over Anti-matter. Matter and Anti-matter may have had different properties which could cause anti-protons to decay faster than protons but the researchers found within quite strict limits that the charge-to-mass ratios are the same. Thus, this latest research has clearly, yet again, confirmed there is absolutely no inequality of Matter over Anti-matter which may have possibly allowed more Matter to survive than Anti-matter, **due to any currently proposed reasons**, after a Big Bang. Scientists need only to accept and believe the predictions of their own theories, the laws of their own physics and the true results of their own experiments to know that the universe could not have begun with a Big Bang.

The COST is based on a Quantum vacuum fluctuation which has become more permanently configured as a virtual proton. Given that a COST produces pairs of virtual particles, one of Matter and one of an anti-particle there is no need for an Anti-COST (Anti-Dark matter). That is, Dark matter (a virtual proton) is both Matter and Anti-matter of the virtual kind.

At the end of the spiraling, magnetic fields of Quasars hot protons and attracted electrons finally escape the spiraling, magnetic fields and balloon out into the cool of Space where protons cool and grab an attracted electron to form the most basic atom, a hydrogen atom. It is fitting that these most powerful objects from near the very beginning of the universe and residing at the center of **the first observable Matter** played a major role in the creation of all regular Matter. This is also both observable and obvious evidence. Plumes of new hot Matter may at times be more like a prolonged or repeated supernova explosion which may last for millions of years. The Natural Model is designed to always maintain a sum-total zero mass-energy universe so any amount of Matter can be produced over extended time. This importantly preserves the laws relating to the conservation of energy.

With the offload of anti-particles the anti-particle-key required to immediately return this energy to the symmetry of Space has now been lost. Hawking Radiation theory says that if a virtual particle and anti-particle pair appears near a Black Hole and the virtual anti-particle succumbs to the Black Hole the surviving virtual particle becomes a regular particle. However, the theory says the process causes the Black Hole to release radiation which takes away a tiny bit of a Black Hole's mass.

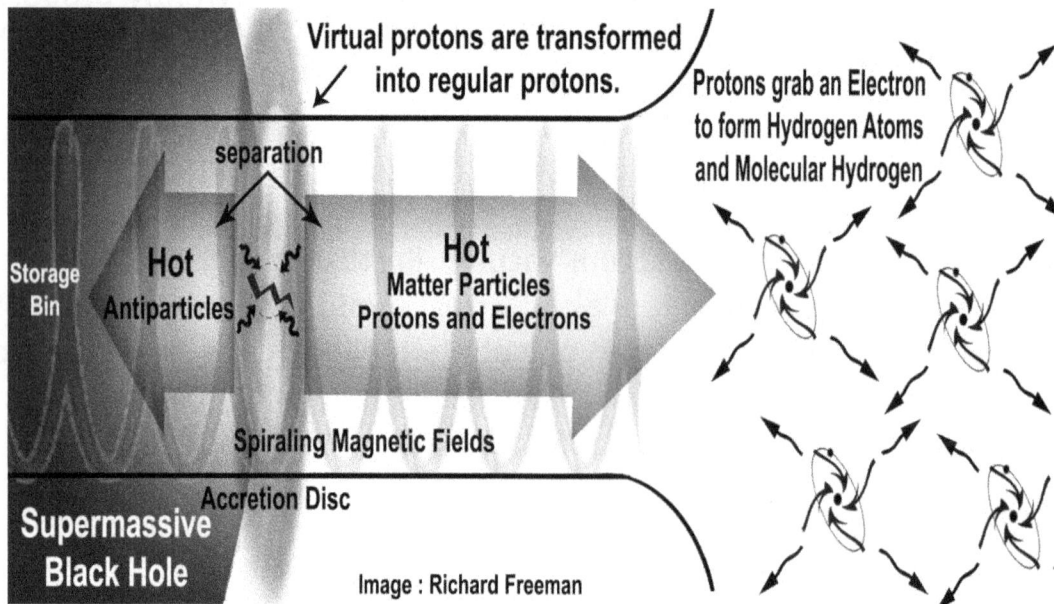

Because our anti-quarks similarly succumb to a Super Massive Black Hole our first Super Massive Black Holes may possibly be required to similarly lose mass as regular Matter is being created. This requires the farthest back in time Super Massive Black Holes to be generally more massive than the closest. The farthest away (back in time) known Super Massive Black Hole is associated with Quasar ULAS-J1120+0641 has a mass of 2,000,000,000 times that of our Sun. The closest, the Super Massive Black Hole at the center of our Mighty Milky Way Galaxy, is by comparison only 4,600,000 times the mass of our Sun. Unfortunately, Galaxies come in many different sizes so this is not a consistent comparison.

Because this process acts independently with a COST unit transforming from a virtual proton into a regular proton, the majority of first atoms created by this model are naturally destined to be mostly hydrogen atoms. This is for the reason that a hydrogen atom has a nucleus of just one proton. A <u>virtual proton can only be transformed into one proton</u> which can only create (golly gosh) a hydrogen atom. Hydrogen atoms can only create stars, however, stars and exploding stars can create atoms of almost all other elements. Meaning the reason as to why the universe first produced mostly simple hydrogen atoms is very clear with this model. Hydrogen today remains by far the most abundant element for the simple reason that the transformation from a COST unit (a virtual proton) to a hydrogen atom was clearly the most direct and easy path. Many of the seven, billion, billion, billion atoms which make up the average person are hydrogen atoms and each hydrogen atom has just one proton and one electron.

Hydrogen within all of the universe's stars continues to first fuse to form helium and then helium atoms fuse into carbon. Many of the elements which we have all been made will be ejected and form a planetary nebula at the end of a star's life. [102] Although this process has been ongoing inside of stars for over 13 billion years the majority of all <u>atoms</u> (90%) remain as hydrogen atoms. The majority of all Matter, now said to be <u>85%</u>, remains as Dark matter (virtual protons). The numbers are compelling evidence of the source of all regular Matter. Virtual protons were transformed into regular protons which grabbed an attracted electron to make hydrogen atoms.

Today research and observations reveal <u>all Galaxies</u> reside within the filaments of the cosmic web which is believed to be made from mostly Dark matter. Studies suggest at least 90 percent of a filament's mass is Dark matter. At junctions where there are <u>larger clumps</u> of Dark matter is where one finds clusters of Galaxies. Why are Galaxies found in the larger clumps? This is where Dark matter became dense enough to cause a Gravitational collapse which formed central Super Massive Black Holes complete with halos of Dark matter. The Super Massive Black Holes provided the power houses to transform virtual protons (Dark matter) into regular protons to make hydrogen atoms to create the stars to build Galaxies. Because Galaxies are found only within these regions this obviously provides remarkable evidence that this modeling is correct.

Has the speed of light slowed over billions of years?

[051] Theories say the speed of light has always been a constant, however, a team of Scientists led by Australian theoretical Physicist Professor Paul Davies of Sydney's Macquarie University has uncovered astronomical data involving light from a distant Quasar which implies the speed of light has slowed over billions of years and may have once been *"close to infinity"*.

Researchers have used the data from actual observations with well-known and respected mathematical formulas. The results showed that the speed of light has slowed. According to the **astronomical data,** at very near the beginning of the universe, the speed of light would have been **close to infinity**. The Scientists say this has far reaching consequences. For instance, theories of how energy cooled to form Matter after a Big Bang allowing the formation of stars and planets may be absolutely wrong. Paul Davies is a much respected Physicist. I'm sure he never set out to find evidence those current theories 'could be completely wrong'. Instead he is respecting the real *'astronomical data' which is telling a different story.* A faster speed of light would obviously violate the physics of Albert Einstein's very successful theory of Relativity which strictly implies the speed of light remained a constant over time. So how is it possible for the speed of light to have <u>both</u> been much faster in the past and <u>always</u> remain the very same constant speed? It <u>requires</u> the modeling of a Natural Universe to make sense of this.

My Natural Universe may well at first had a much faster speed of light corresponding to this actual *'astronomical data and experimental evidence'*, however, this does not signify trashing $E=mc^2$ and the theory of Special Relativity as one may at first believe but it may require a better understanding of **how the universe complies with Special Relativity.** Relativity is protected because of 'expansion speed time dilation' which is directly related to the expansion speed of the dimensions of Space and time. Expansion speed time dilation allows for a much faster rate of time at the beginning of a Natural Universe so as the speed of light would always be observed, if you were there, at the same constant speed as today.

[052] The Cosmic Microwave Background radiation has a near uniform temperature in all directions but how did this come to be? Portuguese Physicist João Magueijo provides an answer. He also proposes that in the early universe, light travelled many trillions of times faster than it does today. These superfast heat-carrying photons then explain how the temperature of the universe equilibrated as fast as it did.

Why the Speed of light may well be slowing: Within a Natural Universe the source of the attraction part of outer Gravity is the seemingly endless outer void of Anti-Space which provides a massive source of negative energy. Because this void would have been much closer at the beginning of the universe it may well have attracted and driven both light and the tiny Gravitational waves which provide entangled dimensions to Space at a much faster rate, as this is how Gravity works closer to its source. (See: Quantum Gravity near the beginning of chapter 13). Since, like photons, the tiny, massless Gravitational waves travel at the speed of light, the speed of light would have at first been faster, however, the passing of the same tiny Gravitational waves provide the passing of time which means time also ran faster. The passing of a tiny Gravitational wave provides units of both Planck length and Planck time which locks light to always travel a Planck length in Planck time due to expansion speed time dilation occurring during the slowing of light. Because time would have also been faster, a faster speed of light in the early universe would still be observed, if you were there, the same constant speed as today. Consequently, the speed of light remains a constant regardless of the speed of dimensional moments. A ramification of this would be a universal slowing of the rate of applying Gravity which, since Gravity is directly linked to the passing of time, would also remain self-observed as unchanging. However, if there had been a Big Bang, applying Gravity at speeds *"close to infinity"* would likely have allowed Gravity to accelerate Matter to speeds which may exceed today's speed of light without the theorized energy of cosmic inflation.

However, if the rate of passing time has been slowing since the beginning of the Universe this slowing will eventually have profound consequences for a Natural Universe. It means a <u>slowing of the vacuum fluctuations</u> of the protons and neutrons which create the tiny Gravitational waves which <u>provide the passing of time</u>. This slowing within protons <u>would not be directly measurable</u> since time itself is now slowing at the same exact rate causing a proton to <u>appear to be eternally stable</u>. However, the <u>vacuum fluctuations</u> will eventually slow to a stop. When the vacuum fluctuations of all Matter slow to a stop <u>all Gravity fades away</u> and passing time naturally stops. Without Gravity all stars and planets fall apart. Without their vacuum fluctuations, atoms and protons and neutrons will have no mass or energy and without a strong force will also fall apart. All particles now default to invisible waves and even dimensional Space-time fades into the distance. Thus, the Natural Universe will likely, long after all stars exhaust their fuel supplies and eons from now, eventually wind down to a definitive end.

A year of today's slow time may naturally equal many years of fast time at the beginning of the universe. Scientists measure the age of the universe in the slower Earth time of today but one year of today's slow time may equal many years of fast time at the beginning of a Natural Universe. Now when the ageing of star HD 140283 is measured in slow Earth time, as you count farther back in time, each year may equal many years of past faster time causing one to alarmingly run out of time. This naturally means, if we accept Paul Davies's research, the universe is likely far older than thought.

[118] The James Webb Space Telescope has recently discovered a cluster of

very distant baby Quasars. The Quasars are powered by Super Massive Black Holes which the Scientists say *"look like they have grown faster than is possible"* (from a Big Bang) and conclude *"something doesn't add up."* Such observations are no problems for a Natural Universe which clearly creates Super Massive Black Holes <u>before</u> it creates Matter or Quasars.

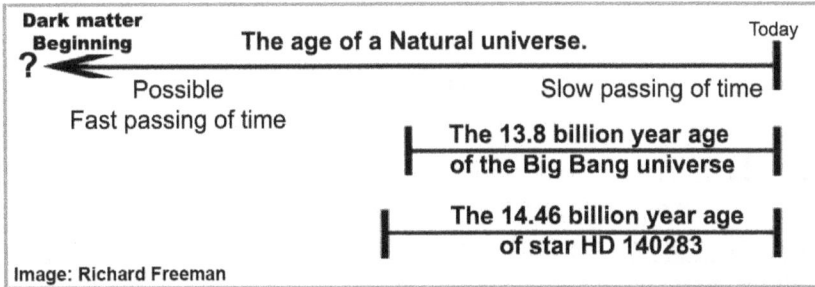

Image: Richard Freeman

All regular Matter within a Natural Universe is created and accumulated over billions of years as opposed to the instant Big Bang universe where all Matter is created within a very tiny fraction of one astonishing second.

Credit: NASA/ESA/G.Bacon,STScI. *An artist's impression of one of the most distant, oldest, brightest Quasars ever seen.* *"Hubble astronomers have looked at one of the most distant and brightest Quasars in the universe and have said that they are surprised by what they did not see: the underlying host Galaxy of stars feeding the Quasar. The best explanation is that the Galaxy is shrouded in so much dust that the stars are completely hidden everywhere."*

This Quasar appears to have the features of a Super Massive Black Hole (SMBH) creating regular Matter <u>before</u> the fledging Galaxy created stars.

213

9 GALAXY FORMATION.

<u>'Fountain Formation'</u>   By Richard Freeman.

Made from <u>Dark matter</u>, a Natural Universe first creates disk shaped <u>Dark</u> Galaxies with central Super Massive Black Holes. The first Galaxies of stars are likely <u>double lobed or elongated</u> which a <u>Dark Galaxy's</u> disk shaped, Gravitational scaffold regularizes as a disk shaped Galaxy of stars. The Big Bang may have created one massive Galaxy with stars spaced throughout all Space. Within a Natural Universe Matter has no long journey to where a Galaxy will form, instead, a Galaxy will actually create its own Matter.

[005] The temperature at the core of Quasar 3C 273 has been discovered to be an astonishing 10 trillion Kelvin which is said to be an astounding <u>100 times hotter</u> than the current theories for Quasars allow. [038] Since this is where my Matter creation cores reside it is a very <u>unlikely coincidence</u> that this is the <u>exact same temperature</u> which is <u>calculated</u> to have **only** existed during the creation of Matter and the annihilation of Anti-matter at just <u>one ten thousandth of a second after the said Big Bang</u>, and provides <u>undeniable evidence</u> of Matter existing here in a state thought to have only existed for a fraction of a second after the proposed Big Bang. Scientists have known for <u>many years</u> that Quasar's jets eject massive amounts of protons and electrons. One proton and one electron make a hydrogen atom and two hydrogen atoms make molecular hydrogen. Very large clouds of hydrogen aligning with these jets are common. [116] Link describes *"a molecular gas fountain in the ancient (early) universe";* the Astronomers have discovered

a Quasar, J2054-0005, powered by a SMBH which they say is *"spewing out molecular gas, the raw material needed to form new stars"*. Hydrogen and molecular hydrogen gas make stars and stars make everything else. I have provided links to evidence of all these things. Now this all provides overwhelming, <u>observable evidence</u> of Matter originating from the poles of a Galaxy's Super Massive Black Hole allowing Galaxies to independently form where they actually created their own Matter. However, is this proof? It truly provides outstanding evidence for me to really believe it is true.

All computer simulations of Galaxy formation I have seen show gas streaming in <u>towards the center</u> of a forming Galaxy, however, actual images from the early universe clearly show gas and radiation streaming <u>away from the center</u> of a forming Galaxy. <u>My Fountain Formation model:</u>

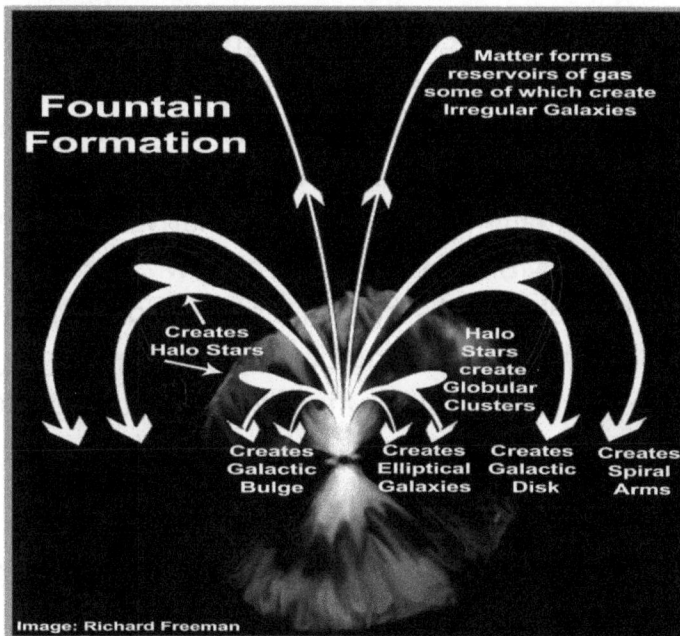

Within large clouds of virtual protons (Dark matter) a Gravitational collapse first forms a Super Massive Black Hole (SMBH) at the center of a <u>disk shaped Dark Galaxy</u>. [114 See: A dark matter disk in our Galaxy]. *An international team of scientists say their supercomputer simulation has exposed the presence of*

the disk of Dark matter within our own Milky Way Galaxy but the rotating dark disk is (today) only about <u>half</u> of the density of the Dark matter of the outer, non-rotating halo. Now ask why the dark disk appears to have lost half its density? Answer; originally <u>denser</u> than the halo, the dark disk has lost density due to its Dark matter being used to create a regular Galaxy's Matter during which much of the dark disk is <u>replaced</u> with a disk shaped, bright Galaxy of stars. **The rotating, disk shaped <u>Dark Galaxies</u>** provide the complete Gravitational Scaffolds essential for stars to <u>quickly reconfigure</u> from early <u>elongated</u> Galaxies to <u>disk shaped</u>, bright Galaxies of stars. The 'unforeseen' early formation of disk shaped Galaxies have been observed by the James Webb Space Telescope. An outer halo, of <u>original density Dark matter</u> (virtual protons), remains today as evidence of how virtual protons were sourced from a central Dark matter disk to create a Galaxy of stars.

Note: The <u>dense</u> dark disk becomes <u>less dense</u> due to Matter production and the outer halo of Dark matter becomes <u>closer</u> to a Galaxy due to both Gravity and a Galaxy of stars growing larger; both of which influence the rotation of Galaxies. This caused the observed rotation of a bright Galaxy of stars to over time gradually speed up, particularly in its outer regions. At the poles of a Galaxy's SMBH hot newly created Matter builds <u>pressure and density</u> until it explosively breaks free, in a succession of long dense pulses, of powerful magnetic fields. The long dense pulses are <u>imperative</u> for they will provide regions of exceptionally high rate of early star formation. One can observe this <u>essential</u> lumpiness in the image below of Galaxy 3C353, tiny spot at center, from when the universe was aged just 1.4 billion years.

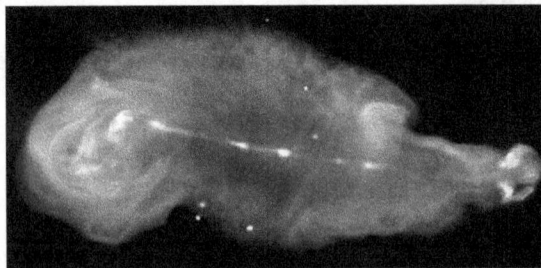

Image Credit: X-ray: NASA/CXC/Tokyo Institute of Technology/J.Kataoka et al, Radio: NRAO/VLA.

First released <u>close</u> to the central Super Massive Black Hole Gravity <u>slows</u> the speeding away Matter which is <u>first directed</u> to the halo region before it falls to a developing, adolescent Galactic plane. Dense clumps cool and create regions of intense star formation. As a Galaxy grows the spiraling magnetic fields grow and carry Matter farther from the central Super Massive Black Hole. [105] <u>Because of the way Matter is released in long dense pulses</u> many young Galaxies may develop <u>correspondingly</u> bright bubble-like lumpy regions of intense star formation. Such regions have amazingly been observed in recent images from the James Webb Space Telescope.

Image Credit: NASA, ESA, and Z. Levay, F. Summers, G. Bacon, T. Davis, and L. Frattare (Viz 3D Team/STSxl. Active Galaxy Hercules A: Visible & Radio Comparison. [096] A great video which begins with an optical image of the Galaxy before the jets are faded in from a image created by a Radio telescope.

[096] Note the <u>extraordinary long</u> jets carrying material <u>well away</u> from the central Galaxy and the <u>observable lumpiness</u> of the released material. The jets which are clearly visible to a radio telescope span a length of over 1,500,000 light years which is way farther than the 100,000 light years across our entire Milky Way Galaxy. The Galaxy, 3C 348 Hercules 'A', is 2.1 billion light years away and is actually a <u>supergiant</u> Elliptical Galaxy. How is it possible for such a massive amount of material to be explosively coming out of the very center of a relatively smaller Galaxy? There is obviously something truly amazing happening here which is creating unimaginable energy the source of which is today not understood. The jets of this Galaxy have now grown too long to now feed new Matter to their own Galaxy;

however, one can clearly observe how Matter and radiation is released in dense pulses. Virtual protons in their Dark matter state are invisible and Scientists know there is many times more Dark matter associated with Galaxies than regular Matter which allows a Galaxy of stars like this to be freely visible while massive quantities of invisible virtual protons are being fed to a central Super Massive Black Hole where virtual protons are transformed into regular protons and expelled along with electrons, radiation and charged particles from the poles of a Galaxy's central SMBH.

[006] NASA's Swift Satellite was launched November 20, 2004. According to the Swift team, the jets appear to be made of protons and electrons, which Scientists say solves a mystery (of what the jets are made from) as old as the discovery of jets themselves in the 1970s. Images from radio telescopes of early Galaxies confirm many of the jets are at first short which is ideal to begin creating the first Galaxies. Matter cools allowing atoms of mostly hydrogen to be created when many of the regular protons grab an attracted electron. Now clear of the magnetic fields the Gravity from both the parent Super Massive Black Hole and it's forming new Galaxy now clearly acts in the correct manner to slow the released Matter which already contains massive dense clumps created from the way it was released. The dense hot blobs of Matter at first expand but as they cool a blob's self Gravity attracts more Matter which directly leads to the creation of stars within the blobs. Creating Matter this way is essential because the blobs both hastened and provided the perfect configuration for creating stars. Much of the cool Matter and many of the already formed stars gradually migrate to the Galactic plane. Migration to the Galactic plane is naturally enhanced by a Galaxy's already existing Dark matter Gravitational scaffold in the shape of a Dark Disk Galaxy. Unlike a Big Bang which sent Matter speeding apart aimlessly in all directions throughout the universe, a Natural Universe introduces created Matter directly to a Dark Disk Galaxy.

With an event of a near uniform Big Bang all Matter for all Galaxies remains associated with the same <u>single</u>, highly inflationary event. With such an event it seems to make common sense that most Galaxies, because they are all moving apart from each other <u>would seldom come together</u>. [053] Scientists today do not really understand why there are Super Massive Black Holes at the center of even the earliest Galaxies. Because this model first creates Super Massive Black Holes and then actually creates its Matter at the poles of Super Massive Black Holes there is no doubt to why the first Galaxies naturally harbor Super Massive Black Holes at their centers. **See:** [113] **'The JWST Has Spotted Giant Black Holes All Over the Early Universe'. The James Webb Space Telescope is discovering an** *"unexpected abundance"* **of Super Massive Black Holes within the very early universe.**

If all Matter had been created by a Big Bang the Space between Galaxies should contain large amounts of primordial hydrogen and helium left over from the Big Bang which is <u>what Scientists believed.</u> [049] However, in the 1970s, X-ray observations revealed intergalactic Space generally contains less than one atom per cubic meter and are the most rarefied (empty) environments in the Universe. The X-ray observations also revealed quantities of <u>metals</u> mixed in with the hydrogen and helium. The metals could only have been created within the stars of surrounding Galaxies. Consequently, the evidence now tells the intergalactic medium is composed of material which had come from <u>within</u> Galaxies and <u>not directly from a Big Bang</u>. This material is now thought to be driven from Galaxies by Galactic winds from either starbursts or active Galactic nuclei.

Within a Natural Universe the central densest regions of large, clouds of virtual protons (Dark matter) began to <u>rotate and flatten</u> before collapsing to form Super Massive Black Holes. The collapsed dense central region creates a Super Massive Black Hole (SMBH) which naturally leaves behind a thick, disk shaped <u>Dark Galaxy</u> with an outer halo of Dark matter. All Matter is created at the SMBH <u>inside</u> of the outer halo which is why a Galaxy is created <u>inside</u> of this halo of Dark matter. [024], [025] and [054] This provided a detectable correlation between the mass of the central SMBH, the mass of the outer halo of Dark matter and often the total mass of created stars. **It is obviously an essential requirement of this model that Galaxies commonly reside within regions richest with Dark matter (virtual protons).**

Before the collapse of Dark Matter (virtual protons)

The collapse of virtual protons provides the Gravitational Scaffolding essential for the early creation of Galaxies.

Dark Galaxy

Outer Halo

Image: Author

A large cloud of Dark matter (virtual protons)

A resulting Gravitational Implosion forms a central Super Massive Black Hole

Scientists should ask; why do Galaxies themselves reside within a relatively hollow center of massive clouds rich with Dark matter? The center of these clouds were hollowed out when virtual protons (Dark matter) first collapsed to form a central Massive Black Hole which then provided the means to transformed many more virtual protons into regular protons to create hydrogen atoms to create stars. Countless protons were lost due to annihilation. Annihilation energy powered Matter creation cores at the poles of Super Massive Black Holes.

Ejected from the poles of a Galaxy's Super Massive Black Holes newly created regular protons and electrons cool and form hydrogen atoms which make stars which are born with the correct Dark matter, Gravitational scaffold complete with outer halos which is essential in order to quickly shape first double lobed or elongated Galaxies into regular types of disk Galaxies. [004] Without the influence of this essential Gravitational scaffold the stars within our Milky Way Galaxy would fly away, or orbit slower, particularly in outer regions. Surely these Gravitational scaffolds should be constructed at precisely the time to be in place to provide the foundations for the creation of Matter. A Natural Model provides the ideal sequence of vital events to create Galaxies. A Big Bang provides no such modeling so it's no wonder Galaxy formation remains mysterious. **Scientists should ask; how was this halo first created and why do Galaxies reside inside them? A Natural Model tells why the outer halo remains today as undeniable evidence or proof if you now prefer, of how Galaxies were** <u>**really**</u> **created.**

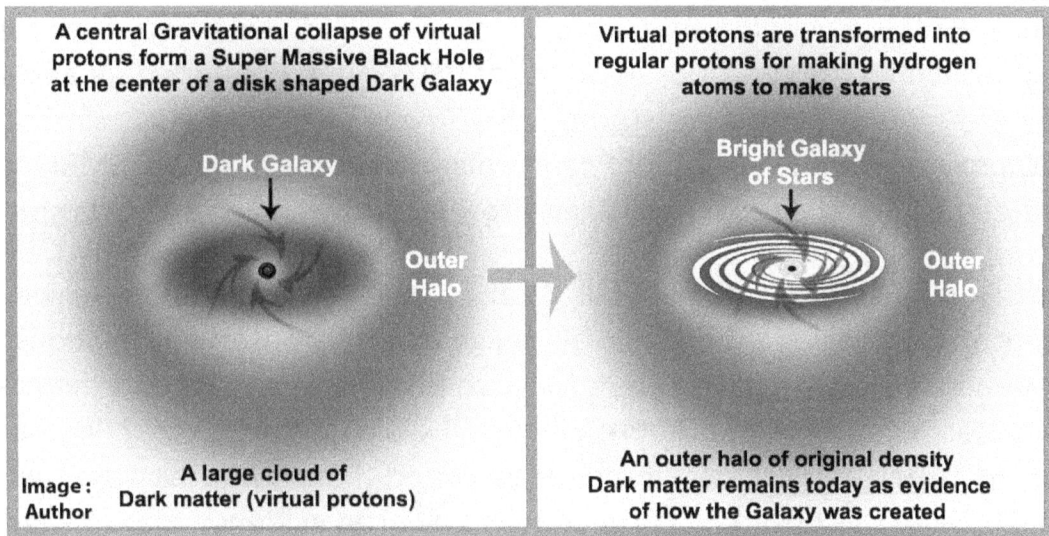

A central Gravitational collapse of virtual protons form a Super Massive Black Hole at the center of a disk shaped Dark Galaxy	Virtual protons are transformed into regular protons for making hydrogen atoms to make stars
Dark Galaxy / Outer Halo	Bright Galaxy of Stars / Outer Halo
Image: Author — A large cloud of Dark matter (virtual protons)	An outer halo of original density Dark matter remains today as evidence of how the Galaxy was created

Naturally, this created a direct link between the size of a Galaxy, its central Black Hole and its <u>halo of Dark matter</u>. Although these links have now been discovered it is a <u>complete mystery</u> to why they should exist. If the universe is created by way of a Natural Model these links will obviously exist.

[024] This clearly provided a correlation of the mass of the Super Massive Black Hole and the size of a Galaxy's halo of Dark matter. Scientists say they have discovered a *"mysterious link"* between the amounts of Dark matter a Galaxy holds and the size of its central Super Massive Black Hole. With a Natural Model there is obviously no mystery. [054] Scientists <u>also</u> say they

have discovered there is an *"unexplained"* direct mass correlation between a Galaxy's central Massive Black Hole's mass and the mass of its central bulge of stars. Scientists ask: *"is this a statistical fluke, or is there a physical reason for the connection?"* No, this is not a statistical fluke. If the universe is created by a Natural Model these links will certainly be expected to exist within today's Galaxies.

[113] **Since a Natural Model actually creates a Galaxy's Matter at the poles of its already established SMBH, adolescent Galaxies will naturally at first harbor very Massive Black Holes in proportion to their forming Galaxy of stars.** UHZ-1 is a very distant Galaxy 13.2 billion light years away. Scientists have combined the JWST observations of UHZ-1 with Chandra data and say they discovered *"something strange the Galaxy's SMBH mass is similar to its Galaxy of stars"*. This model can explain why; the Galaxy's SMBH was created first and the Galaxy will likely grow larger by creating its own stars.

Both Spiral Galaxies and Elliptical Galaxies have a center bulge, however Elliptical Galaxies are basically all bulge so this observed relationship between a Galaxy's central Super Massive Black Hole's mass and the mass of its stars extends to the entire Elliptical Galaxy. Although described as a mystery to Scientists today who have theorized that all Matter was created at a Big Bang this observed correlation is powerful evidence that all Matter was created by way of my Natural Model which creates Matter at the poles of Super Massive Black Holes. The correlation is from when Super Massive Black Holes were created at the centers of large clouds of virtual protons (Dark matter). Consequently, the larger clouds of Dark matter created 'more massive' Super Massive Black Holes and larger clouds of Dark matter naturally provided more virtual protons (Dark matter) to be transformed into regular protons to create hydrogen atoms to form more stars.

The correlation of Matter does not extend to the spiral arms of Spiral Galaxies like our own Milky Way because the spiral arms were created at a later time when spiraling magnetic fields grew longer and carried newly created Matter a great distance, some of which didn't return to the Galaxy. The Natural Model provides two phases for the release of Matter; first an explosively release of Matter almost directly from Matter creation cores or released when spiraling magnetic fields were at first relatively short. The first phase transitions to a second phase when many spiraling magnetic fields progressively grew longer and carried Matter a great distance.

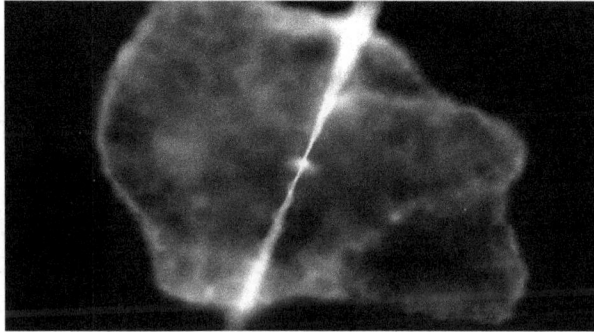

Image credit: M. Kornmesser / ESO. [047] **Credit text ESO.** *The astronomers found that 12 quasars were surrounded by enormous gas reservoirs: halos of cool, dense hydrogen gas extending 100 000 light years* **from the central black holes and with billions of times the mass of the Sun.** *The team, from Germany, the US, Italy and Chile,* **also found that these gas halos were tightly bound to the galaxies.**

[047] The above illustration has been created from observations. Amazingly, evidence of both phases of the release of hot Matter may have been observed in the early universe in this image created by Astronomers using ESO's Very Large Telescope (VLT). The discovery, I believe, is wonderful evidence that Matter was created by way of the Natural Model. The older first phase has formed halos of gas and the younger second phase has developed long spiraling magnetic fields carrying Matter afar. The 12 Quasars from the early universe are over 12.5 billion light-years away and are said to be a major challenge to Astronomer's understanding of Super Massive Black Holes as well as the formation and evolution of Galaxies. However, the discovery agrees very well to a Natural Model. Visually, one may suggest, instead of feeding the Massive Black Hole the hydrogen gas appears to have originated from the central SMBH. The gas has several unbalanced leading edges inferring several explosive releases of Matter.

[091, 059] The recent observed size and shapes of <u>early Galaxies</u> exposed by the James Webb Space Telescope are *'completely upending existing theories'*. Scientists say: ***"The Galaxies are so massive that they conflict with 99% of models representing <u>early</u> Galaxies in the universe, which means scientists need to rethink how Galaxies formed and evolved"***. Unlike a Big Bang, the Natural Universe can begin creating Galaxies <u>before the time</u> of the proposed Big Bang so massive Galaxies existing at this time is obviously no problem. The CMB radiation, the large number of old metal poor stars and Quasar activity peaking in the early universe indicates Matter production within a Natural Universe very <u>likely peaked</u> at a similar time to the said Big Bang. However, unlike a Big Bang <u>a Natural Universe</u> creates all of a Galaxy's Matter for all of its stars **within already created** invisible, Dark matter, Gravitational scaffolds which provided the vital scaffolds to quickly shape young, <u>elongated</u> Galaxies into regular disk shapes. **A whopping 90% of the entire mass of our mighty Milky Way Galaxy is contributed to its Dark matter content which is why this Gravitational scaffold plays such a dominant part in holding on to stars and <u>shaping</u> a Galaxy's regular shape.**

[104, 107] Rather than grow from numerous disfiguring collisions, like theorized from a Big Bang, early Galaxies observed by the JWST are quickly forming as regular disk shapes as they naturally would within a Natural Universe. Young Galaxies are first provided with completed Dark matter Gravitational <u>disk scaffolds</u>, which stars will gravitationally cling to, allowing Galaxies to <u>quickly self-evolve</u> their regular disk shapes. Galaxies are formed independently and are <u>not fixed to a timeline</u> from a Big Bang which makes it very possible for the first Galaxies to form <u>before</u> the proposed Big Bang; all of which explains why the surprising findings from the JWST have been astounding astronomers, by *"revealing that stars and Galaxies were forming and evolving <u>much earlier</u> than anyone had suspected"* and *"<u>thousands of disk galaxies</u> like our own Milky Way were observed in the early universe, where they shouldn't exist"* and *"This implies that most stars exist and <u>form within</u> these Galaxies which is changing our complete* *understanding of how galaxy formation occurs".* When Galaxies create their own Matter their stars naturally <u>form within</u> their own Galaxies.

The shape of the first Galaxies: Unlike a Big Bang, a Natural Universe <u>first creates</u> Super Massive Black Holes (SMBH) so the first Galaxies <u>do not</u> need to create the hypothesized, super stars <u>100,000 times</u> more massive than our Sun to form Semi Massive Stellar Black Holes. The Natural Universe has <u>no need</u> for first Galaxies to collide to combine their Semi Massive Stellar Black Holes to create Super Massive Black Holes and combine their stars to create larger Galaxies. The hypothesized <u>time consuming</u> collisions would have <u>highly distorted</u> the shape of the majority of the first Galaxies which the James Webb Space Telescope has <u>unexpectedly revealed is not true</u>, instead, early Galaxies are commonly observed to be quickly forming their regular disk shapes as they clearly will within a Natural Universe. While some collisions may occur if Galaxies are born very near to each other, Galaxies are generally being <u>moved apart</u> by Outer Gravity.

Quasar-like spiraling magnetic fields had to grow over time. The first Matter is explosively released in corresponding massive amounts almost directly from the Matter creation cores at **each pole of a SMBH, consequently, some of the earliest, youngest Galaxies may correspondingly first form elongated with <u>double lobes</u> harboring a SMBH at the center.** [112] Clearly conforming to this, the James Webb Space Telescope has recently discovered a very distant Galaxy from just 330 million years after the said Big Bang which *"appears elongated, almost like a peanut"*. Others have described *"early galaxies were frequently shaped like surfboards and pool noodles"*. The descriptions clearly match the way a Natural Universe ejects its Matter in opposite directions from the poles of Super Massive Black Holes which logically <u>at first</u> creates an elongated Galaxy which makes no sense from a Big Bang. (See first image Chapter 1). **Note:** When pointed end on <u>directly towards us,</u> a distant elongated adolescent Galaxy will naturally appear as a much more rounded Galaxy intensely forming new stars.

The first protons were explosively released at high speed and <u>very close</u> to the event horizon of a SMBH where the strong Gravity from the SMBH was able to <u>significantly slow</u> many of the speeding away protons so as they could grab an attracted electron to form hydrogen atoms. **Note: Matter and stars which did not have the momentum to remain in the halo region eventually falls to the attraction of the Galactic plane where the <u>resulting</u> lower angular momentum allows the formation of a spheroid shaped elliptical Galaxy or the Galactic Bulge of a disk Galaxy.**

Metals are created inside of stars. A large star suffers a supernova explosion when it begins to <u>make</u> iron which floods Space with metals available for the next generation of stars. When grading stars, metal is anything other than hydrogen and helium. This should have created three generations of stars. **Population III** stars are <u>theorized</u> as being the first stars and should have almost <u>no metals</u> and are made from mostly hydrogen and helium. **Population II** stars are the <u>oldest observed stars</u> and are also relatively metal poor. **Population I** stars, like our Sun, are young or middle age and contain metals such as nickel, carbon, and iron.

There is <u>no observational evidence</u> of Population III stars, nevertheless such stars are theorized as a way of explaining the metal content of Population II stars. This is a sensible theory and if Population III stars ever existed they were likely large, short lived stars and died as supernova explosions which enriched the universe with an early shower of metals. The problem is even the oldest, observed star HD 140283 is a Population II star. <u>Unlike a Big Bang</u>, growing spiraling, magnetic fields, which by <u>carrying Matter afar</u>, provided our Universe with a <u>clear transition</u> from the creation of the old Population II stars observed in Globular Clusters, Elliptical Galaxies and the central Galactic Bulge of Galaxies to the creation of middle age and younger Population I stars observed in arms of Spiral Galaxies and finally to creating the generally younger Population I stars observed within Irregular Galaxies.

[060] Black Hole accretion disks are said to produce small amounts of the

same elements as stars. However, with the Matter creation cores reaching as high as a colossal **10 trillion Kelvin** and during the explosive release of first Matter, like an **exceptionally powerful**, **supernova explosion**, nucleosynthesis can be expected at this time to provide a variety of elements to be ejected from the Matter creation cores. Interestingly, new data from the JWST has revealed the presence of nitrogen in early Globular Clusters. Nitrogen can only be explained by the combustion of hydrogen at extremely high temperatures. This requires temperatures far hotter than any observed star could possibly produce. Scientists suggest the elements may have been created by hypothesize, super-massive, super-hot stars with mass 10,000 times more than our Sun. A Natural Model's Matter creation cores reach **10 trillion Kelvin** which is able to account for the temperatures required to make the discovered nitrogen. Elements are now available to create the first generation of stars as Population II stars as found within Globular Clusters. This is likely why no Population III stars have been discovered and adds additional evidence of where all Matter was created.

Population II stars. What do Elliptical Galaxies, the halos of Galaxies, the Galactic Bulge of Galaxies and Globular Clusters have in common? [055, 056] They all contain mostly old Population II stars which have tiny amounts of metals. Because Population II stars are observed as the oldest they should have all have been created when Matter is first explosively released in a succession of dense pulses, direct from Matter creation cores or from short spiraling magnetic fields. The first release of Matter may be expected to be the most intense due to the densest first supply of virtual protons (Dark-matter. The large numbers of similar aged Population II stars provides evidence that many of these events peaked at a similar time and if close together in an early expanding universe than this is a time which most resembled a Big Bang; be it like numerous Small Bangs.

Following image; Galaxy Centaurus 'A': [097] At just over 11 million light-years away Galaxy Centaurus 'A' contains the closest active Galactic nucleus to

Earth. The plumes have <u>originated</u> from the poles of the Galaxy's central Super Massive Black Hole. Following image, upper mid left: <u>Note</u> how the Matter above the Galactic plane and in the upper <u>halo region</u> appears to be curving back to fall towards the Galactic plane allowing one to actually observe my Fountain Formation model at work. The Matter from the poles of the Galaxy's central SMBH clearly <u>first</u> reaches the Galactic halo region <u>before it slows</u> and begins to fall towards the Galactic plane which is surely why many of a Galaxy's very first stars, which are now the oldest, formed in the Galactic halo and remained here, many within Globular Clusters.

Image Credit: X-ray: NASA/CXC/CfA/R.Kraft et al.; Submillimeter: MPIfR/ESO/APEX/A.Weiss et al.; Optical: **ESO**/WFI. Galaxy Centaurus 'A'.

For a Natural Model this relatively <u>nearby</u> Galaxy displays all of the qualities of an older Galaxy making its own new Matter. Fittingly, clouds of neutral hydrogen are observed in the halo region. Invisible in this image is the radio spectrum, which extend the jets out to at least a million light years. Galaxy Centaurus 'A' may well have had an encounter with a large cloud of Dark matter (virtual protons) which has reignited its Matter creation cores at the poles of its central, 55 million solar mass, Super Massive Black Hole (SMBH). Large plumes of <u>hydrogen</u> flooded with radiation and charged particles are observed with material falling back towards the Galaxy's plane where, amongst older stars, large numbers of new young stars are observed which indicates the Galaxy is receiving a fresh supply of hydrogen atoms.

[097] video time 8.18 Clearly conforming to this, within the swarms of charged particles are several large and revealing clouds of neutral hydrogen atoms much of which appears to be in the Galactic halo region. Neutral hydrogen atoms are considered to be primal first atoms said to have been created shortly after the Big Bang. There is also a significant trace of Halpha which is when an electron of a hydrogen atom falls from its third to second lowest energy level. So what are primal hydrogen atoms doing here together with the charged particles and radiation being ejected from the poles of a Super Massive Black Hole? The answer is visually obvious; they were made here together with the charged particles and radiation! While this evidence is observable the Scientists suggest what they call a *"novel theory"* where the hydrogen, believed created from the Big Bang, rains down and feeds the SMBH's jets which now eject back the charged particles and radiation.

[005, 097] The problem with such a theory is it cannot account for the 10,000,000,000,000 Kelvin temperature measured at the core of a similar system or how the jets of Galaxy Centaurus 'A' mysteriously create radiation of an extraordinary broad spectrum up to very high energies. Such a broad spectrum of radiation may easily be contributed to the creation of new Matter and it's probably a mystery because no one has considered that new Matter is being made here. [098] Centaurus 'A' is also a source of cosmic rays of highest energies. Cosmic rays are said to be high-energy particles of mostly (89%) protons which are nuclei of hydrogen atoms.

Galaxy formation in full: We begin with **forming Globular Clusters**. Because all regular Galaxies have Globular Clusters, which contain many of the very oldest stars, all Galaxies must have begun in a way which allowed newly created Matter to first clearly favor the halo stars of Globular Clusters. Matter had to also be delivered in a way to provide the required directional momentum responsible for the many varied orbits of Globular Clusters. Scientists know Matter from a <u>Big Bang</u> would be expected to <u>first reach the density</u> required to form the very first stars at **the very center** of where a Galaxy will form. So how is it possible to first get Matter denser in far away, outer halo regions to form the very first stars where <u>many</u> of the very oldest stars now reside? This near <u>impossibility for a Big Bang</u> is easy explained by a Natural Universe. Newly created Matter was at first powerfully and explosively released in dense pulses and <u>directed away from the center</u> of a forming Galaxy and <u>towards the halo region</u> so a percentage of Matter as it cooled and slowed <u>reached the halo region first</u> where it naturally created many of the first stars <u>away from the very center</u> of a forming Galaxy. This is why halo stars and Globular Clusters are common in early Galaxies. Globular Clusters and their orbits provide <u>clear observable evidence</u> that Matter was truly created by way of a Natural Model. **The James Webb Space Telescope** has recently discovered a Galaxy just 700 million years after the said Big Bang which has more stars forming in its outskirts than in its center. [105] Sandro Tacchella, an Astrophysicist at the University of Cambridge, UK says; ***"That's surprising because theory suggests the opposite; that early galaxies should have (first) stars forming closer to their centers".*** This discovery is no surprise for a Natural Model.

Globular Cluster M4

Our Sun

Globular clusters, represented here as red dots, are the oldest datable objects in the universe.

Artist's conception of edge-on view of Milky Way
(100,000 Light years)

Image credit: NASA/ESA and A. Feild, of the known globular clusters around the Milky Way. **In this grayscale image Globular Clusters are displayed as 'white' dots. The Milky Way is home to around 150 Globular Clusters.**

By studying this image carefully one can create a roadmap to the source of the Matter which formed the old stars found in Globular Clusters. Where one can observe the Globular Clusters in the previous image is where Matter was <u>first directed to and lingers</u> before it falls to the Galactic plane. Both Elliptical Galaxies and Spiral Galaxies contain Globular Clusters of very old Stars; suggesting the Galaxies had similar beginnings.

Ejected from each pole of SMBHs Matter was <u>first directed</u> to the halo region. Because Globular Clusters are typically the oldest objects in their Galaxy they must have been created from Matter <u>before</u> it fell to the Galactic plane. If Matter traveled <u>too far</u> from the Gravity of the SMBH or an adolescent small Galaxy it failed to fall to the Galactic plane and instead orbited the Galaxy. As the Galaxy grows, its entire mass allows more Matter to fall to the Galactic plane which leaves less Matter in halo regions. Halo stars and Globular Clusters are simply made from large clouds of molecular gas created from a percentage of gas which <u>did not fall to the Galactic plane</u> which allowed these stars to be many of the first stars to form. [057; 058] Like the remnants of a supernova explosion the explosively released material can be expected to have created leading edge shock-waves which accumulated and compressed gas. The compressed Matter formed many of <u>the very first stars</u> which created Globular Clusters. Remaining Matter <u>and stars</u> which could not retain their halo position now fall to the dark disk of the Galactic plane where they create a small Elliptical shaped Galaxy which may become an Elliptical Galaxy or the Galactic Bulge of a Spiral Galaxy. Since all Matter had to first cool the cooling time allowed many of the first stars of the Galactic Bulge to also form at an early time similar to halo stars.

As spiraling, magnetic fields grow they jetted Matter further from the Gravity of the forming Galaxy which allowed released Matter to drift further apart and become more rarefied before it fell to the Galactic plane where it gathered and gained density to form stars. As this trend continued less halo stars were made from the more rarefied gas so Globular Clusters become less common further away from the central area of a Galaxy.

Within a Natural Universe the halo stars were imprinted with the momentum from when Matter was released. When many halo stars gathered together to be tightly bound by Gravity they formed Globular Clusters which orbits were clearly dictated by the angle and the pole from which Matter was released. The jets delivering Matter have been observed to completely change direction. Globular Clusters orbits vary greatly and many actually orbit in opposite directions. Orbits are given similar to how a satellite launched from Earth orbits the Earth. One can visualize this in the previous image on page 230 and the image and 215. The many varied orbits of Globular Clusters provide direct evidence of the source of their Matter.

Globular Clusters and their many varied orbits are obviously difficult to explain from a Big Bang beginning. [093] Computer simulations of forming first Galaxies from gas fail to first form the stars of a Galaxy's many Globular Clusters. Called hierarchical merging, the simulations concentrate on Galaxy collisions and mergers to build a larger Galaxy, thus, solving how large Galaxies quickly grow from the Big Bang. Simulations show the collisions cause Galaxies to develop weird and wild shapes. What several collisions would do to the orbits of numerous Globular Clusters is anyone's guess!

Within a Natural Universe Galaxies are moving apart with Outer Gravity so collisions are rare. [091] This conforms to the said startling observations from the James Webb Space Telescope which *"blindsided theorists and observers alike (who) have been scrambling to explain them"*. [059] *"Scientists are puzzled because James Webb is seeing stuff that shouldn't be there …… the models just don't predict this"*. The problem is the images failed to show the predicted peculiar shapes caused from the theorized collisions and mergers of tiny adolescent Galaxies said to be required from a Big Bang beginning. Instead, the early Galaxies are observed to be *"way more massive than anyone expected"* and have formed as regular shapes with disks and spirals which clearly conform to a Natural Universe where Galaxies are provided with the gravitational means to quickly normalize and grow themselves.

[121] The heading reads; **'Webb telescope spots old, massive galaxies that shouldn't exist'.** The Galaxies are so massive they should not be possible under current cosmological theory. *"It's bananas, these Galaxies should not have had time to form"* said Erica Nelson assistant professor of astrophysics at CU Boulder. The images of Galaxies too massive and matured to have come from a Big Bang obviously <u>conform</u> to a Natural Universe which is clearly able to form first Galaxies <u>before the time</u> of the proposed Big Bang.

Forming Elliptical Galaxies:

Credit: NASA, ESA, R.M. Crockett (University of Oxford, U.K.), S. Kaviraj (Imperial College London and University of Oxford, U.K.), J. Silk (University of Oxford), M. Mutchler (Space Telescope science Institute, Baltimore), R. O'Connell (University of Virginia, Charlottesville) and the WFC3 Scientific Oversight Committee.

Elliptical Galaxies contain mainly old Population II stars with little momentum and outer Globular Clusters of similar old Population II stars. Elliptical Galaxies contain very little excess Matter (gas and dust).

All Matter was <u>directed</u> towards the halo region where many of the first stars will form an Elliptical Galaxy's Globular Clusters.

To crate Elliptical Galaxies the <u>important part</u> is the magnetic fields remained relatively short and Matter was <u>not carried great distances pass</u> star forming regions by long, spiraling magnetic fields. Protons were at first explosively released in dense pulses <u>close to the strong Gravity</u> near the event horizon of the SMBH where Gravity causes Matter to slow and <u>balloon out at a wide angle</u>. The slow, low momentum protons allowed a high percentage of protons to grab an attracted electron to form hydrogen atoms. The <u>resulting</u> low angular momentum of Matter caused the formation of a spheroid shaped elliptical Galaxy or the Galactic Bulge of a disk Galaxy. Consequently, an Elliptical Galaxy has no distant reservoirs of gas to drift back to its Galaxy for ongoing star formation which may create young Population I stars. Instead, nearly all available gas was directed to the forming Galaxy and used for making stars which is why Elliptical Galaxies contain mainly old Stars and is why many were the first completed Galaxies in the universe. [047] In a scenario similar to Galaxy 3C 348, the spiraling magnetic fields may grow too long too quickly causing Matter to be carried out of Gravity's reach for a return to the Galactic plane.

The size an Elliptical Galaxy is naturally dependant on the mass of its central Super Massive Black Hole. Larger Super Massive Black Holes naturally had the ability to create larger quantities of Matter which made larger Elliptical Galaxies. [024] Although said to be a mystery this correlation has been observed to be true. The <u>direct distribution of self-created Matter</u> is also responsible for <u>preserving</u> an original relationship between the amounts of Dark matter (halo) a Galaxy holds and the size of its central Black Hole.

Forming Lenticular Galaxies, Barred Galaxies and Spiral Galaxies:

Credit: Hubble data: NASA, ESA and A. Zezas (Harvard-Smithsonian Center for Astrophysics); GALEX data: NASA JPL-Caltech, GALEX Team, J. Huchraetal. (Harvard-Smithsonian Center for Astrophysics); Spitzer data: NASA/JPL/Caltech/Harvard- Smithsonian Center for Astrophysics.

<u>Spiral Galaxies like our own Milky Way:</u> An outer halo containing Globular

Clusters of very old Population II Stars; a Galactic Bulge also contains primarily old Population II Stars and a Galactic Disk with spiral arms contains many young to primarily middle aged, Population I Stars.

The Globular Clusters and the Galactic Bulge are naturally created first in the same way as with Elliptical Galaxies. As spiraling magnetic fields grow longer they <u>jetted pulses of Matter well away</u> from the Gravity of the central SMBH and the Galactic plane below. Now when Matter is released at the end of longer spiraling magnetic fields Matter has more angular momentum, as it falls to the Galactic plane, to form a Galactic disk which already has a Galactic Bulge. The Galaxy is now called a Lenticular Galaxy.

As spiraling magnetic fields continue to grow Matter is <u>carried</u>, with <u>high momentum</u>, even farther away where it is <u>less slowed</u> by the Gravity from a central SMBH and its young Galaxy. [099] The carried Matter naturally reaches stars forming regions with <u>higher momentum</u> where it forms molecular clouds of gas which are introduced to an already created central Bulge. The higher angular <u>momentum</u> of this Matter provided a higher rotation speed than found in elliptical Galaxies. Matter released from each pole of a SMBH may commonly create two spiral arms; one from each pole. [095] More aggressive jets can now easily create <u>an early Barred Galaxy</u>. Another long pulse of Matter from each pole naturally creates two more arms and **is why** many Spiral Galaxies have two or four main arms. Arms may vary by the amount of Matter being ejected from the poles of a SMBH.

The process may repeat itself to create a Spiral Galaxy with numerous arms. Like a volcano there are few set rules. Because the process ejects Matter from each pole of a parent SMBH it will commonly create an even number of arms. Matter released in <u>varied pulses</u> will have similar affect on the arms. Because all facets of a Galaxy are created by the same basic mechanism when the process remains uninterrupted normal regular Galaxies can form and grow relatively quickly. When the supply of virtual protons (Dark matter) to the Matter creation cores is interrupted or exhausted the jets will die out but may <u>restart</u> anytime a new supply of virtual protons drifts to the SMBH creating new Matter which creates young stars in regions where older stars reside, like Galaxy Centaurus 'A'.

Matter carried afar by extraordinary long spiraling, magnetic fields provide distant reservoirs of gas to slowly drift back and provide a <u>prolonged</u> supply of Matter (gas) for ongoing star formation as found within Spiral Galaxies. A quantity of this far distant Matter will fail to make the journey back to its parent Galaxy and linger in large clouds of gas some of which will eventually form Irregular Galaxies without central Super Massive Black Holes.

How the transition from metal poor Population II stars to metal enriched Population I stars occurred: Jets generally grew longer <u>over time</u> and longer jets naturally released Matter <u>far</u> from where it falls to the Galactic Bulge. Longer jets created the spiral arms of Spiral Galaxies and carried Matter afar to create Irregular Galaxies. The extended time taken for the journey to spiral arms or to regions of Irregular Galaxies created many younger Population I stars which form at a later time to the previously created Population II stars of the Galactic Bulge. [049] Gas for star making in these outer regions is now enriched with metals released from older, large exploding Population II stars within the Galactic Bulge. Metals are carried to these outer regions by Galactic winds.

Forming Irregular Galaxies: Irregular Galaxies contain primarily young stars.

Credit: NASA, ESA and The Hubble Heritage Team (STScI/AURA) Acknowledgment: M. Gregg (Univ. Calif.-Davis and Inst. for Geophysics and Planetary Physics, Lawrence Livermore Natl. Lab.)

One may first ask why do Irregular Galaxies contain mostly young, metal rich Population I stars, why are they irregular shapes and why do they commonly orbit regular Galaxies?

To form Irregular Galaxies the spiraling, magnetic fields have now extended a great distance. Matter is now <u>carried</u> by spiraling, magnetic fields a great distance from the central Super Massive Black Hole. A good example of this is the brilliant jets of radio Galaxy 'Alcyoneus' which are at least 16 million light-years in length and is way more than the 100,000 light years across our entire Milky Way Galaxy. [106] Now Scientist themselves say; the jets from 'Alcyoneus' blast material into Space which ***"includes the building blocks for new star formation"***. Long jets can obviously carry Matter an extended distance pass the outer edge of an existing Galaxy and now eject Matter so far that it is all but out of reach of the Gravity of its parent Galaxy <u>where star formation now slows.</u> The Gravity of its parent Galaxy still slows the outward course of this Matter but does not have the required influence to initiate a return journey.

Having been ejected <u>far and sparsely,</u> Matter has no Gravitational Scaffold provided by a Dark Disk Galaxy to help begin Galaxy formation and merely has its self Gravity to finally clump it together. The process, with Matter having travelled far and sparse, takes an extended time to accumulate the required concentrations of Matter which will form stars. [049] The extended timeframe allowed Galactic winds, generated by either starbursts or active Galactic nuclei, to seed these regions with metals created by Population II stars within the parent Galaxy. Because of the extended timeframe of accumulating Matter, Irregular Galaxies will today consist of stars of different ages including numerous young Population I stars. **Lacking a dark, disk shaped Gravitational Scaffold to evolve within; the Galaxies will develop as odd and irregular shapes.** One would naturally expect, if this process was factual, many of these Irregular Galaxies would <u>today orbit</u> and so be constrained by their larger parent Galaxy, from where their Matter was produced, which is indeed, golly gosh, observed. Naturally, if a large quantity of Matter is ejected millions of light years afar, more <u>oddball</u> Galaxies and even Galaxies <u>poor in Dark matter</u> may now be created. Matter may be ejected so far and sparse that it fails to form stars and lingers within filaments far from a Galaxy's observable stars. On rare occasions a Galaxy may <u>use up most</u> of its Dark matter by creating regular Matter causing it to become very large with little <u>left over</u> Dark matter.

We now have a road map, from the same source, for the development of all facets of all types of Galaxies including rare Dark matter poor Galaxies. In fact, our explanation of Galaxy formation actually fits well to the most common observed types of Galaxies, including their Globular Clusters, which strongly reinforces how Matter is created. If there is an adequate supply of Matter, star forming may begin at any time and in many Galaxies is an ongoing process. Galaxies may occasionally collide and combine to form an amazing variety of different shapes, combinations and sizes.

We have importantly, nevertheless, been able to apply common reasoning to produce the most common types of Galaxies and without difficulty create stars in the order which matches their observed age and matches star formation to the different types of Galaxies to what is mostly observed.

A Mathematical Big Bang. Even when it was smaller than the size of an atom a Big Bang universe is complete with all energy for all time, allowing energy to form all Matter required for forming all Galaxies.

A Natural Universe begins with the tiniest, most minuscule, energy event known to occur within pure nothingness [015]. Beginning with the smallest to grow the largest is surely Mother Nature's way. A Natural Model totally solves the long-standing Matter and Anti-matter asymmetry problem. A Natural Model, by maintaining a sum-total zero mass-energy universe, meets the mandatory requirements of the 'Conservation of energy' law, all of which has evaded the <u>non-exacting</u> science of a Big Bang beginning.

For a Natural Universe the task of producing a universe full of Matter is fittingly distributed amongst hundreds of billions of Galaxies and given a fitting timeframe, not of a second of time at a Big Bang but of hundreds of millions, or billions of years, or whatever amount of time it requires. A Natural Universe is clearly more capable of performing this colossal task than a one-only totally, inexplicable and instant event way smaller than the full stop at the end of this sentence. The concept of 'expansion' itself drives one to easily imagine that an object expanding over time must have at some point, back in time, appeared from an infinitely small point of nothingness. However, if one releases the air from an expanding balloon it will not disappear into an infinitely small point of nothingness.

The notion of extrapolating back in time because the universe is expanding 'mathematically' expresses Matter mysteriously rising from a single, singularity like point at a Big Bang. The mathematics of the Big Bang have worked themselves into a single singularity where the only direction is infinitely smaller, like in a Black Hole, from where there should have been no escape. Within a Natural Universe the energy for creating all Matter is sensibly provided by the massive <u>Gravitational resource</u> of numerous Super Massive Black Holes, all of which are said to harbor 'singularities'. This may wrongly infer the energy for all Matter originated from a single singularity.

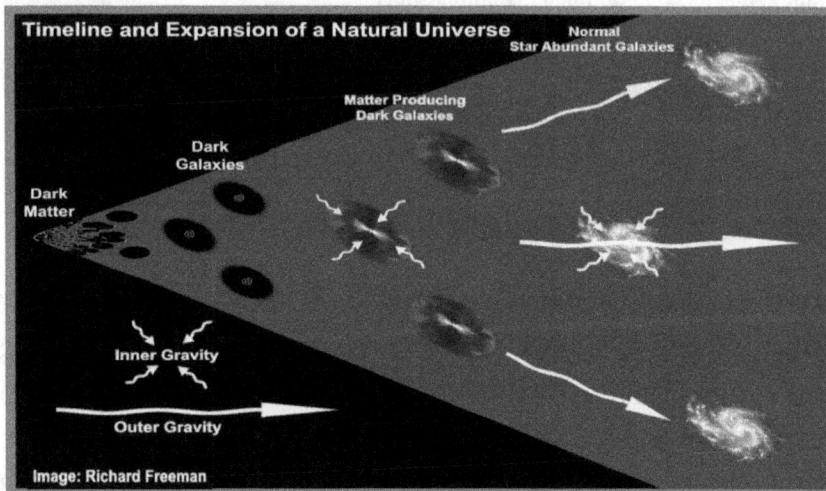

Timeline and Expansion of a Natural Universe

Normal Star Abundant Galaxies

Matter Producing Dark Galaxies

Dark Galaxies

Dark Matter

Inner Gravity

Outer Gravity

Image: Richard Freeman

239

10 DARK ENERGY AND INFLATION.

A Natural Model allows the use of Einstein's General Relativity to explain why Galaxies are accelerating. The Big Bang theory provides no known source of the hypothetical energies for either inflation or Dark energy. My Natural Model makes it clear; Galaxies create their own Matter where they are and have <u>always</u> been accelerating due to **Gravitational Acceleration.** Galaxies are <u>free falling</u> in the direction which the universe is expanding. A <u>free</u> falling object simply <u>falls and accelerates</u> due to Gravity and <u>not due</u> to any forms of hypothetical energy such a Dark energy. When I right click 'hypothetical' and select 'Synonyms' the 2nd word displayed is 'imaginary'.

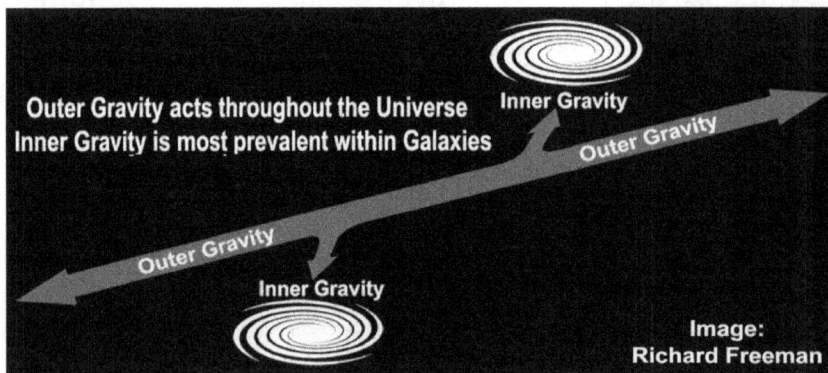

Image: Richard Freeman

The delivery part of all Gravity is primarily driven by the strong attraction part of Outer Gravity. Outer Gravity is always acting throughout the universe and Inner Gravity is always most prevalent within Galaxies. Both Outer Gravity and Inner Gravity operate in unison as one of the same. Galaxies being accelerated by Outer Gravity can be sensibly explained when the source of mass and Gravity is a tiny phase of pure **empty Space** and the universe is expanding into a seemingly endless void of pure **empty Space**. Pure empty Space provides <u>negative energy</u>.

<u>Note:</u> As this is an outer void of empty Space, it <u>contains nothing</u> which may crush our universe. Einstein's Gravity is delivered, in a seemingly back to front way, by entangled dimensions of Space-time <u>originating from Matter within our universe</u>, thus the void can only apply Gravity acting outwards and <u>cannot apply</u> any form of crushing inward Gravity.

The energy of inflation is another hypothetical form of energy used to explain how all the energy for a whole universe of Matter managed to escape from a seemingly inescapable <u>singularity like point</u> at the theorized Big Bang. Like escaping the <u>singularity</u> of a Black Hole, inflation energy required the unfathomable power needed to blast energy and Matter away at speeds <u>faster than the speed of light</u>. No known-to-man form of energy can achieve this. Physicists faced with this dilemma proposed an unknown-to-man form of energy, from a never to be revealed source, caused an extraordinary, powerful pulse of inflation, sending all energy and Matter flying apart at <u>faster than the speed of light</u> which allowed Matter to escape from a <u>singularity like point</u> even if it clearly defied the laws of today's physics. Physicists say they favor the energy of inflation as it appears to explain several mysteries at once, even though no 'direct evidence' of the energy of inflation has yet been found. **Physicists today say they have no idea what caused inflation energy, or why it stopped.**

'Theoretical' physics has a perplexing assortment of different theories for inflation to choose from which are said to be responsible for the initial expansion of the Big Bang universe. I for one have no hope of sorting out Cosmic inflation from Old inflation or New inflation from Cosmological inflation, not to mention Universe nucleation, Slow-roll inflation, Chaotic inflation, Eternal inflation, Stringy inflation, or just plain Inflation. It appears there are so many different bewildering ideas for inflation, mainly because of the difficulty applying these 'same but different' theories to the Big Bang universe and the problems created. This is a common dilemma facing so much of mathematical science, so many different theories, all claiming to be mathematically correct but none may have really, actually occurred. Generally it is thought rapid inflation occurred for just an incredibly small fraction of less than a trillionth of a second at the Big Bang. Driving this incredibly powerful short pulse of inflation was a mysterious and unknown form of unstable energy, which is said to have come from a mysterious, nonexistent source.

My Natural Model provides a small pulse of 'inflation' when magnetic fields broke and explosively released first Matter from Matter creation cores at polar regions <u>near event horizons</u> of numerous Super Massive Black Holes.

Introducing bewildering amounts of purely hypothetical, unknown-to-man forms of energies mysteriously appearing everywhere from a never to be revealed source is not what the Natural Model is about. If I wrote such mind baffling things I'm sure nobody would believe me. Within a Natural Universe all is achieved with the common everyday resource of Gravity. A Natural universe is all about using existing knowledge which is extensive.

The observation that Galaxies are really accelerating away provides truly remarkable evidence that the way the Natural Universe creates its Dark matter, regular Matter, mass and Gravity is truly correct. Within a Natural Universe a tiny phase of pure empty Space at the hearts of Dark matter, protons, neutrons and so atoms provides both mass and Gravity. A Natural Universe is expanding into a seemingly endless void of pure empty Space and pure empty Space provides the attraction part of Gravity, consequently, a Natural Universe has to have, beyond doubt, Outer Gravity which is responsible for accelerating Galaxies. Since the Natural Model does not have the rapid inflation of a Big Bang it has to have Outer Gravity at the very beginning to prevent all Galaxies coming together. Given that our universe is expanding into a seemingly endless void of pure empty Space,

Outer Gravity was there at the beginning and is unlikely to ever go away and remains today responsible for the observed acceleration of all Galaxies.

General Relativity is Albert Einstein's theory of Gravitation. However, General Relativity knows nothing of Dark energy or the energy required for first inflation. Mirroring the predictions of General Relativity both Inner and Outer Gravity are delivered when a <u>stretching</u> causes a <u>warping</u> of the <u>curvature</u> of the 'expansion' of the 'entangled dimensions' of Space-time. Both Inner and Outer Gravity are the same exact phenomenon.

Einstein revealed Gravity was not a true '<u>force</u>' so <u>no additional energy</u> to apply a <u>true force</u> is required; rather Gravity occurs due to the warping of the curvature of the dimensions of Space-time. Consequently, Gravity does not require a massive, unknown-to-man kind of energy to bizarrely appear to provide inflation or another unknown-to-man kind of energy to account for the observed acceleration of Galaxies. <u>Gravity is ideal</u> since it operates without boundaries throughout the universe and simply <u>passively moves objects</u> while they remain at stationary equilibrium <u>at the center</u> of their expanding <u>entangled dimensions</u> of Space-time. We may observe a very far away Galaxy speeding away from our Milky Way Galaxy at <u>faster than light speed</u> but, <u>like ourselves</u>, objects in the far away Galaxy remain <u>stationary</u> (at rest) at the center of **their** <u>entangled dimensions</u> of Space-time. To the far away Galaxy our Milky Way Galaxy appears to be speeding away from it. You remain stationary at the center of your <u>own entangled dimensions</u> of Space, <u>until you yourself exceed the speed of light</u> which is (probably) not possible as it would be like attempting to escape from a Black Hole.

It is well known that Gravity moves whole Galaxies. It is known that Gravity accelerates objects, that is, <u>Gravity does exactly what is being observed with Galaxies</u>. Gravity operates from the beginning of the passing of time so there is never a need for a mysterious inflation phase or for the mysterious Dark energy which mysteriously expands the <u>Space itself</u> between Galaxies <u>but not the Space</u> between stars within Galaxies. When one digs deep with research one finds that science today has no workable source for Gravity, no source for the energy of inflation, no source for Dark energy, no source for the dimensions of Space and no source for passing moments of time.

In the early universe when all Galaxies were at first near to each other Inner Gravity was at first responsible for restricting the acceleration of Galaxies and now as Galaxies are moving away from the Inner Gravity from the nearness of other Galaxies, Outer Gravity is now taking control which is responsible for the now observed and said reacceleration of Galaxies.

Galaxies accelerating provides overwhelming evidence that our universe is expanding into a seemingly never-ending dimensionless void of a lower state of vacuum than the ground-state of dimensional Space-time. The never-ending void of empty Space provides a super-enormous reserve of negative energy which is ideal for the attraction part of Outer Gravity, so much so it drives the delivery part of all Gravity. To accelerate all Galaxies with Gravity obviously requires a massive amount of Gravitational Attraction for which only the Natural Model provides a logical explanation. **Galaxies accelerating in the direction the universe is expanding is also direct evidence that empty Space is truly the source of all mass and Gravity.** Thus, due to Gravity alone, all Galaxies are accelerating as they passively free fall in the directions which the universe is expanding.

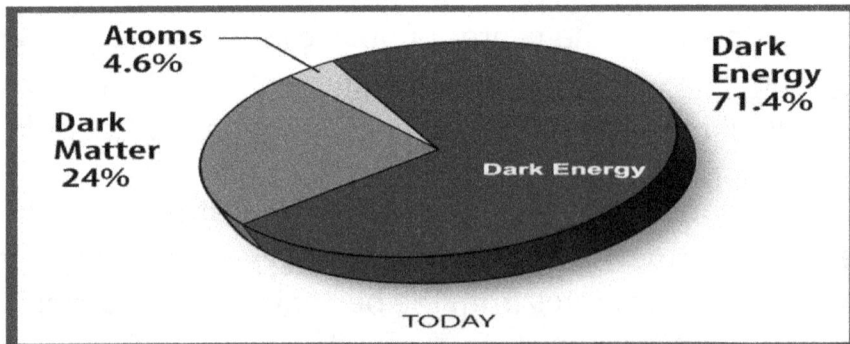

Contents of the Big Bang Universe. Image Credit: NASA.

Note the super-massive amount of energy required to continually accelerate Galaxies with Dark energy. I fail to understand how Dark energy obeys the conservation of energy law! My Natural Model is designed to understandably maintain a sum-total zero mass-energy universe. The conservation of energy law is one of Physicist's most treasured laws but it seems to me Dark energy violates this law and throws it out the window which is why we cannot use Dark energy. There is only one way to accelerate the mass of Galaxies and maintain the conservation of energy law and that is with Gravity. Gravity works passively and not as a growing energy or even as a conventional force which is why it maintains the conservation of energy law. With this modeling Gravity has two parts. The attraction part of Inner Gravity is a tiny negative energy phase of virtual mass which is a phase of pure empty Space and the delivery part of Gravity is provided by tiny Gravitational waves which are positive energy and provide dimensions to Space. Gravity is a combination of both where all negative energy cancels all positive energy leaving sum zero energy 'for Gravity' so as to maintain the conservation of energy law.

Why did Scientists introduce Dark energy?

For many years Scientists were certain that the expansion of the universe had to be slowing. Because Gravity brings objects together, it was believed, with complete certainty that the Gravity derived from all of the Matter in the universe would by now, 13.8 billion years after the initial short moment of the theorized, powerful pulse of inflation at the Big Bang, be slowing the outward speed of Galaxies. There was no doubt since this is exactly what Gravity does; it brings Matter together.

The following is an extract from a NASA Website: **Credit: NASA:**

061 ***Gravity was certain to slow the expansion as time went on. Granted, the slowing had not been observed, but, theoretically, the universe had to slow. The universe is full of matter and the attractive force of gravity pulls all matter together.***

Then came 1998 and the Hubble Space Telescope (HST) observations of very distant supernovae that showed that, a long time ago, the Universe was actually expanding more slowly than it is today. So the expansion of the Universe has not been slowing due to Gravity, as everyone thought, it has been accelerating. No one expected this; no one knew how to explain it. But something was causing it. Eventually theorists came up with three sorts of explanations. Maybe it was a result of a long-discarded version of Einstein's theory of Gravity, one that contained what was called a "cosmological constant." Maybe there was some strange kind of energy-fluid that filled Space. Maybe there is something wrong with Einstein's theory of <u>Gravity</u> and a new theory could include some kind of field that creates this cosmic acceleration. Theorists still don't know what the correct explanation is but they have given the solution a name. It is called Dark energy. Thank-you NASA.

NASA's new James Webb Space Telescope has now confirmed the Hubble Space Telescope's results. However, Scientists say *"What the results still do not explain is why the universe appears to be expanding so fast"*.

<u>NASA's reference to Gravity</u> is made because Dark energy clearly works like Gravity acting outwards. It appears Scientists had two choices; <u>figure out how</u> Gravity acts outwards or insert a bewildering amount (see previous image) of an unknown form of extremely powerful, <u>growing</u> energy into Space to <u>continually</u> act in a way which strangely expands the Space between Galaxies causing Galaxies to accelerate apart. Scientists know the way Galaxies are accelerating mimics how Gravity works. However, with current understanding, for this to be true, the universe would probably be expanding into a sea of solid Matter which is illogical. Thus, Scientists chose the energy idea and because they had no idea what kind of energy this possibly could be or where it came from they called it "<u>Dark</u>" energy.

General Relativity is Einstein's theory of Gravity and is indeed brilliant. My Natural Model, mirroring General Relativity, warps the curvature of the dimensions of Space-time to deliver Gravity. However, General Relativity has no actual source of the attraction part of Gravity or an actual source of dimensional Space-time. General Relativity does not mention Gravitons or Dark energy. Yes, the source of Gravity is said to be mass, however, when one digs deep one discovers the actual source of mass remains a mystery.

When one understands the source of the <u>attraction part</u> of Inner Gravity is a tiny, fluctuating phase of <u>empty Space</u> which exists at the hearts of Dark matter, protons and neutrons and the universe is also expanding into a similar lower state of <u>empty Space</u> the universe will now naturally have two sources of <u>negative energy</u> to provide both Outer Gravity and Inner Gravity. The <u>delivery part</u> of both Outer Gravity and Inner Gravity are created in unison as one of the same. Einstein's brilliant General Relativity can now be used to explain why Outer Gravity is responsible for accelerating whole Galaxies apart while Inner Gravity, the majority of which is from Dark matter, is responsible for holding Galaxies or clusters of Galaxies together.

Einstein's General Relativity describes very well how Gravity is <u>delivered</u> by bending or stretching the curvature of the dimensions of Space and time. However, General Relativity does not tell the very source of Gravity but rather how Gravity is delivered. When one understands how Gravity is derived from Quantum vacuum fluctuations which ties Gravity to the realm of Quantum and how the singularities of Black Holes are merely <u>an area</u> of dimensionless Space rather than an infinitely diminishing point then General Relativity is perfect for predicting the effects of Gravity.

All Matter in the universe is commonly moved about by Gravity. Again, the discovery that Galaxies are currently accelerating like the effect of Gravity, rather than decelerating, really should have told Scientists that their few ideas regarding the <u>source</u> of Gravity are incorrect. The fact that Scientists today cannot explain the actual source of Gravity or the actual source of their Dark energy, both of which influence the movement of Galaxies, should have turned on a few light bulbs. Faced with an observation which obviously did not agree with the predicted and rational 'slowing' after the initial, super-powerful pulse of inflation and energy at the Big Bang, Scientists may have logically questioned if their Big Bang actually happened.

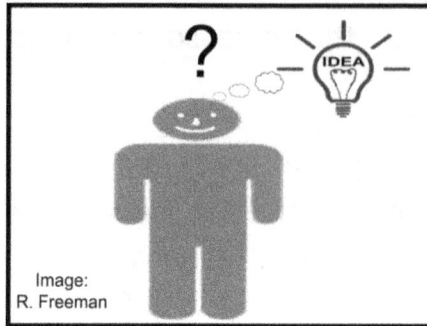

Image: R. Freeman

Instead of abandoning the Big Bang, Scientists decided the best idea was to 'insert' a copious amount of undetectable and totally mystifying sort of 'energy' into the universe to explain this unexpected and inexplicable speeding up of Galaxies which began to show itself approximately five to six billion years ago. When I Google the words 'Dark energy' I quickly see the descriptions; 'Proposed unexplained force', 'Mysterious quantity', 'Perplexing unknown form of energy', 'One of the great challenges of astronomy', 'The existence of Dark energy is still so puzzling', Biggest mystery in the universe' and 'Nobody claims to understand'. Many decades since hypothesizing this puzzling theoretical form of energy, what Dark energy is and where it could have plausibly come from, still remains a complete and absolute mystery.

Image: R. Freeman

All theories for expanding our universe must avoid inflating the Galaxies like the dots which have grown in size on this expanding balloon.

The most bizarre part is the common acceptance of Scientists as 'absolute fact' of something which is astoundingly vague, **has proved not possible to directly detect** and is merely a proposed, hypothesized theory. I truly have <u>great admiration</u> for Scientists but I shudder in my boots when I hear someone in this field portraying purely hypothetical Dark energy being as factual as the Sun rising in the morning sky. Maybe Dark energy came from the same nowhere place the Big Bang came from! Please forgive me for saying this but we fisherman would call all of this 'a can of worms'.

Visualize allowing your car to <u>free</u>wheel and accelerate with Gravity down a steep hill. Your car will accelerate freely without seemingly using any fuel or energy. However, to accelerate your car up a steep hill requires a lot of energy which requires a lot of fuel because your car's engine has to now work against Gravity. Consequently, to <u>accelerate</u> all Galaxies with Dark energy, which is an energy which works <u>against</u> Gravity, would be a truly mammoth task which requires such a mammoth amount of energy that it would now completely dominate the contents of the whole universe.

The problem with using energy to accelerate Galaxies is that, like a car engine, it would be expected to require more and more energy to continually accelerate Galaxies. Eventually this energy will become so powerful that it will become what Scientists call phantom energy which rips all Matter apart and destroys the universe. When the universe was just a billion years old Dark energy is said to have represented just 1% of the contents of the universe. Now the universe is 13.8 billion years old Dark energy apparently accounts for a massive 71.4% of the contents of the entire universe. However, there is one common way to move and actually accelerate Matter without using any <u>additional</u> energy or force and that is with the use of Gravity. Thanks to Albert Einstein, science already knows that Gravity itself is not an actual force. A Natural Universe allows Gravity to **accelerate** Galaxies without a force by allowing the accelerating Galaxies to remain at **'stationary equilibrium'** while length contraction manipulates stretched dimensions of length so as **Galaxies** simply remain at stationary equilibrium within the expansion of their dimensional Space-time. **The answer is not complicated; Galaxies are simply gaining speed as they 'free fall', which requires no enormous amounts of baffling Dark energy.**

An apple falls (moves) from an apple tree to the ground due to Gravity. Apples are made from Matter and Galaxies are made from Matter. Galaxies are made from Matter and are simply moving (falling) and accelerating in the direction of expansion of slightly-warped dimensional Space-time.

Outer Gravity is simply delivered at the level of individual particles at the same moment and in complete harmony as common, every day, Inner Gravity. Outer Gravity should not be considered as an entirely new concept. Here on planet Earth, we observe Gravity acting outwards from the Earth every time the moon passes overhead, giving rise to the ocean's tides. Outer Gravity, being the same exact phenomenon as Inner Gravity, naturally acts in the same exact way, causing Galaxies or clusters of Galaxies to accelerate apart as they 'free fall' in the direction which Galaxies or clusters of Galaxies are moving apart.

To conform to the Big Bang Theory, the speed at which Galaxies and clusters of Galaxies are moving apart should <u>really be slowing</u> which has now been observed to be untrue. There is no mention of Dark energy within the Big Bang Theory. To conform to my Natural Model, the speed at which Galaxies and clusters of Galaxies are moving apart should <u>really be increasing</u> which exactly matches what is observed to be true.

The Big Bang universe cannot have something as 'down-to-Earth' and simple as Outer Gravity because the Big Bang universe of Space and time is nonsensically not expanding into anything, not even empty Space, so it requires a colossal, totally mystifying energy called Dark energy to expand it by <u>accelerating Galaxies from within</u>. Only the concepts of a Natural

Universe can supply a sensible source for Outer Gravity. The source of Outer Gravity is an enormous reserve of <u>negative energy</u> which is a naturally available resource given from a seemingly never ending outer void of empty Space. **Negative energy is sourced as pure empty Space, a state of less energy than the vacuum of dimensional Space-time. Outer Gravity also confirms yet again a Natural Model's source of Gravity and mass; a tiny fluctuating phase of empty Space, called virtual mass, residing at the hearts of Dark matter, protons and neutrons.** It is a very good thing to have a model which seamlessly correlates and <u>marries everything up</u> without the need to resort to introducing unrelated complex theories or colossal unexplainable energies appearing from an unknown source.

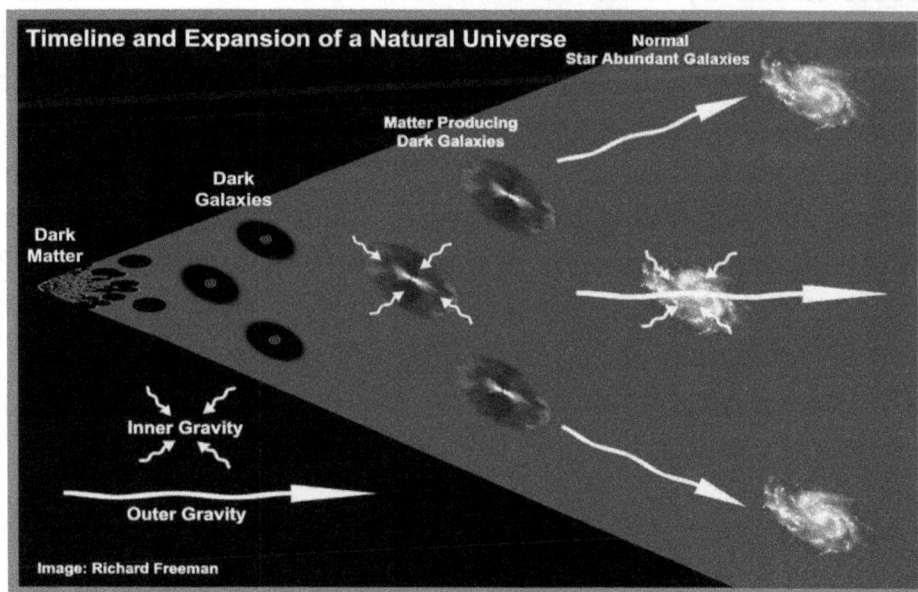

Dark energy functions by opposing Gravity which obviously requires a gargantuan amount of energy while our Inner and Outer Gravity function 'as one of the same' in subtle, complete harmony. Inner Gravity and Outer Gravity are totally combined and delivered in complete unison as just plain every day 'Gravity'. The concept of a Natural Universe cannot avoid having Outer Gravity. At no time were all Galaxies on an inward path of all coming together in a Big Crunch ending. **<u>Outer Gravity avoided a 'Big Crunch'</u> and**

allowed clusters of Galaxies to separate from other clusters. Inner Gravity maintains the structural integrity of Galaxies by <u>firmly</u> holding Galaxies or clusters of Galaxies together. However, every atom within every Galaxy feels a tiny amount of Outer Gravity, which is applied in the direction nearest to the far outer void of Anti-space.

Outer Gravity being driven by the enormous resource of a seemingly never ending far outer void of nothingness allowed Clusters of Galaxies to eventually separate to the point where they were able to partially break free of Inner Gravity from other clusters of Galaxies. Once partially free of the Inner Gravity from other Galaxies, clusters of Galaxies and Galaxies were reaccelerated by Outer Gravity for the simple reason that this is exactly what Gravity does, is exactly how Gravity works and is really just simple common sense. Because this model uses Gravity to accelerate Galaxies, our position in the universe, the distance between Galaxies and the mass of other Galaxies <u>will affect the speed</u> which all Galaxies are moving apart and expanding the universe. Called the Hubble Tension, the different <u>speeds</u> of the expansion of the universe have now been <u>confirmed</u> by the James Webb Space telescope. Scientists now say *"the universe appears to be expanding at bafflingly <u>different speeds</u> depending on where we look"*. The Gravitational expansion of the universe will be further explained near the end of the next chapter which is all about Gravity.

Because Quasars are known to be far more common within our earliest universe, means that Galaxies were much nearer together at a time when Quasars were most abundant. **As a result, one may use the abundance of Quasar like objects to show a time when Galaxies were <u>nearer together</u> and so a time when their combined Inner Gravity was able to slow the effect of Outer Gravity.**

Curious as to whether this was the best way to proceed I began to research if an abundance of Quasar like objects could possibly be tied into the acceleration of all Galaxies over time. After a great deal of research and by studying several of these types of graphs, I was able to draw an overlay showing the abundance of Quasar like objects and the acceleration of all Galaxies but would such a prediction be true? Would the graph show any evidence that the nearness of Quasar like objects to each other in the early universe first slowed the acceleration of all Galaxies.

The correlation is compelling. As can be observed on the 'right side' of my first illustration and in the next illustration where I have copied and pasted the bottom line to show how well the curve correlates.

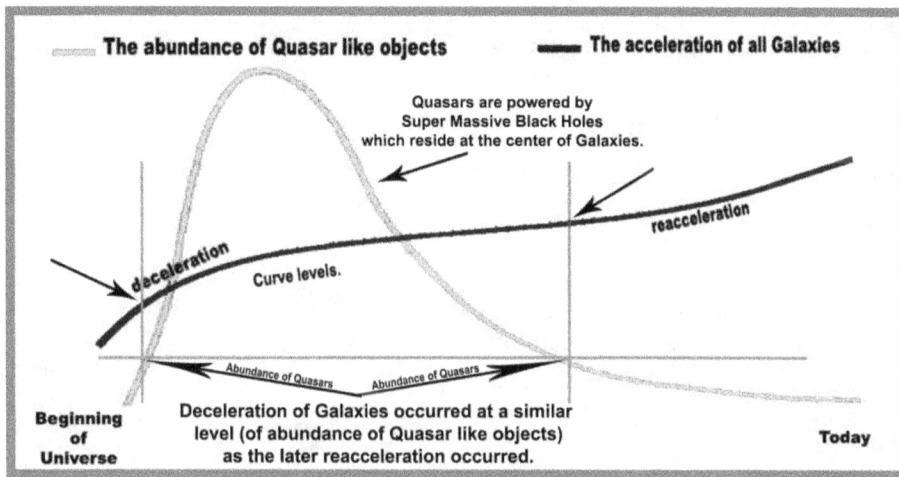

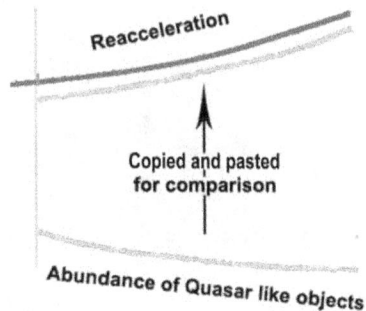

On the right side of this graph one can plainly observe the reacceleration of all Galaxies continuing to **increase** as the abundance of Quasar like objects continues to **decrease**. The correlation at right is also convincing, one can easily observe how well the bend of the two curves correlate.

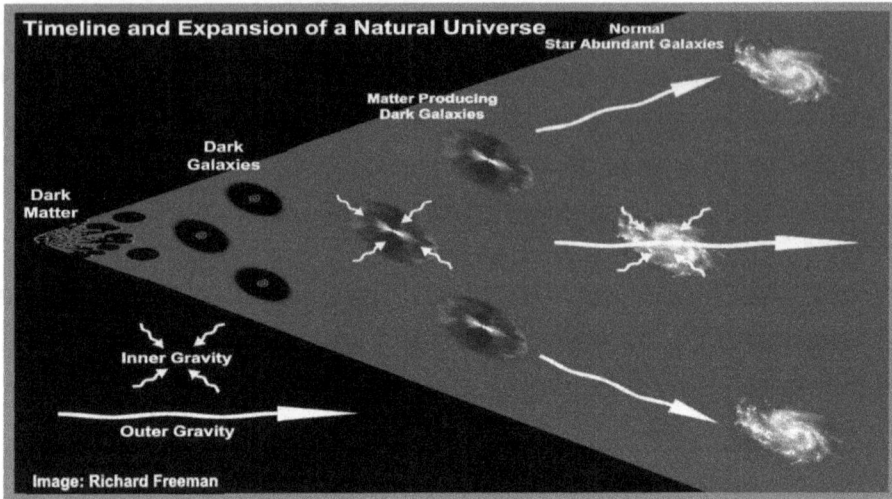

Timeline and Expansion of a Natural Universe

The Natural Universe: Expansion is driven by Gravity acting outwards. The source of all Gravity is fully explained. Both Inner Gravity and Outer Gravity act in unison as one of the same.

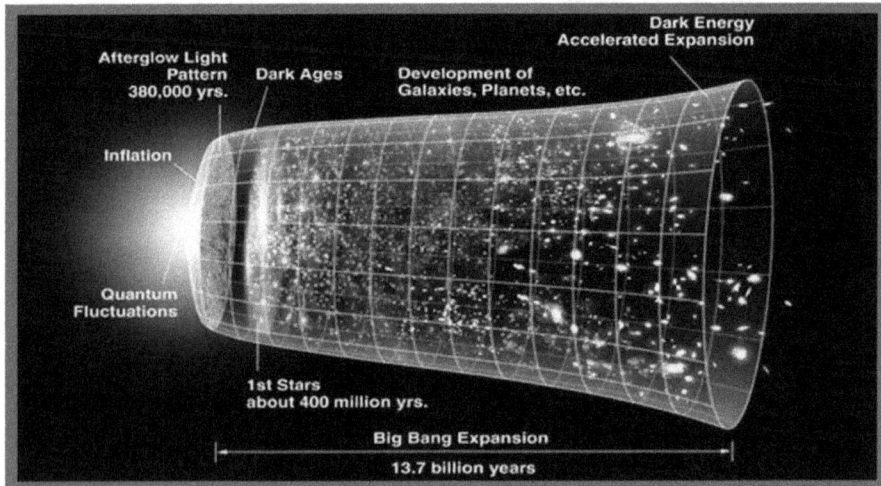

Credit NASA / WMAP science Team

A Big Bang Universe: Expansion is first driven by a mysterious still unknown form of energy called Inflation, followed by a mysterious still unknown form of energy called Dark energy. Dark energy and the energy of inflation have no known source, is why, no one knows what kind of energy created them.

11 THE SOURCE OF GRAVITY.

'Stationary Momentum'   By Richard Freeman.

See chapter 5, page 82, for how Gravity, by squeezing numerous tiny Quantum vacuum fluctuations, gradually arose from empty Space and began building the Natural Universe. For Quantum Gravity see chapter 13.

The source of Gravity is mass and the strong force creates 99% of all mass. Thus, to sensibly reconcile Gravity with mass, the source of the strong force has to also be the source of Gravity. However, the most baffling part is the strong force is about 100 trillion trillion trillion times stronger than Gravity. **So it is thought the source of the strong force could not possibly be the source of Gravity and is likely why nobody knows what Gravity is**. Within a Natural Model the source of the strong force is also the source of Gravity which sensibly reconciles Gravity with mass. However, a Natural Model also has a delivery part for Gravity which has a restraining part working in the opposite direction of a falling object making Gravity many times weaker.

Since both speed and Gravity slow the passing of time, one is required to also clearly show how the model for the passing of time achieves this. *Fittingly, a Natural Model's source of mass and Gravity is a tiny phase of virtual mass which is a tiny phase of totally invisible empty Space supplied by Quantum vacuum fluctuations at the heart of all Dark matter and all protons and neutrons of atoms. The rapidly collapsing and reappearing phase of virtual mass creates very tiny Gravitational waves which actively provide entangled dimensions of Space, including passing time. Virtual mass provides negative energy for the strong attraction part of Gravity and the tiny, positive energy carrying Gravitational waves provide the delivery part of Gravity. The variable delivery part of Gravity allows Gravity to be weakly delivered here on Earth and strongly delivered at the event horizons of Black Holes.* **That is, here on Earth the strong attraction part of Gravity is being very much restrained by the variable delivery part of Gravity.**

A wavelength always maintains a Planck length and the passing of the wave always maintains Planck time. When a tiny, energy carrying Gravitational wave becomes stretched by the attraction part of Gravity the stretching warps the curvature of the dimensions of Space. Since a stretched wave takes longer to pass the passing of time slows but because the wave maintains Planck time one's own time appears unchanging and self-normal. Since the stretched wave covers a greater distance the greater distance contracts to a Planck length which contracts distance and causes an object to fall to the ground. Planck length and Planck time are maintained because the wave's passing of a single 'pulse' of energy is maintained. Thus, an object made from atoms is suspended at the center of very tiny expanding waves of energy. When the tiny waves of energy become stretched in a given direction the object naturally moves in the direction of the stretched waves. The obtained 'speed' times an object's 'mass' provides a falling object with momentum while it remains at stationary equilibrium forever **trapped at the center of numerous very tiny expanding waves of energy.** *This is how my model 'Stationary Momentum' physically provides Gravity.*

In 1687 English Physicist Sir Isaac Newton published a law of universal gravitation in his book Philosophiae Naturalis Principia Mathematica. Then in 1916 Albert Einstein published his theory of General Relativity which correctly predicted a small difference in the motion of Mercury which Newton's law failed to predict. General Relativity also predicted that starlight passing near to the Sun's surface should be slightly deflected which Arthur Eddington confirmed during an eclipse of the sun in 1919. Einstein's theory says that Gravity occurs from mass warping the curvature of Space-time. However, General Relativity fails to tell the physical source of Gravity or the physical source of the dimensions of Space-time. Thus, the actual physical aspects of all parts of Gravity remain as one of the greatest unsolved mysteries of physics. **Hence, everybody knows what Gravity does but nobody knows exactly how Gravity actually physically does it.**

Scientists say Gravity causes Space-time to flow into Black Holes much like water rushing over a waterfall. Since Black Holes are made from the 'Gravity' of the mass of Matter, Space-time very likely also flows into the mass of Matter. If the near light speed dance of a proton's quarks and gluons were to stop, so as they were near motionless, they would still occupy and so disturb the integrity of undisturbed Space-time but would have almost no motion energy to provide a proton's regular mass. No mass here equals no Gravity, consequently, dimensional Space-time for some reason actually flows into the mass of Matter and whatever is creating the near light speed motion energy of a proton's quarks and gluons is somehow, almost certainly, responsible for attracting dimensional Space-time so as it flows into the mass of Matter and actively delivers Gravity.

Decades ago, alone and out of sight of land earning my living on the Pacific

Ocean I considered; what if a realm existed long ago which was so empty that it contained no Matter, no dimensional Space or time, a complete dimensionless nothingness in the form of a deep vacuum which existed for eons before our universe even existed. Now I considered that dimensional Space-time, as an expanding fabric which can be stretched and warped, would flow into such a state of deep vacuum and if one was able to provide a tiny piece of this same pure empty Space to Matter one would solve two things; why dimensional Space-time expands and why Space-time flows into Matter and in doing so delivers Gravity.

Because Gravity is a natural occurring phenomenon the phase of virtual mass, a tiny, lower state of Space than dimensional Space, should also be a natural occurring phenomenon. I originally failed to find evidence of this so I published my first book as the Model was in 2014. I continued my search for <u>many years</u> for evidence of a <u>tiny</u> lower state than ground state of the vacuum of Space. I knew ideally this tiny piece of empty Space should act as a fluctuation so I decided to research Quantum vacuum fluctuations for any evidence that such a phase of empty Space really existed but repeatedly found nothing. My search continued for many years before I discovered more recent research by a team led by Prof. Dr. Alfred Leitenstorfer from the University of Konstanz in Germany. Their research centered on vacuum fluctuations of the electromagnetic field which carries the interactions between electrical charges. I could hardly believe what I was seeing, the image was totally amazing, there it was and I knew exactly what it was; my source of Gravity! I had been searching before it was actually discovered.

[015] Their research has exposed a phase of Quantum vacuum fluctuations where the level of measured noise is *'lower than the ground-state of empty Space'*. With no explanation for why such a tiny state of deep vacuum should exist, the Scientists have referred to this phase of Space as being an *"astonishing phenomenon"*. For me there was no doubt this amazing find of an *"astonishing phenomenon"* was my sought after phase of virtual mass. I

was naturally elated for I had already written and published the model and in my mind, while reading this article, I was seeing direct evidence of my model for Dark matter. The research so <u>closely matched</u> my own modeling of Dark matter that I did not even need to change my basic <u>modeling or even many of my illustrations</u> of Dark matter from my previous book published in 2014. I now had little doubt that my modeling was correct.

I began to realize that the phase of virtual mass needed to mimic a minuscule, rapidly collapsing and reappearing Black Hole which would provide <u>an ideal source</u> of extraordinary tiny Gravitational waves. [002] I much later realized a collapsing and reappearing tiny Black Hole could also provide an invisible source which allows the pressure inside of a proton to peak ten times higher than inside of a neutron star and in doing so solve how a proton was provided with mass. [003] As I further researched I learnt that much of a proton's mass was created by madly exciting and agitating a proton's particles at near light speed; I realized a collapsing and reappearing Black Hole could easily accomplish this. This is what Black Holes do, they give particles near light speed and fluctuations would surely madly agitate a proton's particles.

To provide entangled dimensions to Space the Gravitational waves should ideally be so extraordinary tiny as to provide tiny Planck units, notably Planck lengths, to Space. Because of the <u>extraordinary ability</u> of Gravitational waves to actually change or distort the dimensions of Space I concluded Gravitational waves <u>had to be the only possible contender</u> to actually <u>provide</u> Space with dimensions. Gravitational waves also provide other <u>very special properties</u> which my modeling required. Gravitational waves, like the ringlets from raindrops on a pond, carry energy from the event which created them. This direct relationship to their source provides a tiny Gravitational wave with an <u>entangled</u>, direct link to the event which created it. Energy wise, as these tiny Gravitational waves expand they retain entangled energy of the event. **The tiny unit of energy will allow**

Planck units to remain the same value, derived from one passing unit of energy, regardless of the Gravitational waves being stretched. Because the tiny Gravitational waves are produced by the continuous Quantum vacuum fluctuations of the COST Gravity can be <u>continuously</u> delivered.

[017] <u>Very large</u> Gravitational waves are detected at the LIGO observatory which has two "arms" each more than 2 miles (4 kilometers) long. A passing Gravitational wave from a distant powerful event causes the length of the arms to change slightly. The observatory uses lasers, mirrors, and sensitive instruments to detect these very tiny changes. This is why Gravitational waves <u>are ideal</u> to deliver Gravity; they can <u>alter spatial dimensions</u>.

Commonly, objects which have energy also have mass which prohibit them travelling at the speed of light. An important property of Gravitational waves is, although they carry <u>energy</u> from the event which created them, Gravitational waves are massless which allow <u>Gravitational waves to travel at the speed of light</u>. *Note: This is a vital requirement because Gravity has actually been measured to propagate at the speed of light which means whatever delivers Gravity <u>has to be</u> massless which narrowed the search down to <u>Gravitational waves</u> as they are the <u>only object</u>, besides light itself, which travels at the speed of light.* **When the tiny Gravitational waves reach a source of Gravitational attraction, due to <u>entanglement</u> and speed of light ramifications, the <u>action</u> of Gravity may now instantly occur. (See: Quantum particle <u>entanglement</u> chapter 13).** Gravitational waves pass through objects as if they were not there and <u>propagate forever</u> which is required for delivering Gravity. Gravitational waves also clearly provide a curvature which is required because Einstein has told us Gravity warps the curvature of <u>inseparable</u> dimensional Space and passing time.

However, can Gravitational waves really be stretched and warped by the attraction of Gravity? [017 Ask Ethan: Are Gravitational Waves Themselves Affected By Gravity?] This link provides amazing evidence of how Gravitational waves ***"become definitively stretched"*** and ***"are affected by the warping of space".*** Thus, because Gravity stretches and warps the curvature of the dimensions of Space and the properties of Gravitational waves exactly mirror the <u>many very unique properties</u> required to deliver Gravity it leaves very little doubt, there are no other plausible contenders, tiny Gravitational waves have all of the required special properties for providing the actual dimensions of Space and time and in an unique way which is ideal for delivering Gravity.

But can nothingness which provides <u>negative energy</u> really be used to warp the dimensions of Space-time? This theory says it can. In 1994, theoretical Physicist Miguel Alcubierre wrote a paper which <u>uses negative energy</u> to drive a spaceship by warping the dimensions of Space. [111] His Alcubierre drive contracted the dimensions of Space ahead of a spaceship and stretched the dimensions of Space behind a spaceship. The point here is the Physicist's choice of using <u>negative energy</u> to warp Space by contracting and expanding the actual dimensions of Space. Fittingly, his concept of warp drive is based on a clever utilization of <u>Einstein's General Relativity</u> which provides Gravity by warping the dimensions of Space and time. Warp drive was first proposed by John W. Campbell in his 1957 Sci-Fi novel 'Islands of Space' and later by the Star Trek series. [111b see video] NASA has made progress developing Alcubierre drive theory into a viable proposal.

But there is more! I believe the most compelling evidence, or <u>proof</u> if you now want, that a tiny phase of empty Space <u>truly provides both</u> mass and Gravity is the discovery that Galaxies are accelerating in the direction the universe is expanding. If a phase of pure empty Space provides mass and Gravity and the universe is expanding into a seemingly endless void of pure empty Space the universe <u>will have</u> a massive resource for Gravitational Acceleration which acts in the direction which the universe is expanding. That is, due to Gravitational Attraction, Galaxies are simply accelerating as they free fall in the direction which the universe is expanding and there is absolutely no need for the mystifying and <u>totally, hypothetical</u> Dark energy.

Ideally and like this model superbly provides, the tiny Gravitational waves

are <u>required to originate from the source of mass</u>. The tiny Gravitational waves carry <u>positive energy</u> from the tiny fluctuation and <u>annihilation</u> event which created them which means they are positive and are <u>naturally attracted</u> to the negative charged density of the phase of virtual mass of other COST units. Now the delivery part of Gravity is attracted to the strong attraction part of Gravity and my source of Dark matter was looking good. But I wondered how this mechanism got to be part of Matter. Then I realized Dark matter was a virtual proton. I did not need to get virtual protons into Matter; virtual protons were transformed into regular protons. But where could this happen? There was only one logical answer; at the poles of Super Massive Black Holes and there was no Big Bang.

Because these Gravitational waves are so tiny and expand on a flat plane, like the ringlets from raindrops of a pond, it will require <u>many atoms</u> to come together to build an all around Dimensional Reality from where their combined dimensions of Space are able to smoothly deliver Gravity. A fine grain of salt is said to have around 1,000,000,000,000,000,000 atoms which will surely provide a complete Dimensional Reality. [018] The double slit experiment provides remarkable evidence that this is true. See page 331.

[015] As already expressed, experiments relating to vacuum fluctuations have exposed a phase where the level of measured noise is lower than that of the vacuum state, that is, the ground-state of empty Space, a state described as an *"astonishing phenomenon"*; (Prof. Alfred Leitenstorfer, the University of Konstanz in Germany). The significance of this experiment has without doubt been overlooked, for according to my model the Scientists have made one of the greatest discoveries in history; the actual source of Gravity, the true origins of regular mass and even the origins of the fabric of dimensional Space and time which is surely worthy of a Nobel prize.

[016 false vacuum.] Further evidence of this same tiny phase of a lower state of the vacuum now comes from an entirely different source; the energy state of the vacuum can be calculated from the potential energy of the Higgs field and the masses of the Higgs and top quark; refined measurements and calculations reveals evidence of the existence of a lower state of vacuum than the vacuum of dimensional Space. The discovery is said to be one of the *"biggest unsolved mysteries in physics"* as it means we live in a false vacuum. This mystery is easy to solve; **there is nothing in the tiny phase of virtual mass; it is a tiny phase of true vacuum nothingness while what is now called 'the vacuum' is not a true vacuum but a false vacuum since it contains the positive energy of the dimensions of Space and time. Einstein realized that dimensional Space is not nothingness but needed to be like a fabric which can be stretched or warped to deliver Gravity.** Scientists remain mystified as to the source of the lower state of vacuum. This model leaves little doubt; Scientists are seeing direct evidence of the actual source of the dimensions of Space-time, Gravity and mass.

[011] This extraordinary video called **'Your Mass is Not from the Higgs Boson'**, featuring Professor Derek Leinweber and researchers at the University of Adelaide, **actually reveals empty Space is the source of most of our mass**. For me this video confirmed again my source of all mass. The video is a super computer simulation of the theory Quantum Chromodynamics which describes the action of the strong force within a proton. The video clearly shows how quarks gather around an area of empty Space which is clearly identified as being the source of the strong force which creates almost all regular mass. The conclusion is it is *"extraordinary, because what we think of ordinary empty Space that turns out to be the thing that gives us all, most of our mass"* (99%). Mass and Gravity go together like two peas in a pod; exactly where you find mass you will find Gravity. Since the source of mass is empty Space, Gravity should share the same source of empty Space.

[002] Learning the pressures inside of a proton actually peak ten times higher than inside of a neutron star now confirmed again that the phase of virtual mass mimics a collapsing and reappearing Black Hole. If one ever wanted conformation that the tiny phase of virtual mass acts as a tiny fluctuating

Black Hole then this provides it. [021] **Theories allow a tiny _virtual_ Black Hole such as this to appear spontaneously but only very briefly.** I realized nothing else was likely to provide such astonishing pressure which is astoundingly ten times stronger than inside of a neutron star. I now had confidence my modeling, which began over three decades ago, was truly on the right track. Now if the negative energy phase of 'virtual mass' is the source of the attraction part of Gravity and is also the source of all regular mass, this would clearly and sensibly reconcile Gravity with all mass which is a <u>compulsory requirement</u> which has never before been achievable.

[003] Theoretical Physicists Matt Strassler when relating mainly to the actual results from experiments at the Large Hadron Collider confirms; _"almost all mass found in the ordinary Matter around us is that of the nucleons within atoms. And most of that mass comes from the chaos from the <u>motion-energy</u> of a nucleon's quarks, gluons and anti-quarks, and from the interaction-energy of the strong nuclear forces that hold a nucleon intact. Our planet, our bodies are what they are as a result of a silent, and until recently unimaginable, internal (near light speed) pandemonium"._

A Natural Model provides protons with Quantum vacuum fluctuations which fluctuate to a pure empty Space phase of virtual mass which mimics a tiny, rapidly collapsing and reappearing Black Hole. A tiny, rapidly collapsing and reappearing Black Hole provides an ideal source for the energy vital to create _"unimaginable, internal pandemonium"_ by wildly exciting and agitating quarks and gluons about at near the speed of light and momentarily tightly combine all and so create regular mass. This is exactly what Black Holes do; they give particles near light speed.

[062] John Wheeler was a renowned American theoretical Physicist who once said of Gravity; _"Space-time tells Matter how to move and Matter tells Space-time how to curve"_ Although possibly not intended, this suggests that Gravity is best explained as a two part mechanism. For a Natural Model 'Space-time' should be rephrased to '<u>the dimensions</u> of Space-time'.

A minuscule, rapidly collapsing and reappearing Black Hole will provide an ideal source of extraordinary tiny Gravitational waves. With a Natural Model the attraction part of Gravity is provided by a tiny fluctuating phase of virtual mass. Virtual mass is a <u>strong</u> negative energy phase of deep vacuum empty Space provided by the COST (Dark matter). The delivery part of Gravity is provided by tiny Gravitational waves which provide dimensions to Space and are created by the actual 'fluctuation' of the virtual mass phase of the COST. The <u>variable</u> delivery part of Gravity is fittingly closely modeled on Albert Einstein's brilliant General Relativity. Since the process relies on fluctuations, when the phase of virtual mass is filled by ground-state dimensional Space the virtual mass phase is immediately reproduced by the next fluctuation. This is why Quantum vacuum 'fluctuations' are required so as the flow-in of dimensional Space-time can be maintained.

[017] Attracted to the strong phase of virtual mass the dimensions of Space in the form of tiny, energy carrying Gravitational waves, become <u>stretched</u> which causes both length contraction and time dilation. Length contraction causes an object to fall and time dilation slows the passing of time. This allows dimensional Space-time to flow <u>not only</u> into <u>the mass</u> of Black Holes but also into <u>the mass</u> of all Dark matter as well as <u>the mass</u> of all protons and neutrons to deliver Gravity to all objects made from atoms.

[026, 070] The failure at the Large Hadron Collider to discover any of the most sought after Supersymmetry particles, has also allowed this model to be fittingly and further developed. Physicists have described this failure as a *"nightmare scenario"*, for it means they have *"not made one correct prediction in thirty years"*. Theories for the <u>origins of the mass of particles</u> must now be abandoned or at least require a brand new form of physics.

Why does Gravity <u>fail to work</u> with the Standard Model of particle physics and why did Einstein say, mathematically, Gravity itself is not an actual force? With my 'Stationary Momentum' model Gravity is not conveyed by a force carrying particle belonging to the **Standard Model.** Acting exactly like

Einstein's General Relativity, Gravity is a mechanism which stretches and *'warps the curvature of the dimensions of Space and time'* and in doing so continually repositions an atom's stationary equilibrium position at the center of its <u>own entangled dimensions</u> of Space. **Since a free falling object remains at stationary equilibrium within its Space-time it feels no force acting on it.** But if an atom's <u>stationary equilibrium</u> position within Space-time is <u>blocked</u> by other atoms, its stationary equilibrium position has now moved passed its blocked position. The blocked atom is now in a state of continually *'accelerating'* away from its natural stationary equilibrium position. Like Einstein said, on Earth we feel this acceleration as one 'G' of Gravity. **Because your entangled Dimensional Reality is continually being renewed at the amazing speed of light you continually feel acceleration.**

Thus, when an atom's stationary position is <u>blocked</u> by other atoms an atom is now '<u>forced</u>' away from its stationary position. This is the <u>only kind of 'force'</u>, which is felt as acceleration, applied by Gravity but is the <u>passive force</u> which allows us to weigh ourselves and which causes nuclear fusion by collectively squeezing and colliding hydrogen atoms in the core of the Sun so tightly that four hydrogen atoms fuse into a helium atom and in doing so powers my solar panels and provides me with free electricity.

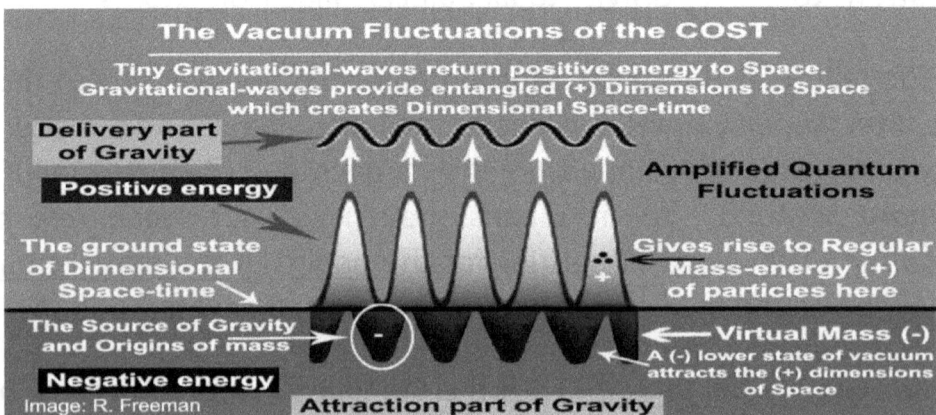

Scientists have theorized a hypothetical force carrying particle, called a Graviton, may be responsible for Gravity. [012] However, almost all <u>theories</u>

containing Gravitons <u>suffer from severe problems.</u> There are no successful theories within physics which use a Graviton and there is no experimental evidence supporting the existence of Gravitons. [110] The theory says when a hypothetical Graviton is exposed to the vacuum of Space it breaks into <u>negative mass and positive mass.</u> Similarly, my phase of virtual mass fluctuates at the vacuum of Space to a phase of <u>negative mass which directly delivers positive mass</u> to particles. Because virtual mass is negative energy it can be called negative mass. Unlike a hypothetical Graviton the 'lower state phase' of virtual mass has been observed in experiments. Inside of a Black Hole <u>all particles</u> are crushed out of existence so how could a Graviton continue to exist where Gravity is strongest? A phase of virtual mass is a phase of nothingness which cannot be crushed out of existence.

The COST fluctuates from a <u>positive</u> energy phase of <u>virtual</u> particles to a <u>negative</u> energy phase of <u>virtual</u> mass. **The positive energy phase is created from the <u>annihilation</u> of pairs of virtual particles. <u>Positive</u> energy is returned to Space in the form of extraordinary, tiny Gravitational waves which provide <u>entangled</u>, dimensional moments to Space which provide a woven fabric or structure to the nothingness of Space; structure which can be stretched and warped to deliver Gravity. A <u>restraining part</u> is what physically causes Gravity to be puzzlingly <u>far weaker than other forces.</u>**

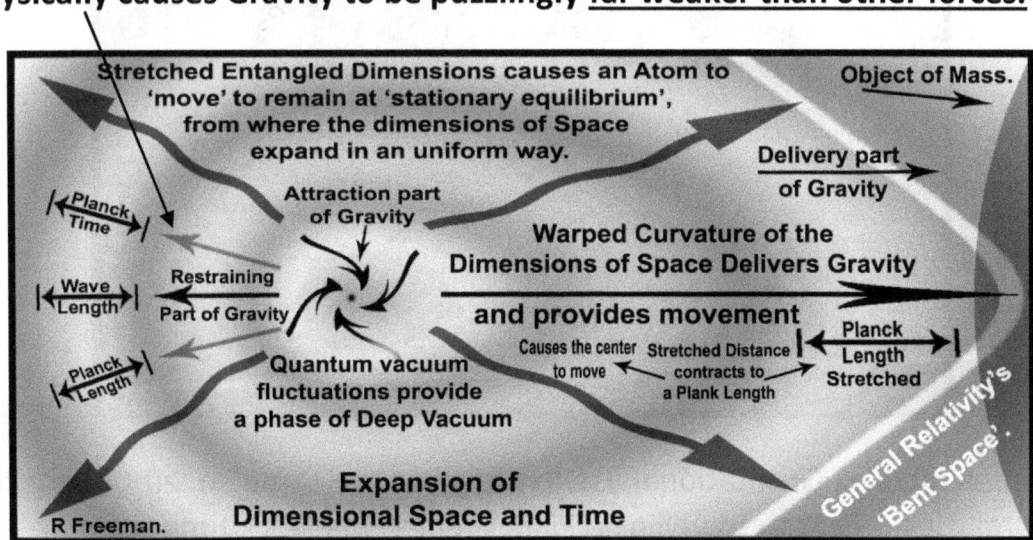

Stretched Entangled Dimensions causes an Atom to 'move' to remain at 'stationary equilibrium', from where the dimensions of Space expand in an uniform way.

Object of Mass.

Delivery part of Gravity

Attraction part of Gravity

Planck Time

Warped Curvature of the Dimensions of Space Delivers Gravity

Wave Length

Restraining Part of Gravity

and provides movement

Causes the center to move   Stretched Distance contracts to a Plank Length

Planck Length Stretched

Planck Length

Quantum vacuum fluctuations provide a phase of Deep Vacuum

General Relativity's 'Bent Space'.

Expansion of Dimensional Space and Time

R Freeman.

The delivery part of Gravity provides movement but has a <u>restraining part</u> which operates in the <u>opposite direction of a falling object</u> causing Gravity to be <u>very weakly delivered</u>. This <u>solves the long time mystery,</u> which has confounded theorists, of why Gravity is so very much weaker than other forces. The <u>positive</u> state, expanding, tiny Gravitational waves which provide entangled, dimensional moments to Space are naturally <u>attracted to the negative energy</u> of the phase of virtual mass of the protons and neutrons of other atoms. Attraction stretches and warps the curvature of the expanding Gravitational waves from other atoms. The tiny Gravitational waves provide Planck dimensions to Space. The distance <u>trough to trough</u> is the wavelength of one tiny Gravitational wave which provides <u>one dimensional unit or Planck length</u> which is the smallest distance allowed within a Dimensional Reality. The passing of the same tiny Gravitational wave provides <u>Planck time</u> which is the shortest possible moment of passing time within a Dimensional Reality. Consequently, like Einstein has said, dimensional Space and the passing of time cannot be separated, thus Space-time. Planck length is the distance which light travels in Planck time.

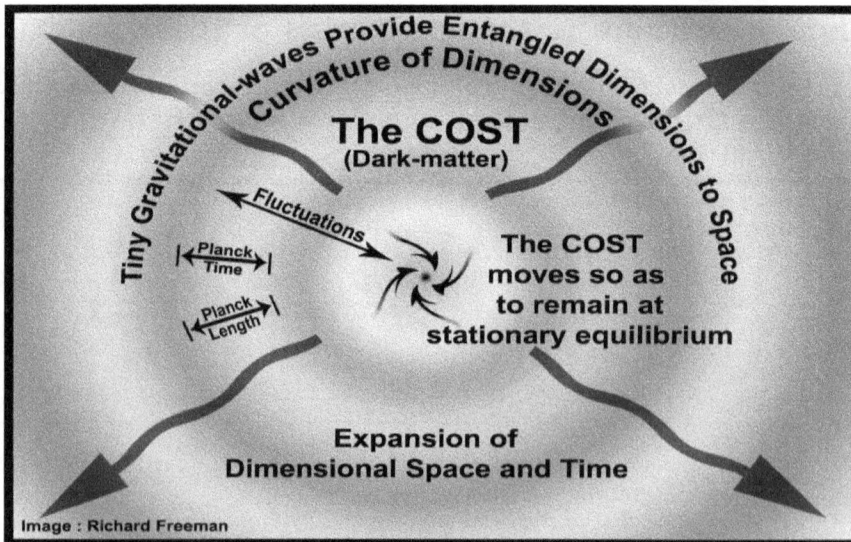

Because the tiny Gravitational waves travel at the same speed of light this naturally <u>locks</u> light to travelling a Planck length in Planck time regardless of the tiny Gravitational waves being stretched or warped and causing time dilation or length contraction. This is the <u>physical</u> process which allows the speed of light to be a constant so as we all observe the speed of light as being the same regardless of our time dilation or length contraction factors.

[017a] Although the energy that Gravitational waves carry is said to be fixed, as they spread out and <u>perpetually</u> travel through Space, Gravitational waves do become smaller but forever retain a vital, single pulse of energy. When a tiny Gravitational wave is stretched towards a tiny phase of virtual mass the stretched wave naturally covers a <u>greater distance</u> but because the stretched Gravitational wave retains the same <u>single pulse of energy</u> from the event which created it the <u>stretched</u> Gravitational wave also retains a Planck length which causes the dimensional distance the stretched wave covers to contract to a Planck length. This is Gravitational length contraction which since there is a contraction of distance moves an atom's stationary position towards other atoms. An atom now <u>moves</u> because its dimensional distance towards other atoms has contracted. This movement is Gravity and is how Gravity manipulates the dimensions of Space by stretching and warping the curvature of very tiny Gravitational waves.

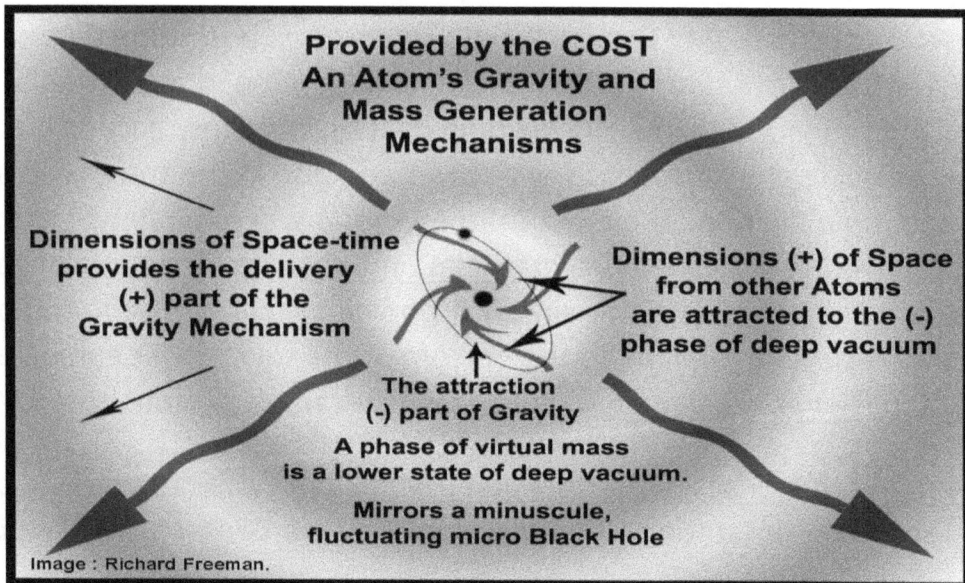

Provided by the COST
An Atom's Gravity and
Mass Generation
Mechanisms

Dimensions of Space-time
provides the delivery
(+) part of the
Gravity Mechanism

Dimensions (+) of Space
from other Atoms
are attracted to the (-)
phase of deep vacuum

The attraction
(-) part of Gravity

A phase of virtual mass
is a lower state of deep vacuum.

Mirrors a minuscule,
fluctuating micro Black Hole

Image : Richard Freeman.

[017] Gravitational waves travel at the speed of light so does not the very act of stretching a Gravitational wave need to exceed the speed of light which is not allowed? Yes, it is rational to suggest the actual stretching needs to exceed the speed of light, however, because the passing of the same tiny Gravitational wave maintains a Planck unit of passing time, as it becomes stretched, it takes longer to pass which slows time and allows the tiny Gravitational wave to be stretched without the act of stretching the wave exceeding the speed of light. **This is exactly why applying Gravity 'is required to slow time' because if it did not, the act of stretching and warping the curvature of a tiny Gravitational wave would exceed the speed of light which Relativity forbids.** This provides confirmation that the passing of the same wave which provides dimensional distance also provides the passing of time. The more a tiny dimension providing Gravitational wave is stretched the more Gravity is applied and the more the passing of time is slowed by time dilation. Because the Planck value of a tiny Gravitational wave is continually maintained, Gravitational length contraction, just like time dilation, is masked from one's self-reality. However, the telling tale that it truly is being stretched is distance contracting which causes an object to fall and the passing of time to slow.

Because a 'stretched' tiny Gravitational wave causes both time dilation and length contraction wherever one finds time dilation one will also have length contraction. Thus, measuring time dilation with very accurate atomic clocks by placing one at the base of a tower and one at the top of the tower will detect time dilation. The clock on the ground where Gravity is stronger will run slower, this is called Gravitational time dilation. The coupled length contraction can be detected by simply dropping a clock from the top of the tower and recording it falling to the bottom of the tower. This can be called Gravitational length contraction. Both of which can be demonstrated.

Why you do not feel acceleration when free falling. When free falling one is really remaining at 'stationary equilibrium', at the center of where one's own <u>entangled</u>, expanding, dimensional moments are at equilibrium. The whole Gravity mechanism neatly <u>acts on the dimensions of Space alone</u>. **NOTE: there is no direct force-like attraction of atoms to other atoms**. A free falling object remains at a position of 'stationary equilibrium' at the center of its expanding, <u>entangled</u> dimensions of Space-time. In a free fall towards Earth this acts like you remain stationary as the dimensions of Space between you and the Earth are being stretched and the stretched dimensional distance is contracting and bringing the Earth to you. The Gravity mechanism now provides a free falling object with what is now best termed as 'stationary momentum'. A stationary object feels no force or acceleration which is why you feel no acceleration when free falling. Hence, I have called my theory of Gravity 'Stationary Momentum'.

This is why one can think of Gravity as being passively sourced as a byproduct of a mass generation mechanism. On page 136 of my copy of Stephen Hawking's 'A Brief History of Time' modeling relates negative energy to Gravitational Attraction and positive energy relates to Matter. The COST fluctuates to a phase of virtual mass which is of a lower state of vacuum than the ground state of dimensional Space. Virtual mass is negative energy for the reason that it is both the source of Gravitational Attraction and sits <u>below</u> the ground-state of dimensional Space.

The COST is transformed from its Dark matter state of a virtual proton into a regular proton or a neutron. The COST provides a fluctuating phase of virtual mass which mirrors a minuscule, rapidly collapsing and reappearing Black Hole which provides the tiny seed of all Black Holes. **Virtual mass is what stays behind to provide mass and Gravity when all particles are crushed out of existence inside of all Black Holes.** Black Holes are one of the most powerful objects in the universe. This is why the tiny phase of virtual mass has the strong ability to agitate quarks at near <u>the speed of light.</u> When the phase of virtual mass appears the quarks of nucleons of atoms slam <u>tightly</u> together and when the tiny phase of virtual mass is consumed by dimensional Space the particles are momentarily free to aggressively rebound. <u>One cycle of this near light speed crazy agitation naturally provides</u> particles with energy which now equals their regular mass by the speed of light squared ($E=mc^2$). Particles have now acquired regular mass-energy in a state where their regular mass cannot be separated from the energy acquired. The particles now have regular mass created by a phase of virtual mass and virtual mass provides the strong attraction part of Gravity. This is why; although they are created in completely different ways, regular mass cannot be created without Gravity. Thus, our source of regular mass and the source of the attraction part of Gravity share the same virtual mass source but Einstein's General Relativity's Gravity is significantly different to regular mass. Sensibly derived from <u>the source of the strong force</u>, the attraction part of Gravity is obviously <u>very strong,</u> but the variable delivery part of Gravity, provided by tiny Gravitational waves, has a <u>restraining part</u> which operates in the <u>opposite direction of a falling object</u> causing Gravity to be <u>weakly delivered.</u> This solves the long time mystery, which has confounded theorists, of why Gravity is far weaker than other forces. See image page 67.

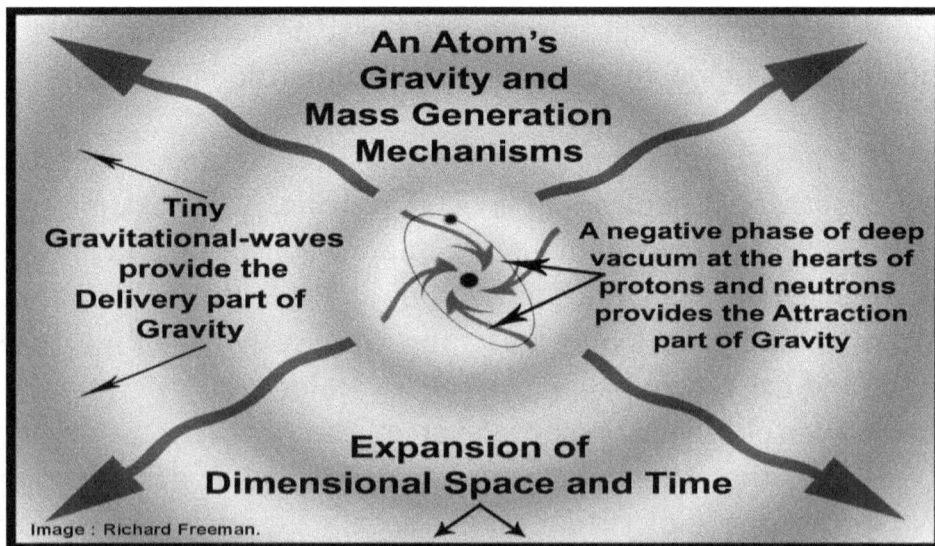

An Atom's Gravity and Mass Generation Mechanisms

Tiny Gravitational-waves provide the Delivery part of Gravity

A negative phase of deep vacuum at the hearts of protons and neutrons provides the Attraction part of Gravity

Expansion of Dimensional Space and Time

Image : Richard Freeman.

Virtual mass is both the source of Gravity and the source of regular mass. A proton's regular mass is positive mass-energy given to its particles, mainly its quarks, from being excited and agitated at near the speed of light by the negative energy phase of virtual mass. Hence, regular mass itself has no Gravity and is different to virtual mass. However, regular mass which has no Gravity cannot exist without virtual mass which is the attraction part of Gravity. Consequently, positive regular mass cannot come into existence without having the byproduct of Gravity. Thus, wherever you find regular mass you will also naturally find Gravity.

The delivery part of Gravity is varied by the proximity and quantity of other COST units phasing to a phase of virtual mass and the proximity of a far outer void of a lower state of empty Space into which all dimensional Space is expanding. Gravity is applied by the amount the stretching warps the curvature of dimensions from zero to 100%. At a Black Hole's event horizon dimensions are so stretched that they are attempting to accelerate an atom to near the speed of light. Inside of a Black Hole the tiny Gravitational waves which provide entangled dimensions to Space cannot escape the COST because to escape they would now need to exceed the speed of light. Consequently, the passing of time stops. Now without entangled dimensions the Space inside of a Black Hole is effectively dimensionless which mirrors a singularity. Even when zero Gravity is being applied by the delivery or movement part of Gravity, a COST continues to retain 100% of its attraction part of Gravity derived from a tiny phase of virtual mass.

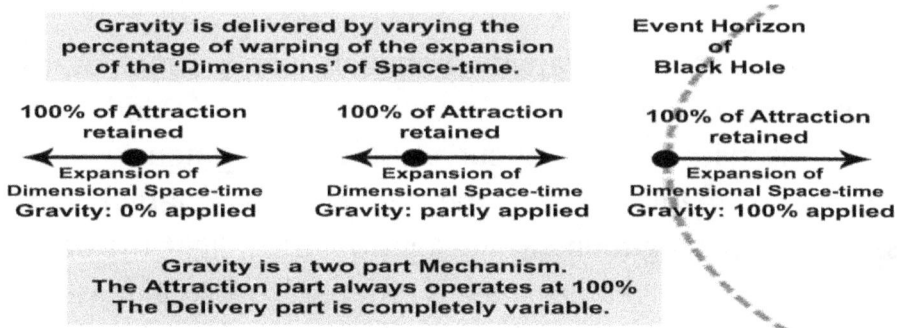

Gravity is delivered by varying the percentage of warping of the expansion of the 'Dimensions' of Space-time.

Event Horizon of Black Hole

100% of Attraction retained   100% of Attraction retained   100% of Attraction retained

Expansion of Dimensional Space-time Gravity: 0% applied   Expansion of Dimensional Space-time Gravity: partly applied   Expansion of Dimensional Space-time Gravity: 100% applied

Gravity is a two part Mechanism. The Attraction part always operates at 100% The Delivery part is completely variable.

Gravity is applied independently by every COST unit (Dark matter,

protons and neutrons) to collectively provide the stretching which warps the curvature of all dimensional Space and time in the vicinity of a large massive object. This allows Gravity to exist throughout all Space-time.

Special Relativity within a Natural Universe: Albert Einstein's Special relativity says that all laws of science shall be the same to all non-accelerating observers no matter their location. This is easily achievable when all Matter (atoms) is always located at the center of expansion of dimensional Space and time. Within our Natural Universe Special Relativity applies to how an atom always remains, <u>unless</u> it is blocked by other atoms, at stationary equilibrium at the center of its expanding **Gravitational waves** which provide entangled dimensions of Space and time. This allows all laws of science to be the same for all observers no matter their location.

General Relativity within our Natural Universe: It is imperative to show exactly how my concept called 'Stationary Momentum' explains the very source of Gravity and at the same time has a delivery part which is well-matched to General Relativity's highly successful concept of 'bent Space'. However it is said General Relativity and 'bent Space' alone cannot explain the precise physical source of Gravity. General Relativity strongly suggests that Gravity is not truly a force but a product of the reality that the curvature of dimensional Space-time is in some way stretched or warped near objects of mass. Albert Einstein has also revealed that the feel of acceleration and the feel of Gravity are both fundamentally the same thing.

When referring to General Relativity Scientists often simply say "mass warps the curvature of Space". While with my Natural Model one may simply say "virtual mass warps the curvature of the <u>dimensions</u> of Space and time". However, regular mass is the energy (mass-energy) created by virtual mass aggressively agitating at <u>near the speed of light</u> the particles of the nucleus. Consequently, regular mass is different to virtual mass.

Within my Natural Universe the curvature of an atom's tiny entangled

Gravitational waves become stretched or warped in the direction of an object of mass and is a trend which also increases nearer to mass. Stretching causes a warping of the curvature of the tiny Gravitational waves which because they provide dimensions to Space naturally causes a warping of the curvature of the dimensions of Space-time.

The following image is a rubber band analogy: Within a Natural Universe Einstein's General Relativity ingeniously describes the position 'A' from where the expanding dimensions of Space-time expand more stretched towards objects of mass and a position which an atom **naturally attempts to avoid** by continually 'moving' to the position "B", where it is able to remain naturally stationary at 'stationary equilibrium' and at the center of where it's expanding waves of dimensional Space and time are at equilibrium. Because an atom remains stationary, at the center of its expanding dimensions, there is no 'force' being applied to a free falling atom. However, if an atom's stationary equilibrium position is blocked by other atoms, a blocked atom is now forced from its stationary equilibrium position. Being 'forced' away from its stationary position is felt as acceleration. This is why we feel one 'G' of Gravity as acceleration.

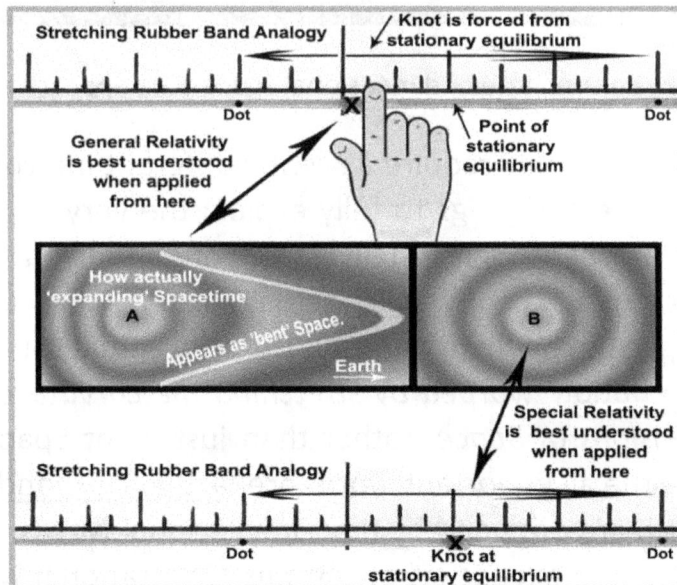

Physicists will tell you that Albert Einstein's General Relativity says the fabric of Space is stretched or warped by Gravity near a massive object. What the physics do not tell the Physicists is that it is not 'Space' which is being bent, stretched or distorted; it is the dimensions of Space in the form of tiny Gravitational waves which are stretched or warped. A Natural Universe provides extraordinary tiny Gravitational waves permeating throughout all Space which provide Space with dimensions. However, Einstein did say that Space was not nothingness but like a fabric. Like a fabric being woven by thread, the raw nothingness of Space is woven into a dimensional fabric by numerous trillions of tiny Gravitational waves which are <u>very fittingly a prediction of Albert Einstein's General Relativity</u>.

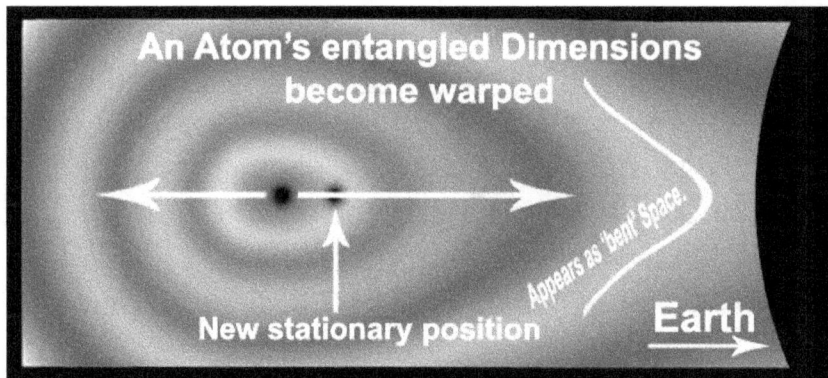

How the warping of the curvature of dimensional Space-time appears as 'Bent Space'.

General Relativity may only require a slightly different interpretation of the theory, or the smallest of twigs to fully explain the very source of Gravity. So yes, in this way, 'bent' Space delivers Gravity just like a Physicists may casually say but it is what <u>physically</u> actually causes the curvature of the dimensions of Space to be 'bent' which is the true actual source of Gravity. I much prefer the notion *'warped by stretching the curvature of entangled, expanding dimensions of Space'* rather than just 'bent Space'. Within our Natural Universe the all important words are '<u>actively expanding, entangled dimensions</u>' which are provided by tiny Gravitational waves. <u>In a seemingly back to front way</u>, the expansion of the tiny Gravitational waves naturally allows Gravity to propagate at its known speed of light.

General Relativity's warped, bent Space alone can precisely predict the amount of Gravity at any point in Space. However, without <u>active expansion</u> of the dimensions of Space, bent Space, arguably, cannot produce the active 'flowing' effects of Gravity. Gravity is a very active phenomenon. Observe an object actively falling to the ground. Since it is the actual dimensions of Space <u>between the object and the ground</u> which are actively moving the object and causing the object to fall is why the dimensions to Space are <u>required</u> to be actively created. The passing of time, which is the fourth dimension, is also required to be actively created by objects so as time can <u>actively pass by at an object's very own unique rate of time dilation.</u> Self <u>entangled,</u> expanding dimensions achieves this as it allows all other Dimensional Realities **<u>to be ignored</u>**, except your own.

An atom as it falls to Earth sees its dimensions of Space expanding evenly and not warped at all. Only when its Gravitational path is blocked by other atoms does the atom feel the full effects of the warping of its dimensions of Space. Nothing much happens when an atom is alone in Space where it can occupy its natural place trapped at 'stationary equilibrium' within its expansion of Space and time. In sharp contrast, lots of meaningful and wonderful things begin to happen when an atom is blocked by other atoms from occupying its most natural position at 'stationary equilibrium' within the expansion of its entangled dimensions of Space and time.

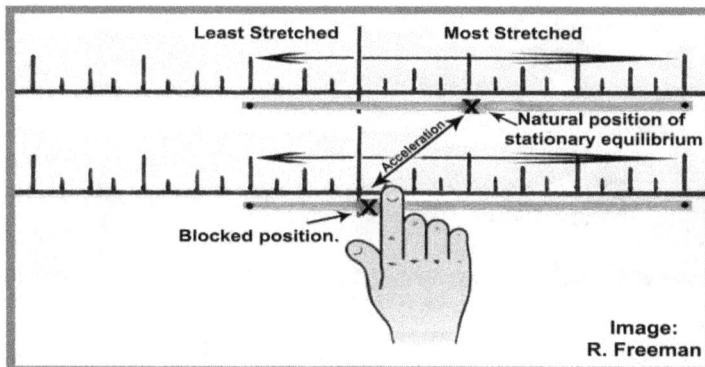

Image:
R. Freeman

Like this strip of rubber band analogy, if the stationary path is blocked by other atoms, like on the surface of the Earth, an atom's stationary position has moved passed its blocked position. Now it is like you slid the knot with your finger and accelerated the knot in the rubber band to a new but blocked position. A blocked atom is now in a perpetual state of accelerating away from its natural stationary position, this is the acceleration revealed by Albert Einstein which we feel as one 'G' of Gravity. The stationary position is <u>continuously being reset,</u> at the amazing speed of light, by the continuous supply of tiny expanding Gravitational waves which is why the feel of acceleration is continuous. Only when its Gravitational path is blocked by other atoms does the atom feel the effects of the warping of its entangled dimensions of Space. Blocked by other atoms its most natural stationary equilibrium position within its entangled, expanding, moments of dimensional Space-time has now <u>travelled passed</u> its blocked position.

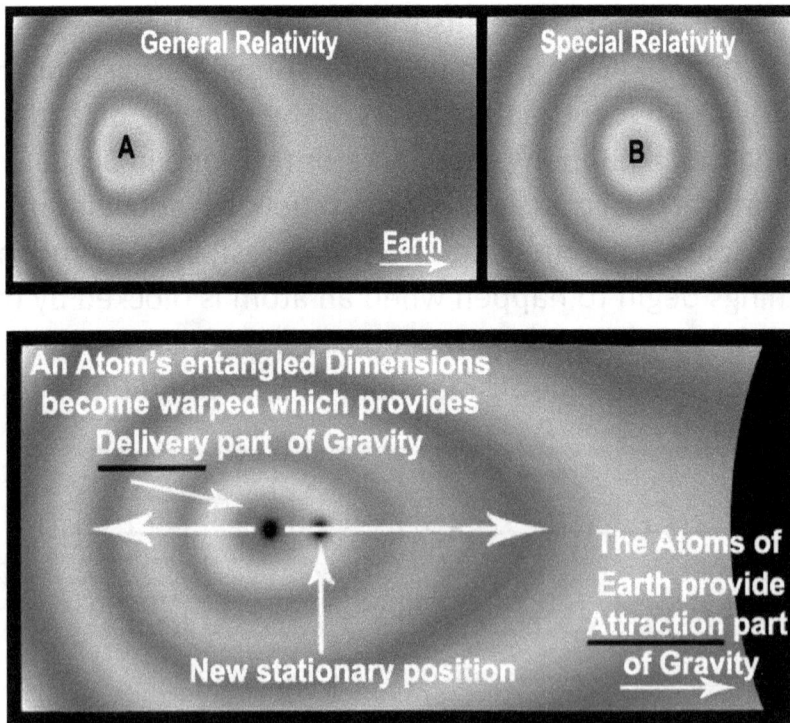

Because the delivery part of Gravity mirrors General Relativity one can understand why bent-Space and General Relativity can calculate the rate of Gravity applied, even though <u>bent-Space alone</u> cannot explain the very source of Gravity which is why science has attempted to explain the very source of Gravity with a hypothetical, elementary particle called the Graviton. The dimensions of Space-time within our Natural Universe are actually **expanding** everywhere at the speed of light which can fully explain the delivery part of Gravity and why Galaxies are accelerating away without resorting to purely hypothetical Gravitons and purely hypothetical Dark energy. There are no force carrying particles such as Gravitons or energy such as Dark energy within Albert Einstein's brilliant General Relativity.

Observed how atoms reside at the center of expansion of dimensional Space-time and how light radiates away in all directions from its source, for example a light bulb. The photons leaving the light bulb simply travel away *(due to Quantum Gravity; see chapter 13)* at the same speed as the dimensions of Space-time from the atoms where the photons were produced, while the light bulb (atoms-of-Matter) remains at the center of their expansion of dimensional Space and time.

Special Relativity says speed causes time to slow and the dimension of length to contract. At the speed of light time completely stops and dimensions of Space contract to a single point. Ask how can the dimensions of Space and the passing of time be speeding away from these street lights so as to appear still or stopped to a photon travelling at the speed of light? <u>Within a Natural Universe</u> this is achievable because the tiny Gravitational waves providing the dimensions of Space and time are radiating away from the street lights at the same speed of light. There can be no other rational conclusion other than each street light is actually residing at the center of <u>expansion</u> of both <u>dimensional</u> Space and time. This is both observable and obvious, which is in itself evidence that all atoms reside at the center of expansion of the dimensions of Space and time. Tiny, expanding Gravitational waves produced by vacuum fluctuations at the hearts of the **nucleons** of atoms provide entangled dimensional moments to Space. Gravitational waves travel at the speed of light. The passing of a dimensional moment provided by the passing of a tiny Gravitational wave provides the passing of time. At light speed <u>one is travelling at the same speed as a dimensional moment,</u> consequently, the <u>passing</u> of the dimensions of Space and the <u>passing</u> of time stops.

This is how <u>both</u> time and dimensional Space stops at the speed of light. Consequently, it is the '<u>dimensions</u>' of Space, and <u>not the nothingness</u> of Space, which cannot be separated from the '<u>passing</u>' of time, hence, <u>dimensional</u> Space-time. Because a photon travels at the <u>same speed</u> as the tiny Gravitational waves which provide dimensions to Space a photon effectively travels within a moment of time from where there is <u>no actual passing of time</u>. Because a photon effectively remains within the same dimensional moment (a tiny Gravitational wave), a photon effectively travels no dimensional distance, even if it travels an astonishing distant <u>within one's Dimensional Reality</u> of dimensional Space and time.

One may ask what of the waves of electromagnetic radiation? Electromagnetic radiation is <u>the 'sum' of all emitted photons</u> from a light emitting object. Gamma rays are at the high frequency end of the electromagnetic, radiation spectrum, visible light sits about midway and low frequency radio waves are at the low frequency end of the spectrum. Electrons orbit the nucleus of atoms. A photon is produced when an electron in an orbit higher than regular orbit falls back to its usual orbit.

During the fall from high energy to regular energy, the electron emits a photon. The emitted photon has a frequency which exactly matches the distance which the electron falls and has the same energy as the difference between the high and low levels which the electron is moving between. This allows high energy and low energy photons to be created. High energy photons create high frequency waves of gamma rays and low energy photons create radio waves. One may think of a single photon as a very tiny part of an electromagnetic, radiation wave. A single photon does not make an electromagnetic, radiation wave. An electromagnetic, radiation wave is a combination of many photons in the same state.

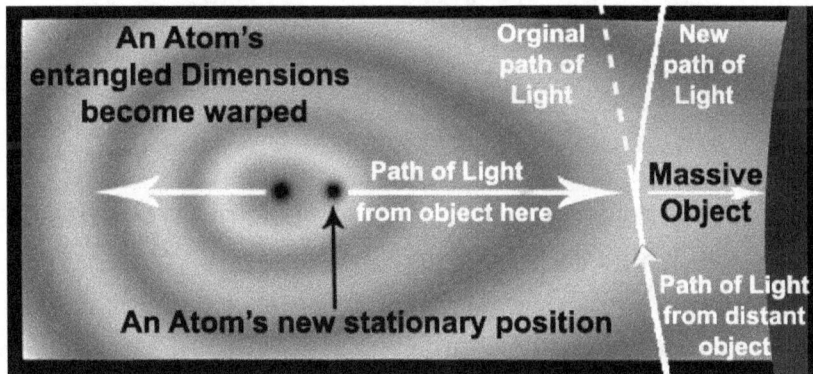

017 **Both light and the tiny Gravitational waves which provide the dimensions to Space are <u>massless</u> which is why both travel at the same speed for the same reason that they are both attracted to the <u>attraction part of Gravity</u>.** (See <u>how easy</u> the mystery of <u>Quantum Gravity</u> is finally solved on page 347). The attraction part of Gravity warps or bends light for the same reason it stretches, warps or bends the tiny Gravitational waves which provide the dimensions to Space. This is how Gravity bends light and bends the tiny Gravitational waves which provide the dimensions to Space. The path of light travelling pass a massive object is bent by a small amount before speeding out of Gravitational reach. **Actually, observing how a massive object bends light <u>provides evidence</u> of how tiny Gravitational waves are similarly stretched and warped while delivering Gravity.**

Within our Natural Universe <u>entangled</u> moments of dimensional Space and time expand away from every atom similar to the expanding circular ripples created from raindrops on a pond. Whether the water is at rest or flowing, expanding circular ripples created from raindrops on a pond act independently, freely expanding through (including the energy of) other expanding circular ripples, almost as if they were not even there.

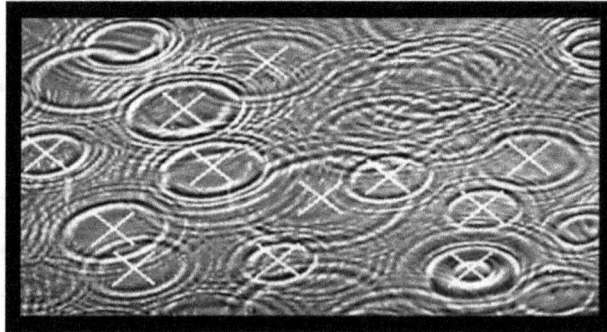

Entanglement is why an atom is not affected by time from other atoms and provides an atom with its own unique rate of passing time which originates from the protons and neutrons within itself. Naturally this means that your very own Dimensional Reality including the passing of time originates from within yourself while all other Dimensional Realities are <u>totally ignored</u>.

My model required a <u>delivery part</u> for Gravity which mirrors Einstein's General Relativity but, the big question has been, **without applying a force,** what can possibly do the heavy lifting of accelerating a massive object like Gravity can? Again, I have turned to Einstein and Relativity to provide the answer. There can be but one answer which is 'dimensions' and a phenomenon known as 'length contraction'. **Because an object remains at <u>stationary equilibrium</u> at the center of its entangled dimensions of Space-time, length contraction can cleverly reposition an object nearer to another object without actually physically moving it.**

When an atom's entangled dimensions become stretched there are less Planck lengths towards a massive object which causes an atom to 'move' to remain at a stationary equilibrium position within its expanding dimensions. A Natural Model delivers Gravity with length contraction. Length contraction is more defined as a prediction of <u>Special</u> Relativity and the warping of the curvature of Space is a prediction of <u>General</u> Relativity. With a Natural Model the attraction part of Gravity stretches and so warps the curvature of tiny Gravitational waves which provide dimensions to Space. Because a stretched Gravitational wave maintains the same '<u>single pulse</u>' of energy as a non-stretched wave a tiny wave maintains a <u>single unit</u> of Planck length. This causes the stretched distance covered by the stretched wave <u>to measured a Planck length</u> so as the stretched distance contracts to a Planck length which causes Gravitational length contraction. [017a] Although the energy that Gravitational waves carry is said to be fixed, as they travel throughout Space, the waves are also said to become smaller but continue to retain a <u>vital, passing single pulse of energy</u>. **The emphasize being a passing <u>single pulse</u> of energy rather than the amount of energy.**

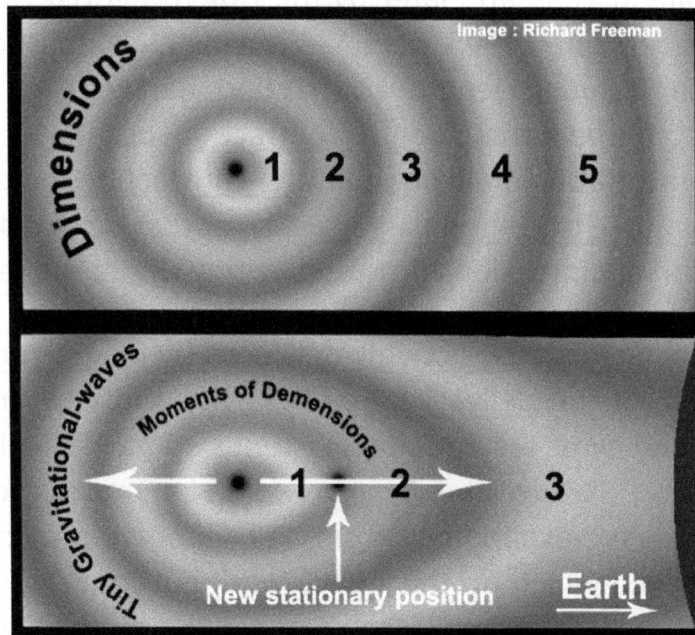
Image : Richard Freeman

To demonstrate <u>Gravitational</u> length contraction, one only needs to let go of any heavy object and watch it fall to the ground.

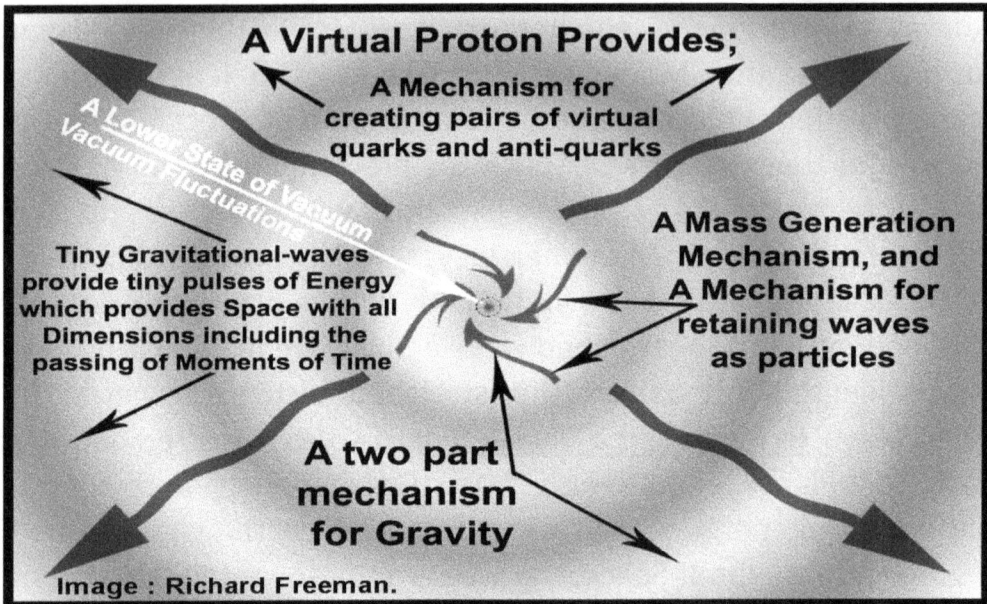

A Virtual Proton Provides;

A Mechanism for creating pairs of virtual quarks and anti-quarks

A Lower State of Vacuum Vacuum Fluctuations

Tiny Gravitational-waves provide tiny pulses of Energy which provides Space with all Dimensions including the passing of Moments of Time

A Mass Generation Mechanism, and A Mechanism for retaining waves as particles

A two part mechanism for Gravity

Image : Richard Freeman.

Length contraction was first proposed by the Irish theoretical Physicist George FitzGerald in 1889 and later by the Dutch theorist Hendrik Lorentz in 1892. It was not until 1905 that Albert Einstein published his Special Relativity which included length contraction as well as time dilation. Einstein has established that length contraction occurs along the radial direction of a Gravitational field and that the contraction increases when an object is nearer to Gravitational mass. My operating mechanism is the active 'expansion' of dimensions provided by tiny Gravitational waves.

All is why Gravity needs to be **actively** delivered by length contraction. This is the mechanism of the delivery part of Gravity and why an object 'falls' towards Earth and why we require Einstein's remarkable Relativity and length contraction to do the heavy work of moving objects. Tiny, Gravitational waves expanding from all atoms provide entangled dimensions to Space which do this heavy work but in a most passive way. The tiny Gravitational waves may stretch or expand but because they retain units of Planck values which in this case is a vital single pulse of energy the distance is consistent of a single Planck length per Gravitational wave.

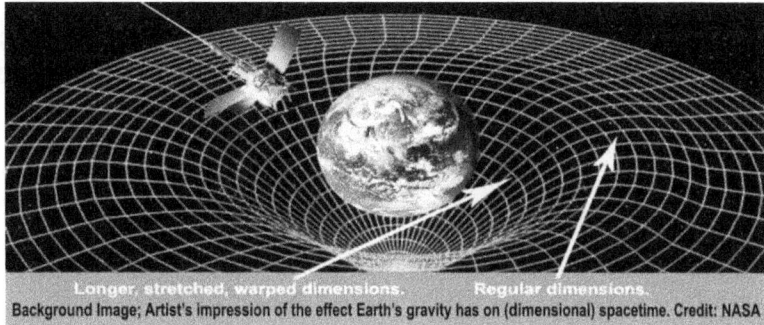

Longer, stretched, warped dimensions.   Regular dimensions.
Background Image; Artist's impression of the effect Earth's gravity has on (dimensional) spacetime. Credit: NASA

Perhaps not intended to show length contraction, in this illustration from NASA we can see how the curvature of dimensions become stretched and warped towards Earth. For the reason that the dimensions of each circular grid of Space are stretched there are fewer dimensions of length or Planck lengths nearer Earth. Images like this are poor at showing how Gravity **actively gives objects movement to fall to any place on Earth.** Although the fabric of Space and time is fundamentally vital for the workings of our universe <u>nobody today</u> knows what Space-time is. The Natural Model tells how Space-time is **actively** made in a way it provides your Space with <u>your own unique **entangled** dimensions.</u> Entanglement allows your time dilation and length contraction to be unique to your own speed and Gravitational position while all other Dimensional Realities are **totally ignored**.

Travelling at the speed of light and like a photon of light the tiny Gravitational waves themselves experience no passing of time or passing distance; in their own time they instantly travel across the universe and travel <u>no dimensional distance</u> in doing so. **This is how they remain attracted to a very distant complete void of Anti-Space which may be many billions of light years away from the perspective of one's own Dimensional Reality. To the tiny Gravitational waves it is as if the void is always right next to them. This is another important property of tiny Gravitational waves which is required for the delivery of Gravity.**

I have called this model for Gravity **'Stationary Momentum',** which seems to contradict itself, however it describes quite well, momentum gained by the mechanism of Gravity. A falling atom gains momentum while it remains stationary trapped and restrained at the center of its entangled, expanding dimensions of Space and time. The following is not a perfect analogy but shows rather well how an object is moved by the delivery part of Gravity while always remaining at stationary equilibrium within its expanding dimensions of Space-time. One can now easily demonstrate my Stationary Momentum model with a knot in a strip of rubber band.

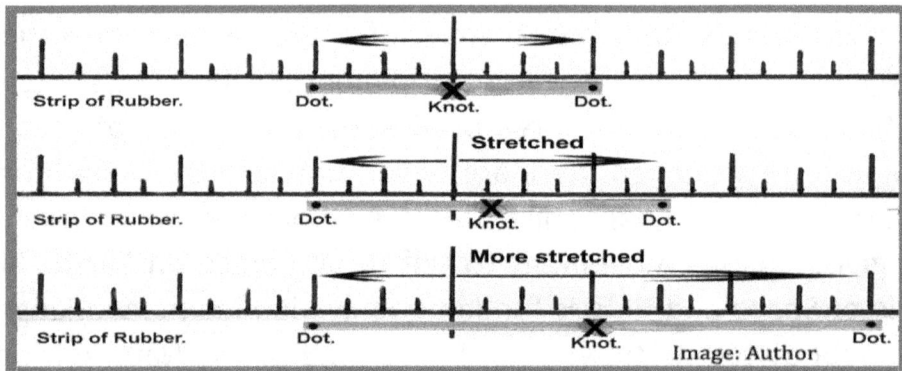

Image: Author

Cut a rubber band and place a reference dot on each end of the strip and tie a knot at the center. First, slowly stretch the rubber strip evenly in both directions. The knot does not move. Now stretch the rubber strip again but a little more in one direction. The knot now moves in the direction of the most stretching which will now correspond to the direction of most mass.

Like an atom the knot stays at the center between the two dots and so does not move from the center of expansion, nor does the knot 'see' the rubber stretching more in one direction. Like the knot in the rubber band an atom, because a stretched dimension retains its Planck value, does not even 'see' its dimensions of Space as being stretched, warped or bent more in one direction. The knot has not travelled 'through' the rubber but has moved in relation to the ruler. The knot has remained trapped stationary within the expanding rubber similar to how an atom remains trapped at 'stationary equilibrium' within its expanding, dimensions of Space-time. *(Now release the least stretched end. Now the knot seemingly instantly* *speeds towards the most stretched end, welcome to how Quantum Gravity works, page 347. Reason; residing without <u>entangled</u> dimensions, tiny Quantum particles <u>ignore</u> the <u>restraining</u> dimensions of Space-time.)*

Within a Natural Universe <u>every pulse</u> of virtual mass and <u>every wave</u> of expanding dimensions speed up and so accelerate an object (atoms) towards Earth. **The obtained 'speed' times an object's 'mass' provides a falling object with momentum while it remains at <u>stationary equilibrium trapped</u> at the center of <u>numerous very tiny units of expanding energy</u>.**

Thus, Gravity causes an object to fall and accumulate momentum and kinetic energy until it decelerates as it impacts with the ground. Gravity surely does not stop or change the way it operates once you are on the ground. While standing on the ground Gravity continues to be applied in the same tiny increments and provide exactly the same momentum. While standing on the ground one is locked into a continuous pulsating loop of acceleration, deceleration and speed providing momentum and relocating in time. Because this all literally happens at the speed of light and in tiny pulses of Planck time this is like driving a car at a constant speed with your foot on both the accelerator pedal and the brake pedal at the very same time and discovering your brakes are on fire. Gravity is continuously providing acceleration and providing speed (within Space-time), which continuously replaces the momentum that has been continuously expelled as heat from kinetic energy, as you continuously impact and decelerate into the ground. You feel this continuously impacting and decelerating into the ground as 1 'g' of Gravity or as acceleration from your stationary position.

Resulting impact heating from Gravity being applied in pulses may possibly play a part in accounting for a portion, of the Earth's inner heat. Particles free falling towards Earth simply accumulate this same kinetic energy, by gaining momentum, until they impact with the Earth. [063] Research from Japan at the KamLAND collaboration, shows radioactive decay of elements like uranium, potassium, and thorium only accounts for about half of our planet's inner heat. The Scientists say we are for now left to wonder where all that other energy is coming from. Similarly, NASA's New Horizons' probe recently accomplished a fly-by of Pluto. Thought to be frozen and inactive the team has been stunned to discover that Pluto and Charon have geological activity taking place. The continually expelled, as heat, kinetic energy from all of a planet's particles continuously impacting, with every pulse of Gravity, may play a part in creating a planet's inner heat.

Merging Inner Gravity with Outer Gravity.

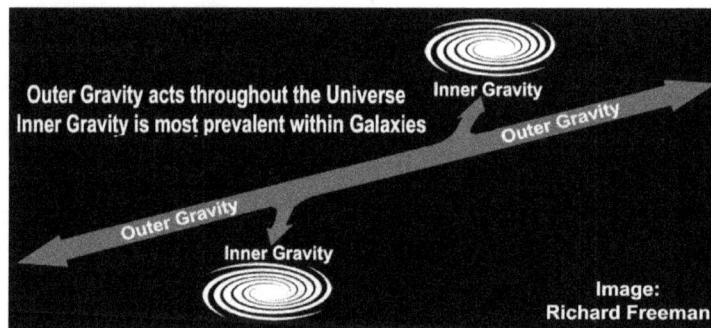

Image:
Richard Freeman

The delivery part of all Gravity is primarily driven by the attraction part of Outer Gravity. Gravity is only actually delivered when the curvature of tiny Gravitational waves, which provide entangled dimensions to Space, become <u>warped</u> from being stretched towards the attraction part of both Inner Gravity and Outer Gravity. Outer Gravity is always acting throughout the universe and Inner Gravity is always most prevalent within Galaxies. Both Outer Gravity and Inner Gravity operate in unison as one of the same. The source of the attraction part of Gravity is <u>negative energy</u> sourced from pure empty Space. For <u>Inner Gravity</u> this is a tiny phase of virtual mass which is a phase of deep vacuum **empty Space** produced by the COST's Quantum vacuum fluctuations. While the source of the attraction part of <u>Outer Gravity</u> is a seemingly, never ending far <u>outer void</u> of a similar lower state of **empty Space** which the dimensions of all Space-time are attracted to and so are expanding into. This understanding of the source of the <u>attraction part</u> of Gravity allows Einstein's General Relativity to be used to sensibly explain the observed acceleration of all Galaxies.

Because of its seemingly infinite size, the very distant outer void has the ability to provide an enormous source of negative energy to drive the <u>all around expansion</u> of dimensional moments from all atoms and deliver an amount of <u>warping of the curvature</u> of the tiny Gravitational waves which delivers Outer Gravity. It obviously requires a very <u>massive resource</u> to accelerate all Galaxies; a seemingly infinite size void of **completely empty Space** may easily fulfill this requirement by providing an enormous reserve of negative energy. The observation that Galaxies are accelerating away provides overwhelming evidence of Outer Gravity. **Outer Gravity is indeed overwhelming evidence that a lower state of the vacuum of empty Space is the source of all Gravity and mass.** Because a Natural Universe is expanding into a void of deeper vacuum empty Space it <u>cannot avoid</u> having Outer Gravity. Thus, Galaxies were always going to be accelerating due to <u>Gravitational Attraction</u> and a Natural Universe is a universe which has no use for the impossible to explain hypothetical Dark energy.

The clearest evidence that the universe really has a center and an outer edge is the observation that Galaxies are accelerating apart. Outer Gravity can only be explained with the understanding that our universe truly <u>has a center and an outer edge.</u>

Scientists will tell you their Big Bang universe today and even when it was the size of a pea has always had no center or outer edge which is irrational or at least sounds irrational. The Natural Model provides tiny Gravitational waves eternally expanding from all of the protons and neutrons of our atoms which provide entangled dimensions to Space. Consequently, despite <u>the universe having a center and an outer edge, we all correctly observe ourselves</u>, no matter where we are, to be located at the center of our own universe of eternally expanding <u>dimensional Space-time</u>. This has allowed Scientists to deduce that mathematically the universe has no true center or outer edge, rather, wherever you are; you are stationary at the center of an eternally expanding universe of dimensional Space and time which has no outer edge. However, when Scientists talk of the expanding universe they are generally referring to Galaxies moving apart and are rarely referring to the expansion of the dimensions of Space and time.

An atom's expanding, entangled dimensional moments of Space and time become stretched and slightly warped in the direction nearest to a far outer void of vacuum nothingness now called Anti-space. Unless an atom is at the very center of the actual, whole, combined universe, all dimensions expanding from all atoms are slightly more stretched and so warped in the direction of universal expansion which delivers Outer Gravity to all objects.

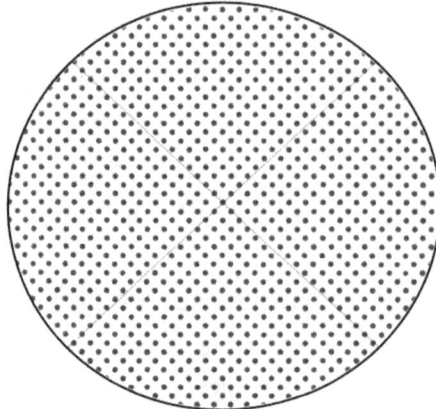

Notice how, except at the very center, every dot is positioned where the outer edge of the image is nearer in one direction. Now imagine every dot is a Galaxy. This warps by stretching the expanding dimensions of Space of every atom and every COST unit of Dark matter in the direction nearest to the outer edge which provides Outer Gravity. If all dots were simply moved outwards at the same speed one would develop an expanding 'hole' at the center which at this point of time is wrong. If we were to move a dot from near the center outwards it becomes nearer to the dot ahead, now the dot ahead needs to move outwards at a greater speed to provide true expansion, all of which continually snowballs as we move closer to the outer edge. With Gravity, the dots which are closer to the 'moving away'

outer edge will naturally move at a <u>greater speed</u> than the dots near the center. This happens for the reason that, being <u>nearer</u> to the outer void is like being <u>nearer</u> to a massive object. This is naturally <u>different</u> to the current concept of Galaxies moving apart from each other because the Space itself is expanding between Galaxies and in a similar way to how raisins are moved apart by the expanding dough while baking a raisin bun. Obviously, with the raisin bun idea all Galaxies will move apart by a more consistent rate which is not what is observed. **Within a Natural Universe the tiny, expanding Gravitational waves, which provide the dimensions of Space, travel <u>through Galaxies and their Matter</u> as if they were not there and the relatively void like regions between Galaxies simply become larger due to Outer Gravity moving and accelerating Galaxies apart.**

<u>Note:</u> **The outer void <u>contains nothing</u> which may crush our universe. Einstein's Gravity works with the (entangled) dimensions of Space-time originating from <u>within our universe</u>, thus the void can only apply Gravity acting outwards and <u>cannot apply</u> any form of crushing inward Gravity.**

Outer Gravity provides the overall expansion throughout our home universe and operates because all Gravity is naturally strongest near its source. When using Gravity to expand a universe of Galaxies the mass of other Galaxies and our own position within the universe will naturally affect the observed <u>speed</u> of Galaxies. Our Milky Way Galaxy is most likely, in <u>one direction,</u> nearer to the outer edge of the universe and there is most likely, in <u>all directions</u>, an inconsistent amount of mass from Galaxies existing between our Milky Way Galaxy and the outer edge. Galaxies formed wherever Dark matter was most dense so <u>are not bound</u> to a more uniform Big Bang beginning. In regions with few Galaxies is where Galaxies are less bound to the Inner Gravity of other Galaxies which gives Outer Gravity the authority to further <u>increase the speed</u> of Galaxies, that is, the varied distance between Galaxies will also naturally affect their speed and is why all Galaxies generally accelerated as they moved apart from each other about five to six billion years ago. Faster speeds in the direction of the shortest distance to the outer edge of the universe can also be expected.

Consequently, primarily driven by Outer Gravity all Galaxies will speed away at different rates depending on where we look. **This is not what one would expect from Dark energy expanding the Space between Galaxies but is understandable when Gravity is responsible for <u>the speed</u> at which Galaxies are accelerating and expanding the universe.** Called the Hubble Tension, the different <u>speeds</u> of the expansion of the universe was noted in 2019 with measurements made by the Hubble Space Telescope and have now been <u>confirmed</u> by the James Webb Space telescope. A Natural Model uses Outer Gravity which can sensibly explain why Scientists now say *"the universe appears to be expanding at bafflingly <u>different speeds</u> depending on where we look"*. Consequently, instead of the <u>Space itself</u> expanding between Galaxies due to hypothesized Dark energy, Galaxies are simply free falling and <u>accelerating,</u> in the direction of expansion, due to Gravity which is obviously exactly the way Gravity works.

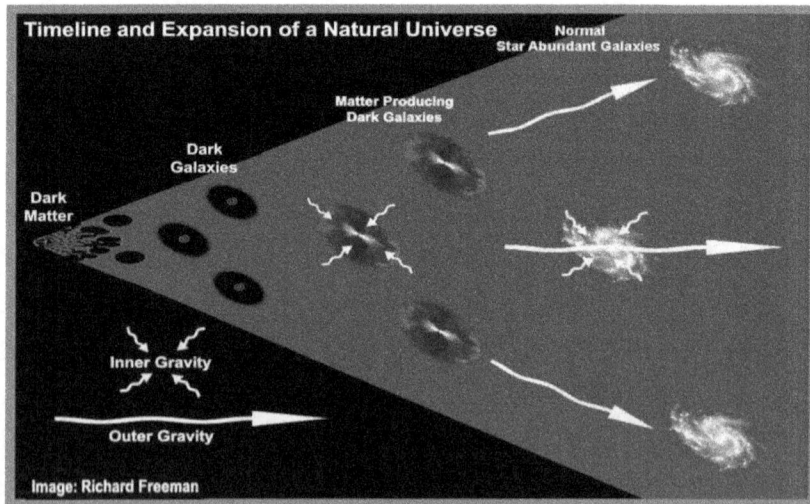

With this modeling, acceleration is at first restricted in the early universe due to the Inner Gravity from the close proximity of other Galaxies. The said reacceleration phase; Galaxies once breaking free of being in close proximity of other Galaxies, were then more freely accelerated by Outer Gravity. This modeling has no use for the bothersome and purely hypothetical Dark energy, Galaxies are simply falling outwards. Not all Galaxies broke free of the Inner Gravity from the close proximity of other Galaxies and associated areas of Dark matter. These Galaxies formed as clusters of Galaxies which now speed apart from other clusters of Galaxies.

All Dark matter and all atoms within a Galaxy are exposed to an amount of Outer Gravity. Outer Gravity, being the exact same phenomenon as Inner Gravity, naturally acts in the same way, causing Galaxies to speed apart at an accelerating rate as they 'free fall' in the direction which the dimensions of Space are expanding most stretched. An object 'falling' to Earth is exposed to expanding dimensions of Space-time, which are stretched towards Earth, plus some added 'stretching' in the direction of the expanding universe. Only the concepts of my Natural Universe can fully and sensibly explain why Galaxies are naturally speeding away at an accelerating rate which obviously matches the characteristics of Gravity and can do so without the use of a totally mysterious, mind-boggling powerful, enormous amount of Dark energy. All that is required is a seemingly endless outer void of nothingness which I have called Anti-Space.

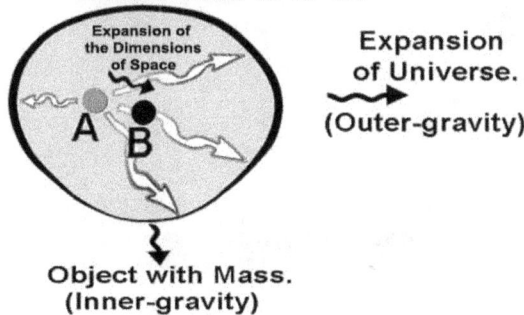

Merging Inner Gravity with Outer Gravity
To remain at its natural position at stationary equilibrium within it's expanding Dimensions an Atom 'moves' from 'A' to 'B'.

Expansion of the Dimensions of Space

Expansion of Universe.
(Outer-gravity)

A B

Object with Mass.
(Inner-gravity)

The far outer void of <u>empty Space</u> provides a <u>massive resource</u> of negative energy allowing the positive dimensions of Space-time to be naturally attracted to the void. It is easy to understand that the Natural Model actually requires Galaxies to be accelerating apart unless they are close to each other which is observed to be true. A Big Bang model has no reason for this unforeseen acceleration of Galaxies. In an honest attempt to explain the unexpected results of observations that Galaxies were accelerating and not slowing as theorized, Scientists have attempted to add to the universe a colossal amount of a most baffling form of undetectable energy and call it Dark energy. However, the Scientists have no known source for their Dark energy. Whereas Outer Gravity, being the exact same phenomenon as Inner Gravity, naturally acts in a similar way, causing Galaxies or clusters of Galaxies to accelerate apart as they 'fall' in the direction which their dimensions of Space-time are stretched and warped. Thus, today's accelerating Galaxies are simply free falling. Inner Gravity and Outer Gravity are seamlessly combined and delivered as one of the same.

Black Holes and Super Massive Black Holes:

If you were free falling feet first into a stellar Black Hole the attraction part of Gravity becomes very strong and so quickly that your expanding dimensions become much stretched in the direction you are falling and squeezed around your body.

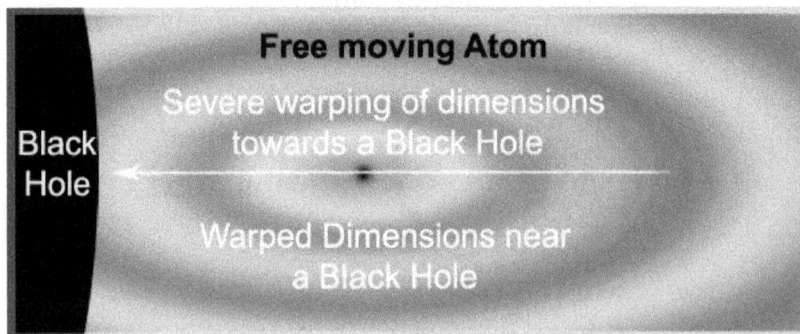

This now stretches you head to feet and squashes you from the side in a way which Scientists call spaghettification. Because of its large size a Super

Massive Black Hole is said to be less severe.

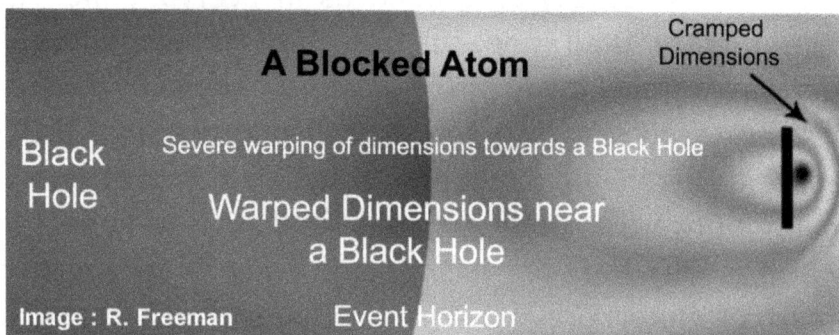

Should you remain stationary as if you were blocked from falling through the event horizon of a stellar Black Hole the severe warping of your expanding dimensions now causes Gravity to squash you way more flat than a pancake. Being stretched into spaghetti or squashed into a pancake are good reasons to stay away from stellar Black Holes. Since an atom's passing dimensions retain the same Planck value one's own passing of time remains self-normal. Within a Black Hole the tiny Gravitational waves which provide dimensions are now unable to escape from COST units. This is because the Gravity is so great that to do so the tiny Gravitational waves would need to exceed the speed of light which is prohibited by Relativity.

Inside a Black Hole a COST unit's fluctuations are unable to express their tiny Gravitational waves and can now only express the attraction part of Gravity. Because there is now no passing of entangled dimensions there is now no passing of time and because there are no expanding Gravitational waves to provide Space with entangled dimensions all Space is now effectively dimensionless. Where Space is now effectively dimensionless is where everyplace is effectively in the same nowhere place. This is where we have a mathematical singularity. The singularity can only be dimensionally measured from the outside. Since the inside is without entangled dimensions the inside measures dimensionally zero.

Because photons and Gravitational waves normally speed away in all directions one may suggest that a form of super, strong Gravity found within Black Holes is actually all around us in everyday life. At first this sounds a little crazy until one realizes, *'this is why both light and Gravitational waves travel away at the incredible speed of light, which is of course the same speed at which light travels one way into a Black Hole'.* Light speeding away in all directions provides compelling evidence of Gravity acting outwards. Both photons and Gravitational waves are massless which is why they are easily sped away in all directions by the <u>attraction part of Outer Gravity</u>. This is also evidence that dimensional Space-time is expanding into a never-ending void and is why we have Outer Gravity responsible for the acceleration of Galaxies. Even though the tiny Gravitational waves which provide dimensions to Space carry <u>energy</u> from the event which created them; the tiny Gravitational waves have no mechanism for a phase of virtual mass so they are massless which is why they can travel at the same speed as light. See page 121, 122 and **347**.

[064] To measure the speed of Gravity Scientists made an observation when the planet Jupiter passed nearly in front of a bright Quasar on September 8, 2002. The light of this Quasar allowed Scientists to measure the bending of the light caused by Jupiter's Gravity. Scientists concluded that <u>Einstein was right</u> and Gravity propagates at the speed of light. It is now known that the speed of Gravity is equal to the speed of light by an incredibly precise amount. The speed of both Gravity and light has to be exactly the same to conform to Einstein's General Relativity. To propagate at <u>the speed of light,</u> like light, Gravity should be delivered by the tiniest increments. Because Gravitational waves propagate at <u>the speed of light</u> this provides further <u>compelling</u> evidence that the delivery of Gravity is provided by very tiny, expanding Gravitational waves caused from a fluctuating phase of virtual mass at the hearts of protons and neutrons. The virtual mass phase acts as a very minuscule Black Hole which is in a fluctuating state of being renewed and collapsing which is ideal for creating very tiny, entangled, Gravitational waves which provide entangled dimensions to Space.

Nothing can escape a Black Hole because to do so it would need to <u>exceed the speed of light</u> which is prohibited by the laws of Relativity. This has prompted many to ask; how can Gravity, which propagates at the speed of light, get out of a Black Hole? <u>Short answer: it can't</u>! However, <u>everything can get into a Black Hole</u>. Consequently, like light, Gravitational waves cannot escape from <u>inside</u> of a Black Hole which means Black Holes <u>cannot</u> self-express the delivery part of Gravity. Black Holes can <u>only express</u>, in the most powerful way, <u>the attraction part</u> of Gravity which attracts the tiny Gravitational waves from all objects outside of a Black Hole. This enables Black Holes to powerfully stretch and warp the dimensions of Space, provided by tiny Gravitational waves, of all Space-time in their vicinity which for all objects outside of a Black Hole allows Gravity to work normally and very much the same as described by Einstein's General Relativity.

Inside of Black Holes all atoms are shredded. Protons, neutrons, quarks and even electrons cannot survive the massive Gravity. Black Holes are heavy with mass and Gravity but can also be thought of as being <u>very</u> <u>lightweight</u> since all 'heavy' particles' have clearly been crushed out of existence. Not even quarks survive to excite and agitate at near light speed to create regular mass as we know it but by some <u>unknown</u> and unfathomable means this <u>very same mass</u> and its Gravity is <u>said</u> to remain behind. It makes absolutely no sense to say what was clearly providing this mass and Gravity now <u>does not exist</u> but the Black hole continues to provide mass and Gravity. Is it little wonder Scientists say they do not understand what goes on inside of Black Holes. A Black Hole has no regular particles to provide it with the heaviness of the regular mass-energy which we daily associate all objects having. When there are no particles left what remains to provide mass and Gravity? <u>Only a Natural Model</u> can provide a clearly correlated and viable answer. The mass which remains behind is virtual mass and is provided by <u>exactly the same</u> tiny phase of virtual mass which creates the regular mass <u>within us all</u> as the mass-energy of our protons and neutrons within our atoms. Being a tiny phase of <u>complete</u>

nothingness, virtual mass cannot be crushed out of existence.

A Black Hole by warping the dimensions of Space from all objects <u>outside</u> of a Black Hole dictates the position of objects outside of a Black Hole which, in turn, positions the Black Hole relative to objects outside of a Black Hole. However, without a delivery part of Gravity, a Black Hole itself cannot directly 'feel' other Black Holes. In spite of that, a Black Hole often acquires an <u>enormous accretion disk</u> the volume of which dominates a Black Hole, effectively <u>providing</u> a Black Hole with a delivery part of Gravity which <u>feels</u> another Black Hole's Gravity. Thus, Matter outside of Black Holes may steer two Black Holes together and cause two Black Holes to orbit each other by each being trapped within the very warped dimensions from all Matter outside of a Black Hole. As Black Holes <u>are not</u> readily attracted to each other by their own means this may solve why stellar Black Holes, for some unexplained reason, rarely combine to form intermediate Black Holes. Because all Black Holes are made from the same negative energy they are also unlikely to be attracted to each other by the attraction part of Gravity.

Because intermediate Black Holes are mysteriously, extraordinary uncommon is reason why Stellar Black Holes <u>do not commonly combine</u> to form intermediate Black Holes and intermediate Black Holes did not combine to form the first Super Massive Black Holes. In general, stars of mass more than 25 times of the Sun will suffer an explosive end when their <u>cores</u> are believed to ultimately collapse to a Black Hole which is why there should be <u>millions</u> of Stellar Black Holes within our Milky Way Galaxy. It is thought Stellar Black Holes should readily combine to form intermediate Black Holes; thus, intermediate Black Holes should be common. However, intensive searches for intermediate Black Holes have implied they are, curiously, <u>extraordinary uncommon</u>; only a Natural Model can explain why.

Quantum Gravity, for tiny Quantum particles, will be addressed at Chapter 13. <u>For given reasons</u>, Quantum Gravity uses only the strong attraction part of Gravity and completely ignores the dimensions of Space-time.

12 PASSING TIME - THE GREAT REGULATOR.

[068] *"Time is in our very souls, isn't it"? (Professor Paul Davies).*

Gravity makes Gravitational waves; with a Natural Model the source of Gravity at the hearts of protons and neutrons create extraordinary tiny Gravitational waves which actively provide entangled dimensions to Space. In 1916 Albert Einstein predicted Gravitational waves in his theory of General Relativity which is all about Gravity.

Why the Natural Model uses the passing of extraordinary tiny Gravitational waves to actively provide the passing of moments of time:

[017] Gravity warps and so distorts the curvature of the dimensions of Space-time and Gravitational waves also distort the dimensions of Space-time. Gravity propagates at the speed of light and Gravitational waves also propagate at the speed of light. Gravity propagates forever and Gravitational waves also propagate forever. Gravity propagates through everything as if it was not there and Gravitational waves also propagate through everything as if it was not there. There can be no doubt Gravitational waves have the very unique properties required to actually provide dimensions to Space and provide the delivery part of Gravity. But to provide dimensions to Space they would need to be extraordinary tiny and to deliver Gravity they would need to originate from within Matter both observable and dark. Because dimensional Space cannot be separated from the passing of time the extraordinary tiny Gravitational waves have to

provide all dimensions to Space including the passing of time.

To provide dimensions to Space my Natural Model creates extraordinary tiny Gravitational waves at the very hearts of all protons, neutrons and Dark matter. The distance across a wave provides Planck length and the passing of a wave provides Planck time. Planck length and Planck time are the smallest possible units of length and passing time within a Dimensional Reality. The tiny Gravitational waves travel at the speed of light which allows your passing of time to 'tick' pass at the amazing speed of light.

The passing of time allows us all to live out our daily lives and be in charge of every move we make; however, today the source of time itself remains a total mystery. How can it be that something which we rely on to exactly measure the duration of any event, down to the finest of small margins, is also astonishingly, totally variable to the point that it can completely stop? [065] For instance, the satellites for GPS navigation would be completely useless if the rate of time difference between the accurate rate of time of the GPS satellite clocks and the accurate rate of time at the user location was not taken into account. Without adjusting the satellite clocks for this rate of time difference it is said your GPS would lose accuracy within about two minutes and errors will accumulate by more than 10 km per day.

How can this mysterious mechanism be so uniquely affected by both speed and Gravity and seemingly remain totally unaffected to a self-observer? How can two clocks of two different observers run at different rates when both clocks are actually working flawlessly? How does the rate of passing time always vary by the exact amount required to allow the speed of light to always be self-observed at a constant speed? Why is the actual physical process which allows this to take place a complete mystery? My Natural model being a physical reality model is designed to unravel everything.

There is no universal clock which began ticking at the Big Bang. Scientists know that the passing of time is unique to the relative Gravitational position and relative speed of each individual atom in the universe but what kind of mechanism could be responsible for uniquely adjusting one's very own rate of time to one's very own speed and Gravitational position? When your very own dimensions of Space-time begin within yourself this becomes physically possible and rationally explainable.

A beginning by way of a Big Bang has failed to provide a clear understandable physical mechanism for the source of the passing of time and has not provided a physical mechanism for time dilation and has not provided a physical mechanism for length contraction. Consequently, beginning with a Big Bang means nobody can 'rationally' explain these things for there is no known physical way to rationally provide these vital mechanisms by way of a Big Bang other than with mathematical equations. Physics and mathematics are indeed very powerful and useful tools.

Within our Natural Universe time has two facets, an instant and a passing moment. A passing moment will not fit into an instant. Passing moments of time are provided by COST units to protons and neutrons and so atoms as a continuous rhythm of passing moments of time. We are made from atoms so we all reside within this traditional, observable realm of passing moments of time. Albert Einstein's Special Relativity and General Relativity excel within the dimensional, realism realm of passing moments of time. Observation or measurement requires the passing of moments of time.

With a Natural Modeling, an instant relates to photons travelling at light speed and all other unobserved Quantum particles regardless of speed and is where time between events is instantaneous regardless of the continuous rhythm of passing moments of time and dimensional Space provided to atoms. Observation within an instant is not allowed for the simple reason an instant is so short it will not fit into a passing moment and a passing moment is too long to fit into an instant. There is a far reaching hidden realm of wonders within a dimensionless realm, where time is reduced to an instant, allowing a particle to reside without the constraints of dimensions or passing moments of time. An instant may be of any length, the shortest of the short or stretch across the length of the entire universe. Hidden within this unobservable realm are particles dressed as waves, silent energies beyond imagination where separation by passing time and dimensional Space is irrelevant. Quantum Theory revels within the hidden, dimensionless realm of an instant. Many of the particles and energies of the Quantum world regularly cross over from an instant to our dimensional, realism world of passing moments of time.

Although Scientists today will say they do not know what time actually is, Special Relativity does in fact offer a marvelous insight into the way time operates, which can help steer us to the source of time. **Albert Einstein's theory of Relativity tells us that time is <u>woven together</u> with the three dimensions of Space forming a four dimensional Space-time continuum.**

Special Relativity tells us that time cannot be separated from (dimensional) Space. This understanding that Space and time cannot be separated and that the expansion of dimensional Space is permanently coupled to the advancement of time, <u>provides the first vital key</u> in allowing an understanding of exactly how we experience the passing of time. Somehow, the dimensions of length, width, depth and time **are all** physically **woven together as one of the same. It is enshrined in theories, I read it over and over again; (dimensional) Space and time are as one.** <u>Because they cannot be separated</u>, when one experiences a moment of passing time one is actually experiencing the 'event' of the passing of a 'moment' of expansion of <u>the dimensions</u> of Space. It is now easy to reason that the dimensions of Space and the passing of time are physically delivered by one of the same and which is always advancing.

Time dilation relates to Gravity and the speed of an object. Importantly, the ramifications of Special Relativity allow the understanding that the passing of time is balanced by design. Balanced means that all the laws of physics, the speed of light and time are required to be the same (balanced), from the current viewpoint of all observers, no matter which 'direction' one is travelling or observing. Balancing the passing of time is easily achievable when Space and time is advancing from you at the same 'balanced' rate in all directions, or in other words, if a person is permanently located at the center of expansion of dimensional Space and time. The point here is being located at the center of expanding dimensions. As time is considered to be the fourth dimension and we are located at the center of the expanding dimensions of Space, the source of the passing of time should come from the center of our own location, that is the actual source of the passing of time should come from within ourselves, and like Professor Paul Davies says *"from our very souls"* but how could this be so?

However, Einstein's Special Relativity does not provide all of the clues, for we will also be required to consider the full ramifications of the 'true results' of Quantum entanglement experiments. These results relate to a prediction from the Quantum world which Einstein had objections to. By combining the predictions of Special Relativity with the results of Quantum entanglement predictions and experiments we will hopefully expose, with little doubt, the hidden true workings of time itself.

The masterful way my Natural Universe operates is the rational way it allows the 'speed' of expansion of inseparable dimensional Space and time to change or to be at any random rate without any self-observed changes to the speed of light or to the strength of Gravity, changes which may have unfathomable consequences for chemistry and fundamental forces.

[051; 052] Reasoning and observational evidence relating to the speed of light imply the dimensions of Space expanded quicker in the beginning of the universe delivering a faster speed of light, faster Gravity and quicker time. Because time ran faster, the speed of light and Gravity appeared, at that time, the same as today. Time running faster within our early universe can explain why science is discovering that objects within this early period developed at a faster rate than their theories allow. A ramification of time running faster in the beginning of our universe is that the universe is now much older than the said 13.8 billion years. This is because the said 13.8 billion years is measured in the now much slower Earth time without taking into account 'expansion speed' time dilation. One year of today's Earth time may possibly equal many years within the early universe. **Naturally, this would mean that the universe is very much older than it is said to be.**

Thus the 'speed' of light and the 'speed' and 'strength' of Gravity is not a random quantity which happened against insurmountable odds to allow our universe to exist the way it does. All is superbly fine tuned and strictly regulated by time, which is in turn regulated by the 'expansion' rhythm of the dimensions of Space-time, all of which silently and efficiently deals with the many random rhythms which may drastically upset the workings of our universe. It is of no problem if one is sitting stationary on Planet Earth as it rotates in Space and orbits the Sun at 108,000 kilometers per hour. Nor does it matter that our solar system orbits the center of our mighty Milky Way Galaxy at 720,000 kilometers per hour, while our whole Galaxy itself is speeding through dimensional Space and time at an astonishing 2.5 million kilometers per hour. For regardless of all of this movement, the wonderfully simple configuration of the COST provides a haven for all Matter (atoms) to always remain 'stationary' at the center of expansion of the dimensions of Space and time from where time superbly regulates so as all can commonly appear self-normal and time appears unchanging for all.

A Moment of Time: When we measure time in hours, minutes and seconds what is it that we are really attempting to measure? [066] If one asked a Physicist what Space or time actually is, the answer may be a shock because they really have no idea. Amazingly, it seems we do know how to measure time even though we do not know exactly what it is that we are measuring.

The all important profession of Timekeeper is allocated to the shortest or quickest events. The word 'moment' effectively refers to the shortest possible period of passing time which is Planck time. Because Space and time cannot be separated it is sound logic that one moment of passing time equals the 'event' of one moment of expansion of the dimensions of Space, which equals one unit of Planck time which is said to be the shortest possible period of passing time. Meaning the shortest and conveniently 'regular' event which we daily experience is almost certainly the passing of an 'outgoing' expanding 'moment' of expansion of dimensional Space-time which has crowned itself the universe's Timekeeper.

The Future, the Present and the Past:

The future is spontaneously arriving and becoming the present and the present is spontaneously departing and entering into the past. The present only exists by the smallest of margins or for just one fine moment.

I like to think of the passing of moments of Space and time as like the passing of frames of a movie film. A movie may have 150,000 frames which flash by your eyes at a regulated 24 frames per second. The present is that one frame which you are viewing at a present moment. The future are all of the frames or moments rolled up on the 'future reel' and the past are all the frames or moments rolled up on the 'past reel'. The movie may be speeded up or slowed down but we will still experience the full time of the 150,000 frames. Time is like the regulated movement of the frames and we all simply reside within a passing reference frame which we expressed as time. Within my Natural Universe the first truly regulated passing of time began with the creation of the first vacuum fluctuations of COST units.

Thanks to the brilliance of Albert Einstein, we know that in spite of one's relative movement or the movement of the source of light, that the speed of light within the vacuum of Space is identical for all observers. The ramification of the speed of light being a constant is the passing time has to be variable. All observers will experience the time of say, one hour as exactly one true and complete hour but one true and complete hour is again like our movie film; it can be speeded up or slowed down giving different observers shorter or longer hours, even though each hour for all observers remains one whole and complete, exactly the same one hour. **How can one's own personal time, the time one feels from within, be so finely locked to one's own personal speed and Gravitational position? How can this be locked so as one personally always observes the speed of light at the same constant speed?**

To establish units of time we have long used the rotation of the Earth and the Earth's orbit around the Sun. In rounded terms one orbit of the Sun is equal to one year which is divided into 365 rotations of our spinning Earth. One rotation is a day which is <u>rounded</u> to 24 hours. Each hour is divided into 60 minutes and each minute is divided into 60 seconds.

One second can be divided into one thousand milliseconds which equal one million microseconds or one billion nanoseconds. At a meeting in 1967, the International Committee of Weights and Measures announced the following classification: *"The second is the duration of 9,192,631,770 periods of the radiation corresponding to the transition between the two hyperfine levels of the ground-state of the caesium-133 atom."*

Clocks normally use a natural period of oscillation to 'smooth' and regulate a notion of time, which allow perceived moments or seconds of time, to pass smoothly and very regularly from a moment of time, to a new moment of time. Clocks may use the oscillation of a pendulum, a Quartz crystal, or in the case of an atomic clock, a Caesium-133 atom.

Clocks do not actually measure time itself, clocks may measure the number of times something oscillates and then we simply record these regular, reoccurring oscillations or events as time.

A pendulum may oscillate a second one way and another second the other way; or in the case of Big Ben which has a pendulum weighing 300 kg and a length of 4 meters and a period or oscillation of 2 seconds in each direction.

Despite the fact that this actually measures how many times a pendulum may have oscillated, we attach the notion of time to these oscillations by simply expressing their oscillations in seconds.

A wrist watch may use a Quartz crystal with an oscillation of 32,768 cycles per second. We attach the notion of time to these oscillations by simply calling them time. The superb accuracy of an atomic clock requires the frequency of a Caesium-133 atom at 9,192,631,770 transitions per second.

The very, one thing that all of these clocks have in common, whether the oscillation of a pendulum, a Quartz crystal, or in the case of an atomic clock, all simply count <u>the number of</u> a series of regular and predictable oscillations or pulses to measure units of a notion of the passing of moments of time. **Important; clocks count the regular rate of single pulses rather than the amount of energy driving a pulse.** All clocks measure the passing of <u>individual oscillations or transitions</u> rather than a smooth and continuous event. It does not matter if it is the energy driving Big Ben's 300 kg pendulum or the energy driving a tiny wrist watch, to tell time clocks count each single pulse or <u>oscillation</u>. It is now very likely that we are required to measure time in this way because time itself is also a series of continuously passing 'single' pulses of energy which <u>regulates</u> all within our universe. This regulation is essential to avoid changes to the speed of light or to the strength of Gravity; changes which may have unfathomable consequences for the chemistry and fundamental forces which allow our universe to exist. Relativity tells us dimensional Space and time <u>cannot be separated</u> which surely means dimensional Space itself, like time, is delivered as individual oscillations or tiny <u>passing pulses of energy</u>.

Within a Natural Universe passing time itself is a tiny Planck scale <u>energy</u> carrying wave which flashes pass at the amazing speed of light at the Planck rate of 10,000,000,000,000,000,000,000,000,000,000,000,000,000,000 passing Planck times per second. Planck time is the time taken for light to travel a Planck length. A Planck length is the distance across the wave.

What is time? How can time be regarded as a dimension like the dimension of length? Actually, the dimension of length provides the next real clue of exactly what time is. This is for the reason that the dimension of length is variable with speed most similar to the way passing moments of time are variable with speed. Special Relativity tells us that speed causes the passing of time to slow (time dilation) and the dimension of length to contract (length contraction). So much so, that at the speed of light, the dimension of length contracts to zero and the passing of time stops. This is clearly telling us that the passing of time is somehow <u>directly linked</u> to the dimension of length which <u>creates distance to Space</u> and if one traveled at the speed of light in any given direction one will **clearly catch up with, so as it stops** in relation to one self, whatever it is which is providing the passing of inseparable dimensional Space and time.

If the passing of time can be slowed to a stop like this, the <u>passing</u> of time is most likely <u>a real physical like object expanding from ourselves at the speed of light.</u> Actually, the speed of light narrows the search for the source of the passing of time to the <u>only two things</u> which travel at the speed of light; light itself or Gravitational waves. As they will need to provide <u>all dimensions</u> to Space, which can be stretched and warped to deliver Gravity, very tiny Gravitational waves are the only really possible known contender which travel at the speed of light, deliver Gravity and clearly provide <u>tiny passing transitions of energy</u> to provide passing time.

<u>Planck length</u> is the smallest distance allowed within a Dimensional Reality and <u>Planck time</u> is the shortest possible moment of passing time within a Dimensional Reality. Special Relativity says the passing of time and the dimension of length are both variable due to time dilation and length contraction. Special Relativity also says, <u>regardless</u> of length contraction or time dilation factors, the speed of light always remains a self observed constant. The mechanism which provide these <u>variable factors</u> to the dimension of length and a passing of time must operate within <u>perimeters</u>

so as the speed of light is always <u>self-observed</u> as unchanging regardless of time dilation or length contraction induced by motion or Gravity.

How is it possible for the same strict perimeters to be <u>attached and locked to yourself</u>? Answer; being <u>entangled</u> with your very own dimensions of Space is why you always self observe the speed of light at the same constant speed. The speed of light being a constant provides outstanding evidence we are all entangled with our very own dimensions of Space which allows all other dimensions of Space to be <u>totally ignored</u>.

Because <u>time cannot be separated from dimensional Space,</u> the <u>passing</u> of the <u>dimensions of Space</u> should also be the <u>passing of moments of time</u>. Time is said to be the fourth dimension so the <u>passing</u> of time should be a <u>passing</u> of a spatial dimension. It is now beginning to become obvious what time itself is. The passing of time can be no more than the 'event' of a passing wave of dimensional Space. Because it provides the smallest allowed dimensional distance the source of the passing of time should be the passing of a unit of the dimension of length. Within a Natural Universe the dimension of length is the wavelength, trough to trough, of a very tiny <u>energy carrying</u> Gravitational wave which equals a Planck length. Since the physics told Einstein dimensional Space <u>cannot be separated</u> from the passing of time means there can be little doubt that the passing of the same tiny Gravitational wave also provides a Planck unit of passing time.

A Natural Universe has a series of regular and predictable oscillations from vacuum fluctuations at the heart of all protons and all neutrons of all atoms, providing atoms with both regular mass-energy and Gravity. Physics require that the energy from vacuum fluctuations is returned to Space. We return much of this energy to Space in the form of tiny Gravitational waves which will be ideal for providing Space with dimensions which are <u>naturally entangled with a tiny pulse of energy from their source.</u> This changes dimensionless Space into expanding dimensional Space which is entangled with energy from its source proton or neutron and so all atoms. This also means that when these tiny Gravitational waves which provide entangled dimensions to Space become warped or stretched the dimensions of Space are similarly warped or stretched. The all important word is 'entanglement'. The tiny Gravitational waves are clearly born with and so spontaneously entangled with a tiny <u>pulse of energy</u> from the event which created them.

Special Relativity tells us when the passing of moments of time slows the dimension of length becomes contracted. Time and the dimension of length share a common and seemingly unbreakable bond which is required to be locked to the speed of light. The speed of light in a vacuum is 299,792,458 meters per second regardless of your own speed. Physics, theories and experiments all confirm it so but science lacks a rational, explainable physical mechanism which makes is so. So how is it that the <u>distance</u> that light travels in <u>one second of time</u> is always 299,792,458 meters <u>regardless</u> of your variable time dilation factor or your variable length contraction factor? Both speed and Gravity <u>slow</u> the rate of time but one's self-rate of time always appears self-same regardless of the amount of time dilation induced by your speed or Gravitational position. Only when comparing one's own rate of time with that of an observer with a different time dilation factor is a discrepancy exposed for both parties.

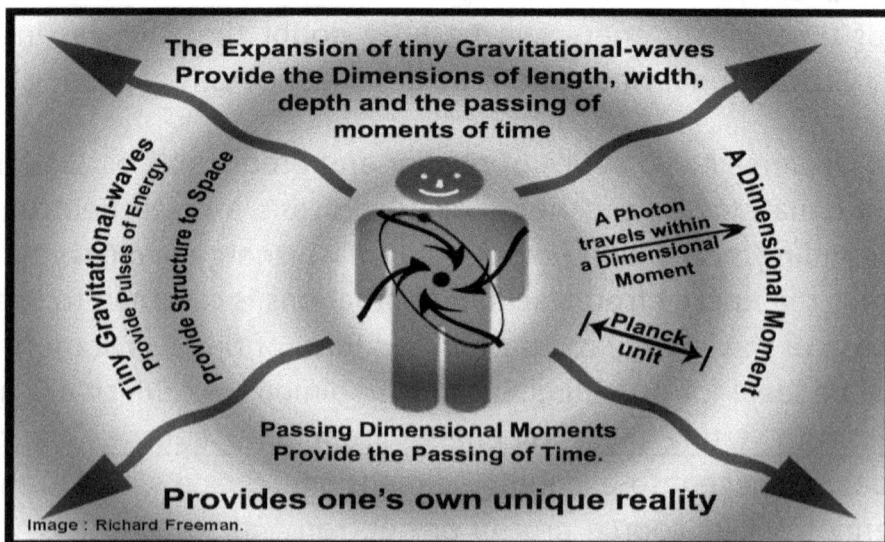

The Expansion of tiny Gravitational-waves Provide the Dimensions of length, width, depth and the passing of moments of time

Tiny Gravitational-waves Provide Pulses of Energy

Provide Structure to Space

A Photon travels within a Dimensional Moment

A Dimensional Moment

|← Planck unit →|

Passing Dimensional Moments Provide the Passing of Time.

Provides one's own unique reality

Image : Richard Freeman.

310

For the speed of light to be a constant and the passing of time variable there needs to be an unbreakable personal bond between the distance of 299,792,458 meters and the passing of one second of time. One should ask; what provides <u>this unbreakable personal bond to the speed of light?</u> Answer; tiny Gravitational waves <u>providing dimensions</u> and expanding from all protons and neutrons of all atoms at the <u>same speed of light</u>. ^{017 see article} Gravitational waves <u>are actually affected</u> by the stretching and warping of the curvature of Space-time and obey the rules of time dilation. Within a Natural Model tiny Gravitational waves <u>actively</u> provide the dimensions to Space which become stretched or warped to provide time dilation and length contraction which occurs due to both speed and Gravity.

The key point in allowing the speed of light to be a constant is the tiny <u>Gravitational waves</u> which provide <u>dimensional distance</u> to Space travel at the <u>same</u> speed as a <u>photon</u> of light. For light to be <u>always self observed</u> as travelling the same constant speed, one must be able to <u>self-lock</u> a set dimensional distance traveled by a photon to a set amount of passing of time. This is achievable when tiny Gravitational waves begin from the hearts of all protons and neutrons of the atoms from which we are made. Within one's Dimensional Reality, <u>regardless of being stretched</u>, the distance across a tiny Gravitational wave provides a Planck length and the passing of the same tiny Gravitational wave provides a Planck unit of passing time. Because a photon travels at the same speed as a tiny Gravitational wave a photon effectively travels with the <u>pulse of energy</u> provided by a tiny Gravitational wave. Within your <u>entangled</u> Dimensional Reality, since light travels a Planck length in Planck time, light is now locked to always travel one Planck length, the distance trough to trough of a tiny Gravitational wave, in one Planck time, the passing of the same tiny Gravitational wave, regardless of the wave being stretched to provide length contraction or time dilation. This allows you to always observe light to travel 299,792,458 meters in one second <u>regardless</u> of your own <u>variable</u> time dilation factor or your <u>variable</u> length contraction factor. **This is how**

the speed of light is locked as a constant within one's Dimensional Reality.

Planck length, denoted ℓ P, is a unit of length that is the distance light in a perfect vacuum travels in one unit of Planck time. This model explains why this is so, why there is Planck length, why there is Planck time and explains precisely what they physically are. A Planck length is one unit of the dimension of length which is the trough to trough (wavelength) of a tiny Gravitational wave in the direction of expansion. This naturally becomes the smallest possible dimensional size for anything within a Dimensional Reality which is equal to the Planck Length, which is about a millionth of a billionth of a billionth of a billionth of a centimeter across.

Because the tiny Gravitational waves, regardless of the amount of energy, always retain the same vital 'single pulse value' from the event which created them they retain the same single Planck value so as the speed of light always remains a constant within different dimensional realities. The point being a single pulse value, like a clock, derived from the energy of a passing wave. Stretching an entangled Gravitational wave will naturally change one's dimensional, spatial reality by way of time dilation and length contraction. Each tiny Gravitational wave is required to have a one pulse value of one unit of dimensional Space or a Planck unit of Space-time whether stretched or non-stretched. **This naturally conforms to the way time dilation is known to occur, one second remains a full one second regardless of time dilation caused by stretched out tiny Gravitational waves which take longer to pass.**

Thus, regardless of the passing rate, fast or stretched slow, of a dimensional moment provided by the energy of a tiny Gravitational wave, a photon is locked to travelling a Planck length (the distance trough to trough of the tiny Gravitational wave) in Planck time (the passing of the tiny Gravitational wave). This is also why dimensional Space and the passing of time cannot be separated, hence, Space-time. Because the tiny Gravitational waves are generated from the protons and neutrons of the atoms within our bodies, the passing of time is <u>locked to one's own self-observed</u> constant speed of light. The key here is time dilation is provided by tiny Gravitational waves retaining Planck values while being stretched by variable amounts, this allows all observers to self-observe the speed of light from all sources as a constant. If a photon was not locked to a self-dimensional moment, which provided both <u>Planck length of distance to Space and the passing of Planck time</u>, the speed of light would not be self observed as a constant.

The very same tiny Gravitational waves naturally provide the <u>all important physical structure</u> to the nothingness of empty Space, structure which can be stretched, warped or bent to deliver Gravity, passing time, time dilation and length contraction. Gravity is delivered by stretching the same tiny Gravitational waves. One may wish to suggest that because Gravitational waves travel at the speed of light that stretching such a wave, like pulling on it to stretch it, would require the stretching to <u>exceed the speed</u> of light which is not allowed. Because the <u>energy value</u> is provided by the event which created the Gravitational wave the energy of a stretched or non-stretched Gravitational wave <u>remains a vital single pulse constant.</u> This preserves the Planck length value which causes the perceived greater <u>dimensional distance</u> of a stretched Gravitational wave to contract to the regular <u>dimensional distance</u> of a Planck length. This is called length contraction which is a prediction of Relativity. Since the Planck length value is maintained, the stretching mechanism is not required to exceed the speed of light. A <u>stretched</u> wave takes <u>longer to pass</u> which <u>slows</u> the passing of time. So <u>rather than exceed the speed of light the distance contracts to maintain a Planck length and the passing of time slows</u>. This allows light to always travel the distance of a Planck length in a Planck time.

Now one <u>cannot apply Gravity</u> with length contraction caused from stretching a tiny Gravitational wave <u>without slowing time</u> because to do so without slowing time the act of stretching would exceed the speed of light. Although this is my physical modeling of time and Gravity it is really a clever ploy of Relativity. **This is why time dilation is a fundamental property of Gravity and how my Natural Model shows why Gravity slows time and shows exactly what passing time is. Without knowing exactly what time is one cannot physically show why Gravity, when it stretches and so warps the curvature of the dimensions of Space, slows the passing of time.**

Because a Gravitational wave is effectively occupying dimension<u>less</u> Space each tiny Gravitational wave can only retain the same self-value of <u>one pulse or unit of dimensional Space</u>. Consequently, the dimension<u>less</u> Space which <u>the Gravitational wave occupies</u> cannot be dimensionally counted because it is dimension<u>less</u>. Similarly, the dimension<u>less</u> Space towards a distant object cannot be counted because it is dimensionless. Mother Nature knows nothing of kilometers or miles, centimeters or inches or even Planck length or Planck time, she only knows that there are these extraordinary tiny entangled pulses of <u>energy</u> between objects made from atoms. Consequently, this modeling means that '<u>smaller</u>' than the Planck constant can only be characterized within an indefinable realm of dimensionless Space. This is where Quantum Theory allows actions which are not allowed within an <u>entangled</u> Dimensional Reality.

Within a Dimensional Reality only the number of dimensional units at the rate of one dimensional unit per Gravitational wave can be counted. One dimensional moment provides one self-moment of passing time and one self-moment of a dimension of length. If there are less stretched dimensional moments (Planck units) towards a distant object there is less dimensional distance to a distant object and the object is dimensionally closer. A stretched dimensional moment takes longer to pass which slows the passing of time. Because each dimensional moment retains the same self Planck value the passing of time <u>appears self-normal</u> and unchanging, however, self-normal can be different within different Dimensional Realities. So how does this all work within different Dimensional Realities?

Travelling in time; Fast Mary's spaceship is travelling to an Earth like planet orbiting a star which for Stationary Bob on Earth is 100 light years away. If Mary's spaceship travels at 99.99999999% the speed of light it will take her just over 100 Earth years to arrive but only 7 seconds of her own self-observed time.

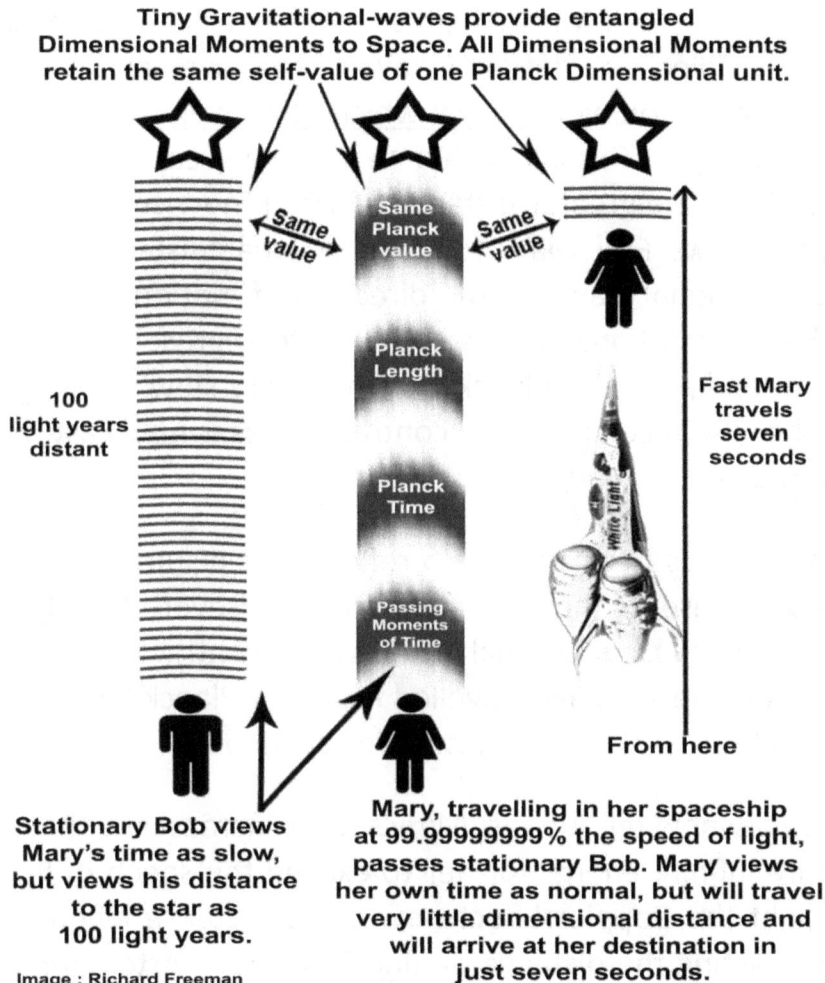

Tiny Gravitational-waves provide entangled Dimensional Moments to Space. All Dimensional Moments retain the same self-value of one Planck Dimensional unit.

100 light years distant

Same value

Same Planck value

Same value

Planck Length

Planck Time

Passing Moments of Time

Fast Mary travels seven seconds

From here

Stationary Bob views Mary's time as slow, but views his distance to the star as 100 light years.

Image : Richard Freeman

Mary, travelling in her spaceship at 99.99999999% the speed of light, passes stationary Bob. Mary views her own time as normal, but will travel very little dimensional distance and will arrive at her destination in just seven seconds.

Time dilation shrinks Fast Mary's travel time from 100 years to just 7 seconds. Although not really possible due to mass increasing for anything made from atoms, if Fast Mary was able to increase the speed of her spaceship to 100% of the speed of light, Mary's dilation factor is now 100% and Mary will now instantly arrive the very same instant she leaves and, like a photon, travel zero dimensional Space or time in doing so. Today it is not possible to explain how this <u>physically</u> happens, other than to say that this is a ramification of the successful theory of Special Relativity.

Since this is a physical reality model it is required to physically explain how this occurs. Because Fast Mary's high speed is nearly the same as her tiny, dimensions providing Gravitational waves, the waves escape from her, in the direction of motion, <u>very slowly which effectively stretches the waves over a great distance</u>. This allows her to travel a great distance during the escape of a wave, the greater distance of which now contracts to a Planck length which allows Fast Mary to travel a great distance with <u>very few</u> escaping Gravitational waves in the direction of motion. Since the passing of one of her Gravitational waves continue to represent the passing of a Planck length and a unit of Planck time her 'very slow' time <u>remains self-normal</u> but her distance to the star contracts. **This is how Fast Mary travels a great distance within very little passing of time.**

The stretched waves <u>behind her</u> also reflects the distance she has actually traveled and represents the distance which has severe length contraction due to the stretched Gravitational waves which contract to a Planck length. The end result is she will have travelled far fewer Planck lengths of distance and her travel time is reduced to just seven seconds.

Special Relativity now tells us Fast Mary's mass increases to the point that it would require infinite energy for her to exceed the speed of light. At near the speed light the Doppler Effect now causes intense length contraction which is contracting the distance <u>behind Fast Mary</u>. Attempting to exceed the speed of light <u>the intense length contraction</u> is now the same as applied by intense Gravity from near the event horizon of a Black Hole. I will call this <u>relativistic</u> Gravity which, like attempting to escape a Black Hole, prevents Fast Mary exceeding the speed of light. Where one finds relativistic Gravity one can be expected to find relativistic mass. Approaching the speed of light, due to Mary's relativistic mass, it now appears Mary is becoming her own Black Hole from which she cannot escape by going faster since, exactly like attempting to escape from a Black Hole, it would require an impossible amount of energy to do so.

The length contraction from stretched dimensions <u>behind Mary</u> are clearly pulling against Mary in exactly the same way as if Fast Mary was attempting to escape the massive Gravity of the event horizon of a Black Hole.

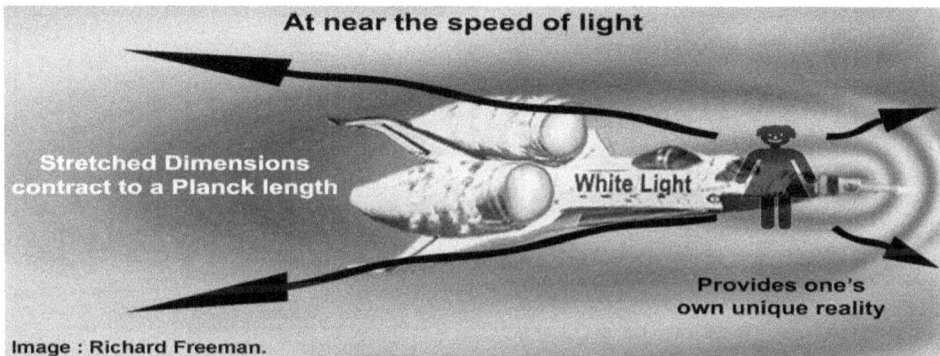

At near the speed of light

Stretched Dimensions contract to a Planck length

White Light

Provides one's own unique reality

Image : Richard Freeman.

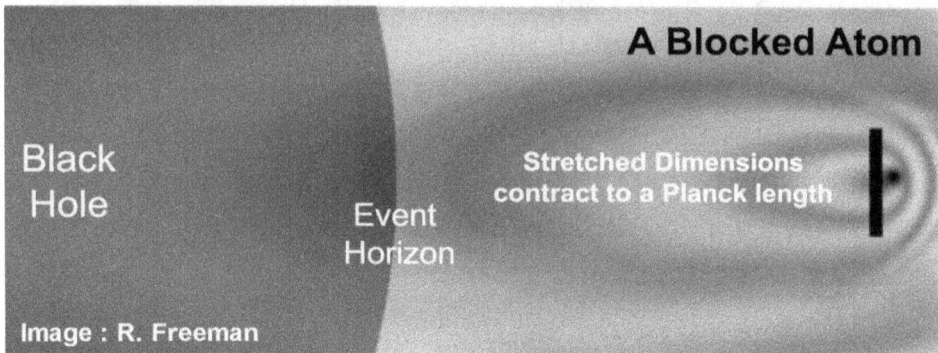

A Blocked Atom

Black Hole

Event Horizon

Stretched Dimensions contract to a Planck length

Image : R. Freeman

If one was stationary at the event horizon of a Black Hole one would effectively have high speed into the flow of the dimensional Space which Scientists say flows into a Black Hole. Now Mary cannot exceed the speed of light for the exact same reason she cannot escape from a Black Hole which is a very neat correlation which only a Natural Model provides.

(Following image) Notice how the influence of Gravity, which slows time increases towards a Black Hole, produces a curve similar to the Lorentz factor curve for speed caused time dilation. The graphs suggest relativistic Gravity is <u>most noticeable at speeds approaching light speed</u>. Within a Natural Universe from a stationary position near a Black Hole there is a stretching of the dimensions of Space towards the Black Hole which is <u>very similar</u> to Fast Mary's near light speed time dilation which means Fast Mary's very slow rate of passing of time occurs for the <u>same reason</u> as being stationary and enduring the massive Gravity close to a Black Hole.

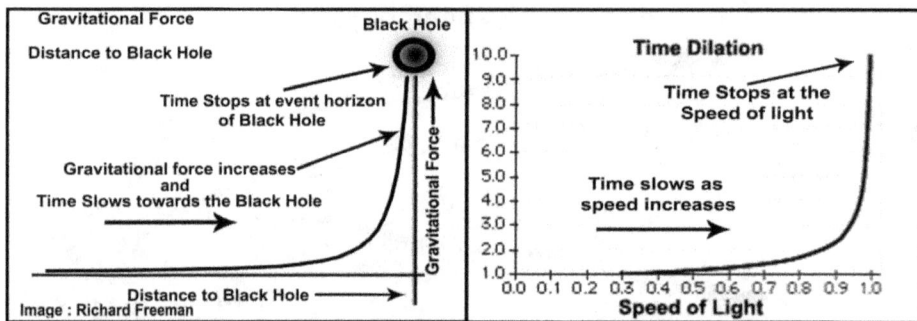

Image : Richard Freeman

How can Einstein's Relativity say Fast Mary now observes the length of her spaceship and her passing of time as normal? [017a] The amount of energy which Gravitational waves carry is fixed as they travel perpetually through space. Each tiny Gravitational wave maintains a vital <u>single pulse of energy</u> from the event which created the wave which for Fast Mary <u>normalizes</u> her *'fewer and slow'* dimensions by always retaining and delivering single Planck values of both length and passing time allowing Fast Mary to always observe the length of her spaceship and her 'slow' passing of time as normal. Relativistic Gravity causes Mary's stationary <u>equilibrium</u> position within her highly distorted dimensions to now be far behind her placing on her the same stress as if she was stationary near the event horizon of a Black Hole. Conforming to Special Relativity, Fast Mary naturally continues to remain at the center of her expanding, entangled dimensions from where the planet which she is travelling to now appears to be speeding <u>towards her</u> at near the speed of light. When Mary attempts to go faster than light, the <u>drag from relativistic Gravity</u> causes her to request far more energy than her powerful Anti-matter twin stellar-drives can provide.

Thus, Fast Mary's journey will take her just 7 seconds of her own time, however, for Stationary Bob back on Earth his entangled dimensions are escaping from him at a far greater rate so Bob ages and dies before Mary arrives at her destination. For Stationary Bob 100 Earth years pass by as do for anyone on the Earth like planet which Fast Mary has travelled to.

Now imagine Fast Mary on board a super fast train travelling at <u>half the speed of light</u>. Similar to when onboard her spaceship, Fast Mary has <u>fewer</u> but stretched passing dimensional moments of Planck length and Planck time. Mary places a mirror on the floor of the train and switches on an above light. <u>Fewer dimensional moments</u> allows Fast Mary to observe the light travel a shorter distance than Stationary Bob standing at the station.

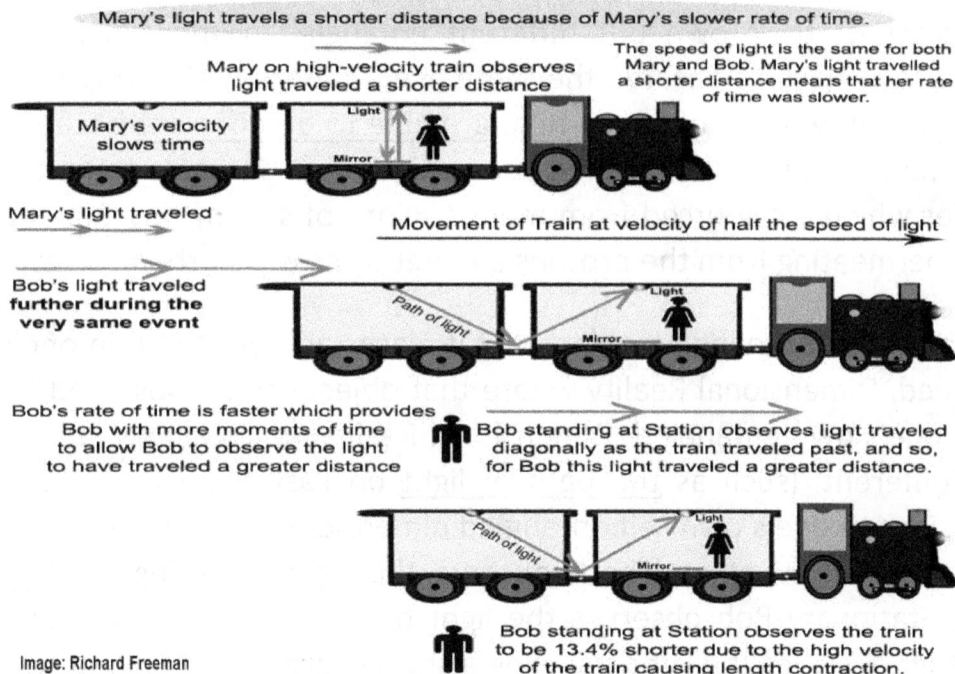

Mary's light travels a shorter distance because of Mary's slower rate of time.

Mary on high-velocity train observes light traveled a shorter distance

The speed of light is the same for both Mary and Bob. Mary's light travelled a shorter distance means that her rate of time was slower.

Mary's velocity slows time

Mary's light traveled

Movement of Train at velocity of half the speed of light

Bob's light traveled further during the very same event

Bob's rate of time is faster which provides Bob with more moments of time to allow Bob to observe the light to have traveled a greater distance

Bob standing at Station observes light traveled diagonally as the train traveled past, and so, for Bob this light traveled a greater distance.

Image: Richard Freeman

Bob standing at Station observes the train to be 13.4% shorter due to the high velocity of the train causing length contraction.

Because light from <u>all sources</u> is locked to a <u>dimensional</u> Planck moment, provided by a tiny Gravitational wave, light can only travel the <u>same dimensional distance within the same dimensional moment</u> regardless of whether a tiny Gravitational wave is stretched or normal. If the very <u>same</u>

event has more moments of passing time in a different frame of reference or <u>within a different Dimensional Reality</u> as with Stationary Bob, it also has more dimensional moments which allows light to be observed to have travelled a greater dimensional distance to match more moments of passing time. This allows different observers such as Fast Mary and Stationary Bob to observe light <u>during the same event</u> to travel different dimensional distances at the <u>same</u> constant speed. Mary and Bob can see each other but each observes the light travel a different dimensional distance during the same event. Because Mary's fast train has less dimensional moments <u>Stationary Bob also observes</u> Mary's train as being shorter due to length contraction.

This is all achievable because both Fast Mary and Stationary Bob reside at the center of their own but different <u>entangled</u>, Dimensional Realities which allows each to observe the same event differently. During the event <u>the actual dimensions of Space-time are able to be different</u> because both Mary and Bob reside within their own unique and <u>entangled</u>, Dimensional Realities which are sourced from many trillions of sets of tiny Gravitational waves permeating from the protons and neutrons within themselves.

So what does observing an object do? It places an object within one's own entangled, Dimensional Reality where that object can be observed 'as it is' within one's own <u>entangled</u>, Dimensional Reality which allows an object to have 'different' (such as <u>the path of light</u> on Fast Mary's train carriage) dimensions to one's own self entangled dimensions. Although they observe the very same event and both observe the light travelling <u>at the same speed</u>, Stationary Bob observes the light traveling a greater distance and Fast Mary observes the light traveling a shorter distance. For light itself the photons travel zero distance in zero time.

Hence, light cannot travel at a faster or slower rate because each expanding <u>dimensional moment</u> from both Fast Mary and Stationary Bob has the same 'dimensional Planck value' as in both 'distance value' and 'time value'.

Consequently, for the speed of light to be a constant, a passing moment is required to remain of the same dimensional value regardless of how fast or slow it is actually expanding, passing or being stretched. Like a full glass of water, drink it slow or fast, there is only a glass full of water to drink.

Tiny Gravitational waves expanding from all atoms provide a unique **clock, which regardless of the pace it operates at, allows all to self-advance at the exact, same self-observed rate.** Because both the tiny Gravitational waves and photons are massless they are both similarly attracted to both the phase of virtual mass nothingness within protons and neutrons and the far outer void of Anti-space nothingness. Because Photons keep pace with the tiny Gravitational waves photons are effectively 'carried' by the same expanding dimensional moments provided by the tiny Gravitational waves.

Actually, Mary's speed now slows the speed of light exactly like Scientists at first anticipated but because time slows as well, like Einstein correctly postulated, the speed of light is always observed at the same constant speed. Regardless of time dilation or length contraction, this model allows all moments of time and moments of length to always provide Planck values so as the speed of light is always self-observed at the same constant speed. Scientists at first thought one's speed should affect the speed of light, however, no matter how the experiments were conducted all failed to show any difference to the speed of light. The speed of light remains self unchanging because the tiny, Gravitational waves, which are expanding from you, are responsible for both the passing of time and for providing dimensions to Space. Now light will always travel a Planck length in Planck time regardless of one's own time dilation or length contraction factor.

Albert Einstein said (dimensional) Space and time are inseparable. Because the exact same tiny, Gravitational waves provide both the passing of time and dimensional distance to Space both passing time and dimensional distance are inseparable. This is naturally mandatory to allow the speed of light to be a constant and allows all the laws of physics, chemistry and fundamental forces within the universe to be regulated by the passing of time so as the laws of the universe, including Gravity, all operate at exactly the same self-observed rate no matter where you are in the universe. Time is indeed the masterful regulator. But if nearly every object in the universe is operating at a different rate of time will not everything get out of sync with dire consequences? No matter how fast or slow the vacuum fluctuations of the protons or neutrons of atoms may be fluctuating we simply use rounding to represent 24 hours for one rotation of the Earth or 365 days for one orbit of the Earth around the Sun. We are all part of the same universal movie and the movie is of the same length whether the individual frames are moving fast or slow.

The passing of time allows the <u>continuous</u> light speed <u>renewal</u> of your entangled dimensions to Space which allow your time dilation and length contraction factors <u>to be renewed</u>, at the amazing speed of light, to exactly match your speed and Gravitational position in the universe so as all appears self normal and unchanging. The Quantum vacuum fluctuations of a COST are responsible for causing the tiny, light speed Gravitational waves which provide atoms with entangled passing dimensional moments. Thus, the final source of passing moments of time is the frequency of the beating heart of all protons and neutrons of all atoms, which is provided by their COST unit's Quantum vacuum fluctuations. **Is there any direct evidence that this is all true?** Recent experiments relating to Quantum vacuum fluctuations by Prof. Dr. Alfred Leitenstorfer and his team from the University of Konstanz without doubt can now reinforce this concept. They have now shown that the vacuum fluctuations relating to light can be measured directly. [067] The following is a totally astonishing, small snippet from their latest amazing research. '*The team at the University of Konstanz has now explored the quantum states of light and vacuum fluctuations, as well as their interplay <u>with time</u>. They have discovered they can <u>redistribute fluctuations from one moment in time to another moment in time</u> and have found that this <u>change in the flow of time is</u> **directly related to the change***

in fluctuations'. [119] **The Scientists have also revealed the dimensions of Space and time *'behave absolutely equivalently'* in their experiments.**

[119] The Scientists by manipulating a Quantum vacuum fluctuation have actually measured and were able to alter and even stop the flow of time and as Einstein has said, (dimensional) Space and time cannot be separated so the team has actually witnessed the flow of the dimensions of both Space and time. This truly remarkable evidence obviously directly aligns with the modeling already expressed in this book as well as in my previous book. By understanding the Natural Model one can rightly elevate this experiment to the significance which it now deserves. That is, the Scientists have likely actually witness the very source of the passing of moments of the actual dimensions of Space and time, thus, this would now be **one of the most astounding discoveries ever made; the source of dimensional Space-time.** Over the last two decades I have come a long way with this model but I still get a sense of real satisfaction when I discover yet another piece of new, cutting-edge science which so precisely reinforces this model.

[069] Additional, reinforcing evidence is provided by experiments which accelerate protons to near the speed of light. **Note**: For example; the 'fluctuations' of gluons appearing and disappearing out of the vacuum **slow down** and the glow of gluons last longer the closer the proton gets to the speed of light. Scientists say this is a direct effect of time slowing. According to my modeling the Scientists are actually observing the source of time which are the vacuum fluctuations given to protons by the COST.

The all important, entangled mechanism provided by the COST cleverly entangles and so locks one's own unique time to one's surrounds and allows one to view the surrounding universe as it appears within one's own personal rate of time within one's personal, Dimensional Reality. All the protons and neutrons of whole atoms have inherited this mechanism for a rhythm of a passing of time from their COST (Dark matter) host.

One may ask; how can the passing of time be uniquely <u>entangled</u> with one's self? The mechanism for the passing of moments of time is individually and uniquely 'entangled' with every COST unit residing at the hearts of all protons and neutrons and so with every atom. Time dilation, length contraction and the speed of light being a constant all provide the strongest, <u>direct evidence of self entanglement with dimensions.</u> Entanglement occurs because the tiny Gravitational waves which provide entangled dimensions to Space are spontaneously entangled from birth with the energy of the event which created them. When your speed or Gravitational position in the universe slightly changes the tiny Gravitational waves also very slightly change by becoming stretched or warped. This allows your Dimensional Reality to be continually renewed at the amazing speed of light. Because of self entanglement with one's own dimensional moments the speed of light is always self-observed at the same constant of 299,792,458 meters per second. **Note: <u>Entanglement</u> prevents another observer's rate of time interfering with one's own. That is, <u>all other Dimensional Realities,</u> belonging to all other observers, are <u>totally ignored.</u> Understanding this will, at chapter 13, lay the foundations to finally provide a clear and rational understanding of Quantum Theory.**

Thus, an atom's tiny Gravitational waves being superimposed throughout dimensionless Space infuse Space with an object's entangled dimensions. Entanglement effectively changes an object's dimensionless Space into dimensional Space-time. The predictions of Special Relativity require this superimposing of dimensions throughout dimensionless Space to be 'unique' to one's self-speed and Gravitational position within the universe. The <u>frequency</u> of the Quantum vacuum fluctuations which provide the tiny Gravitational waves is derived from one's own self-speed and Gravitational position within the universe. Entanglement allows uniqueness. Entanglement just means each proton and neutron of every atom acts independently creating its own tiny Gravitational waves. Entanglement provides a person's 'unique' spatial, Dimensional Reality. We are all provided with many trillions of tiny entangled <u>energy carrying</u> Gravitational waves which <u>collectively</u> provide our own entangled dimensions to Space, all of which begin within the Quantum world, rapidly expanding at the speed of light to provide one's everyday world of dimensional realism.

At the speed of light or within the Gravity of a Black Hole time dilation and length contraction is 100% which causes the passing of time to stall and dimensional Space to contract to a dimension<u>less</u> point. This occurs within a Black Hole because a COST's dimensions providing tiny Gravitational waves would be required to escape from it at faster than the speed of light which is not allowed. If dimensions and passing of moments of time was a universal constant of all Space, entanglement would not be required and the passing of time would be a universal constant. Special Relativity clearly tells us this is not so; we all truly live our lives within our own unique time.

Entanglement provides us all with our own totally unique time which can be uniquely varied as with time dilation. Our bodies have no problem if our legs are moving quicker than our arms for the reason that compared to the speed of light the differences are minuscule and in everyday life are mostly insignificant. Our bodies, for all <u>rounding</u> reasoning, combine all of our atoms different realities as one rounded and smooth Dimensional Reality.

One may have first wished to believe that the four dimensions of Space, which include the passing of time, fluently flows throughout the universe in a way that allows time and dimensions and coordinates to proliferate at a universally fixed rate for everything within the universe. Einstein's Special Relativity and experiments clearly tell us that this is not so. Passing time and the dimension of length are variable. Every person, all life, all plants, all rocks, all pebbles on the moon, all stones on Mars and all things made from atoms throughout the universe have their own unique time and own unique, spatial reality within a coordinated system. To allow this to occur all atoms within a Natural Universe have been provided with their very own individual, entangled mechanism which provide a continuous rhythm for the passing of time and unique dimensions to their Space.

[068] Renowned Physicist, Professor Paul Davies once said while commenting on the subject of the mystic of time; *"The point is that time is rather special. It's different from other physical quantities because we feel it inside ourselves. Time is in our very souls, isn't it? Time is something that is inside us, and that is why I think people are so fascinated with the nature of time".* Within our Natural Universe what Professor Paul Davies says is literally true. We all truly have our very own individual built in internal clocks which provides us all with our own unique and entangled rhythm for the passing of time. Our time as we know it truly originates from within ourselves. Like the way our body functions are synchronized with the rhythm of night and day, all things have been regulated and coordinated by the rhythm of passing waves of dimensions of Space and time. The Natural Model clearly shows why <u>passing time is caused by a physical mechanism</u> which causes both the speed of an object and the delivery of Gravity to physically slow the passing of time which allows the speed of light to be a constant and provides all atoms with regular cycles of mass and energy. If the source of the tiny waves providing the dimensions of Space were to cease, than Gravity ceases to be delivered, light ceases to be a constant, atoms would have no mass or energy, the beat of time ceases and even the universe itself would have never advanced from dimensionless empty Space.

It is easy to understand how the continuous expansion of extraordinary tiny Gravitational waves, by providing dimensions to Space, allow everything to advance in continuous, micro increments. Because the tiny Gravitational waves are expanding away at the speed of light, the 'arrow of time' can only move in the direction of advancement. **As one tiny Gravitational wave passes a new tiny Gravitational wave immediately takes its place causing one moment of time to pass as a new moment of time immediately takes its place which allows your unique Dimensional Reality to be continually renewed at tiny Planck scale and at the amazing speed of light.**

Because this exact same mechanism provides the 'delivery' part of Gravity, this now reconciles time with the predictions of General Relativity and being at the center of this mechanism reconciles time with the predictions of Special Relativity. Because the tiny Gravitational waves are born at the scale of the realm of Quantum Theory they extend Relativity's practicality down to where it meets the tiny scale of Quantum. Within the Natural Universe neither General Relativity nor Special Relativity could predict what they do without atoms being at the center of their own entangled mechanism which provides entangled dimensions to Space. This provides a physical understanding of how the reality realm of Relativity is set in motion at the tiny scale of Quantum vacuum fluctuations which <u>seamlessly integrates the predictions of General Relativity and Special Relativity</u>.

We began this model within a dimensionless void of empty Space called Anti-Space. Unfortunately, without continuous expansion of dimensional moments, Anti-Space has no passing of dimensional moments but I still sometimes think of it like this; if you were within Anti-Space, watching from afar when our dimensional universe began, you may also see how it ended but do not blink because you may miss it completely. Within Anti-Space-time our universe may exist for just one very stretched instant and be so short lived that it barely even exists.

For all sense of reasoning a photon of light never ages. The remarkable telescopes of today can detect light from <u>near the beginning of the universe</u>. In <u>photon time</u> this light arrived here on Earth at the very <u>same instant</u> it departed its source. However, in <u>photon time</u>, at the instant this light departed its source, the Earth and our whole solar system did not exist. This is difficult for the human brain to grasp, however, Mother Nature found a way, she created amazing, dimensional, spatial realities which allowed billions of years to pass by, <u>thus stalling that instant of time</u> until the Earth, you and I were made and that mighty telescope was built and finally the light from billions of years ago entered the lens of the telescope.

Because your center of expansion (of dimensional Space and time) continuously travels with you, whichever direction one has motion and <u>like the arrow of time</u>, one must travel in the direction of the advancement of the dimensions of Space and the advancement of time. Because you can only travel in the direction of the advancement of a dimensional moment you can only travel in the direction of the advancement of time which is why speed in any given direction will slow the advancement of your dimensional moments which slows your advancement of your passing of time. Your dimensional moments are expanding at the speed of light. Should you reach the speed of your expanding dimensional moments your dimensional moments can no longer advance in the direction of travel and your passing of time stops. This is why the passing of time stops at the speed of light and is why a photon of light travels within an instant of time.

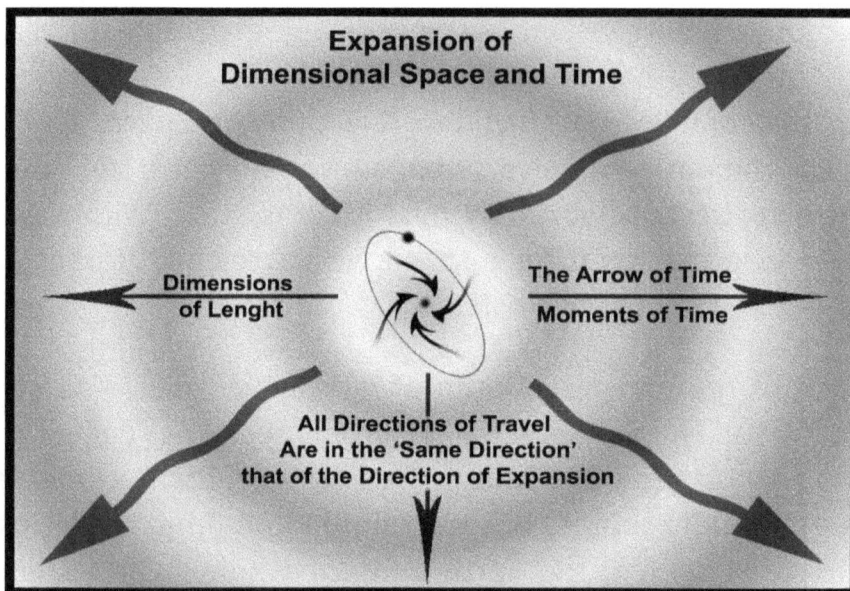

Expansion of Dimensional Space and Time

Dimensions of Lenght

The Arrow of Time

Moments of Time

All Directions of Travel Are in the 'Same Direction' that of the Direction of Expansion

Time dilation and length contraction are like two sides of the same coin. Both are only observable from the perspective of another observer within a different <u>frame of reference and residing within a different Dimensional Reality</u>. Dealing with the many intricate ramifications of the speed of light, length contraction and time dilation within <u>different frames of reference</u> is by nature mathematically complex, however, Einstein found a way. When calculating time dilation Physicists make use of the Lorentz factor which expresses a noticeable trend that time dilation is small at low speeds and dramatically increases near the speed of light, the Lorentz factor allows predictions within different frames of reference.

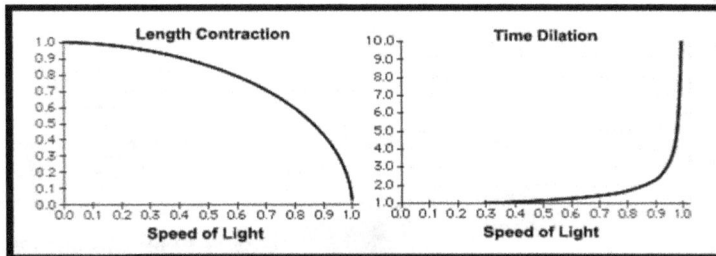

The Lorentz factor is by which time, length and mass changes for an object which is moving at speeds near to the speed of light (relativistic speeds).

With a Natural Model, regardless of time dilation and length contraction, dimensional Space and passing time are locked together as one and cannot be separated, hence Space-time. [119] Consequently and like this experiment has revealed, the dimensions of Space and passing time **behave absolutely equivalently.** To remain a constant, light must always travel a Planck length in Planck time. For example, if light was allowed to travel two Planck lengths in Planck time the speed of light would double which is not allowed.

We all reside within our own entangled, Dimensional Reality and within our own unique dimensional universe. Things may appear different within a different Dimensional Reality because they are actually, really, truly dimensionally different within a different dimensional universe. In everyday life here on planet Earth these differences are insignificant since nearly all objects we observe have entangled Dimensional Realities which are almost identical to our own. However, near to the Gravity of a Black Hole or at velocities near the speed of light these differences become significant which is only apparent when observing or measuring an object which is residing in a state of Space which is significantly dimensionally different than one's own state of Space.

Because Fast Mary's stretched and warp dimensions all normalize to Planck units Fast Mary's <u>slow time</u> allows her to view everything as normal. However, Stationary Bob, with his <u>faster</u> passing of time, may now observe Fast Mary's spaceship, which is residing in a state of Space which is significantly dimensionally different than Stationary Bob's own state of Space, as being short and stubby. This is because over the '<u>same amount</u>' of <u>Stationary Bob's time</u> Mary's spaceship now has fewer but stretched tiny Gravitational waves providing it with less time and fewer Planck lengths.

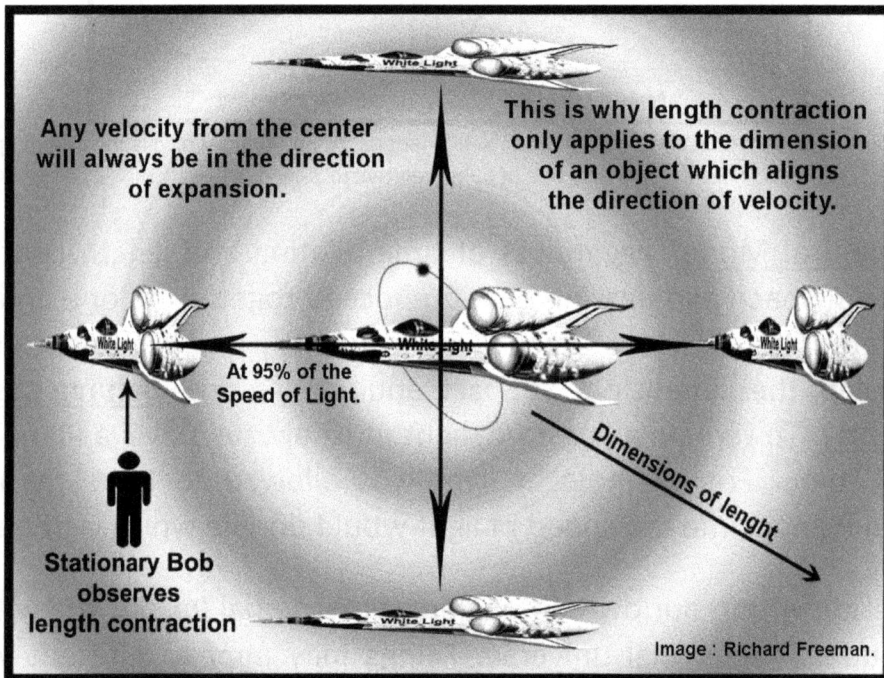

Image : Richard Freeman.

Because all Planck lengths retain the same Planck length value the stretched measurement or distance covered by each stretched Planck length contracts to a Planck length which is why Stationary Bob observes the length of Fast Mary's spaceship contracted so as it is short and stubby. For Fast Mary the stretched Planck lengths take longer to pass so there are fewer but stretched dimensions to a distant star. Because a stretched dimension retains the same self-Planck-value as a non-stretched Planck dimension Fast Mary's time slows and the dimensional distance contracts.

Length contraction is revealed when observing or measuring an object residing within a significantly dimensionally different state of Space than one's own. Please take note how Stationary Bob observes Mary's spaceship as being short and stubby for it will help to understand wave and particle duality in the following chapter.

Building an all-around Entangled Dimensional Reality:

Single virtual protons (Dark matter), and regular protons and neutrons trend to self-reside with a two dimensional Reality; the dimension of length and the passing of time. A complete four Dimensional Reality including the passing of time is built by many protons, neutrons and so atoms coming together, only then do atoms rise fully above the weirdness of Quantum Theory and become fully locked to the realism of Einstein's Relativity.

Gravitational waves expand on a comparatively, thin, flat plane more like the surface of a DVD disk so how do they build an all-around entangled Dimensional Reality? To build an all-around expanding, complete Dimensional Reality one would have wanted the tiny Gravitational waves to expand more like a succession of expanding spheres, however, Gravitational waves are not like that. Because Gravitational waves expand like ringlets from rain drops on a pond this model will naturally require numerous atoms to come together to permanently provide a complete four Dimensional Reality from which there is no escape.

But is there any direct evidence that there is actually a required threshold which needs many atoms to come together to build an all-around entangled Dimensional Reality? Results from double slit experiments confirm that single atoms are not locked out of peculiar, Quantum behavior where particles act as waves and somehow reside in many different places at the same moment. Scientists have discovered there is a size or threshold where objects begin to gain a complete Dimensional Reality from where they can no longer express the peculiar behavior from Quantum-scales.

No other model can precisely explain the exact cause of this known threshold which provides more amazing evidence that this modeling of entangled dimensions is correct.

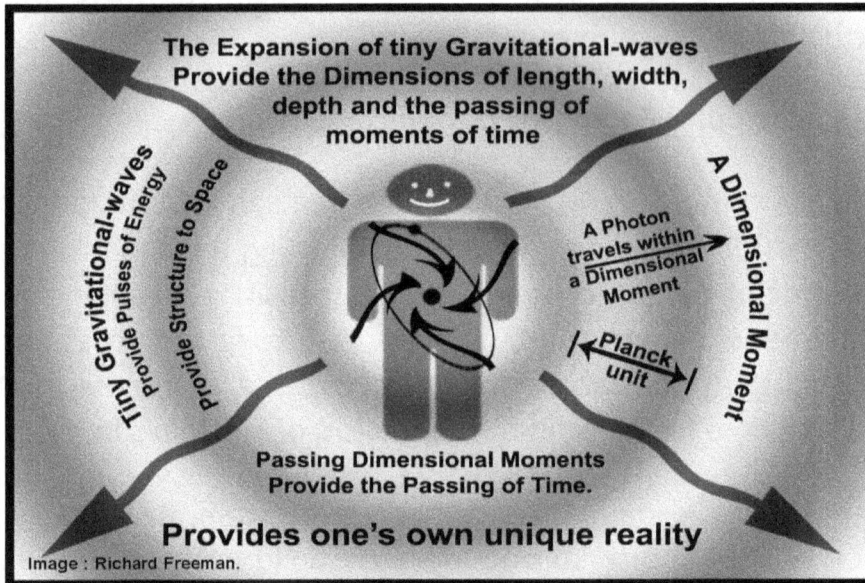

The Expansion of tiny Gravitational-waves Provide the Dimensions of length, width, depth and the passing of moments of time

Tiny Gravitational-waves Provide Pulses of Energy

provide Structure to Space

A Photon travels within a Dimensional Moment

A Dimensional Moment

Planck unit

Passing Dimensional Moments Provide the Passing of Time.

Provides one's own unique reality

Image : Richard Freeman.

[018] The Scientists say it is <u>puzzling as they do not understand</u> why this threshold should exist at the scale or size of an object which has a few thousand atoms. Attempting to obtain answers Scientists <u>have searched for this threshold</u> where Quantum physics cross over to Einstein's classical physics of Relativity by experimenting with different arrangements of particles, in the form of tiny molecules and exposing them to the <u>double slit</u> experiment. As the size of the object increases it becomes more difficult for an object to pass through <u>both slits, at the same time</u>, by displaying wave properties. The largest objects which have now sneaked through both slits by showing wave properties have been molecules of around <u>2000 atoms</u>. Although it may require an object made from just a few thousand <u>atoms</u> to fail the double slit experiment it may require millions of <u>atoms</u> to collectively build a complete, four Dimensional Reality like which we are all permanently locked into and where it is absolutely <u>impossible</u> to be in many places at the same moment of time. **Note the above use of atoms.**

This is how the <u>double slit</u> <u>experiment</u> has provided amazing evidence, or proof if you now prefer, that protons and neutrons and so atoms are indeed <u>entangled</u> with their own dimensions of Space and time which are <u>actually sourced from themselves</u>. That is, without atoms being <u>entangled</u> with their <u>own self created</u> dimensions of Space this experiment could not have provided the result which it has, since, there would be <u>nothing to</u> <u>accumulate</u> and combine to provide a completed Dimensional Reality.

<u>Building a complete entangled Dimensional Reality:</u>

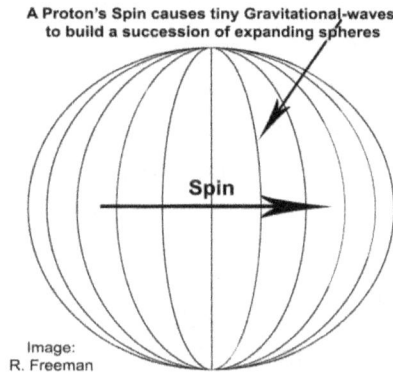

A Proton's Spin causes tiny Gravitational-waves
to build a succession of expanding spheres

Spin

Image:
R. Freeman

All illustrations I have seen of Gravitational waves show Gravitational waves expanding on a thin, flat plane like ripples from rain drops on a pond. To provide all around dimensions the tiny Gravitational waves need to expand more like a succession of expanding spheres. Protons and neutrons <u>have</u> <u>spin</u> which if spin is not in line with the tiny Gravitational waves may achieve this expanding spheres effect. However, spin alignment or not, since the Gravitational waves are so tiny, just a Planck length thick, ideally, it requires many atoms to come together so as their tiny Gravitational waves are collectively expanding in all directions and without gaps. If spin <u>is</u> <u>in line</u> with the Gravitational waves it will naturally require many <u>more</u> <u>atoms</u> to come together to build a robust Dimensional Reality.

Only when many atoms come together does an object gain a complete all-around robust, four dimensional, spatial reality complete with the fourth dimension of passing time. Of course, an object as small as a tiny grain of salt has a complete and robust Dimensional Reality and is always located at a dimensional position which can be precisely calibrated. A fine grain of salt is said to be made from around 1,000,000,000,000,000,000 atoms which all of the <u>spinning protons and neutrons</u> of all of these atoms will surely <u>build a complete robust all-around dimensional web of reality,</u> a succession of expanding spheres, from where there is no escape from residing at a 'certain' position at a certain moment of time within a coordinated system.

Double slit experiments have provided indisputable, <u>direct evidence </u>that a dimensional, classical reality is <u>not a given thing provided by simply being within Space</u> but is a commodity which <u>has to be built by an accumulation of atoms</u>. Scientists should ask; what property of atoms could possibly build a Dimensional Reality in this way? The configuration of tiny Gravitational waves, **<u>entangled</u>** with their source, easily fulfils this requirement by easily explaining the results of experiments. The finding from double slit experiments <u>provides powerful evidence</u> that tiny Gravitational waves, as expressed with this model, do really provide one's own Dimensional Reality. Clearly, a Dimensional Reality complete with inseparable passing time and dimensions which can be manipulated by Gravity are not a given thing from residing within Space. If a Dimensional Reality was given from merely residing within Space all particles would surely behave the same exact way which, as Quantum Theory and experiments clearly reveal, they do not. The experiments firmly reinforce the concept that a complete robust, Dimensional Reality is only achieved <u>when many atoms come together</u> in a configuration which would allow <u>tiny Gravitational waves</u> to build a robust, inescapable, dimensionally woven web of reality.

Many trillions of tiny Gravitational waves expanding from the many trillions of protons and neutrons of our own atoms collectively build one's unique Dimensional Reality from where there is no escape from being at the center of. There could hardly be <u>any other commodity</u> capable of providing and building this <u>observed threshold</u> and which clearly matches the results of double slit experiments. **The tiny Gravitational waves are clearly <u>required to be entangled</u> with their source so as a Dimensional Reality is unique to an object's speed and Gravitational position and <u>is not</u> a given thing by simply being within Space.**

[018] All is why the double slit experiment has provided amazing evidence, or proof if you prefer, that this modeling is truly correct. A tiny phase of virtual mass which mimics a tiny collapsing and reappearing Black Hole is <u>ideal</u> for creating tiny Gravitational waves which are <u>ideal</u> for providing dimensions to Space. The same phase of virtual mass is also ideal for strongly binding and agitating a proton's particles at near light speed to create regular mass and as it mimics a tiny collapsing and reappearing Black Hole is an ideal source of Gravity. [011; 015; 016;] The same lower state phase of virtual mass has been clearly exposed in experiments but no one knows what it is and has been separately labeled by Scientists as being an extraordinary, astonishing phenomenon which is one of the biggest unsolved mysteries in physics.

Four dimensions from just one prime dimension:

Because dimensional Space and time cannot be separated, somehow, Space has dimensions of length, width and depth plus the passing of time and **all are physically woven together as one of the same. It is enshrined in theories, I read it over and over again; Space and time are as one, hence, Space-time.**

Within a Natural Universe the dimensions of length, width, depth and the passing of time are <u>all derived from one of the same</u> which is why, like Einstein has said, (dimensional) Space and time cannot be separated.

Although some proposed theories for the universe require as many as a mind twisting 26 dimensions to build a complete <u>dimensional</u> universe, my Natural Model requires <u>just one dimension</u>. A complete all-around four dimensional, spatial reality is actually built from just one primary dimension which is the dimension of length. The key thing which allows this is dimensions expand from the center of all Dark matter, protons and neutrons. As previously explained the dimension of length is derived from tiny Gravitational waves expanding from all spinning protons and neutrons and so from all objects made from atoms. The dimension of length is the wavelength, trough to trough, of one of these very tiny Gravitational waves. This distance is the smallest possible measurement within a Dimensional Reality which is called a Planck length. The passing of one of these tiny Gravitational waves provides the passing of a dimensional moment which provides the passing of a moment of time which is the smallest possible duration of passing time and is called Planck time. So now we have two dimensions, the dimension of length and the dimension of passing moments of time derived from the dimension of length.

Trillions of these tiny Gravitational waves expanding from an object collectively build a complete all-around, four dimensional web of reality. In doing this, the dimensions of length naturally reach every facet of every object to provide the dimensions of both of what we perceive as width and depth. This is why Planck calculations are not able to expose or express a Planck width or a Planck depth. Now all four spatial dimensions of length, width, depth and passing time are derived from the dimension of length.

Evidence that four dimensions are provided by the dimension of length: Einstein's theory of Special Relativity says the length of objects moving at relativistic speeds undergoes a contraction of the **dimension which aligns with its motion.** Any dimension, regardless of length, width or depth, at any angle across an object which aligns with motion is the dimension affected by length contraction because this is the direction which the tiny Gravitational waves become stretched.

Waves becomes stretched because if Fast Mary is travelling at nearly the speed of light and the tiny Gravitational waves are travelling at the speed of light the waves depart Fast Mary very slowly in the direction of motion. Due to Mary's high speed a wave is now released, in the direction of motion, over a greater distance which 'effectively' stretches the wave over a greater distance. Because the wave maintains Planck length the <u>distance travelled while the wave is released</u> is now contracted to a Planck length which causes length contraction in the direction of motion. Because the slowly released wave takes longer to pass it slows Fast Mary's passing of time but because the wave retains Planck time Fast Mary observes her own passing of time and the length of her spaceship as normal, however, she has a shorter distance to travel. This is similar to how Gravity is delivered.

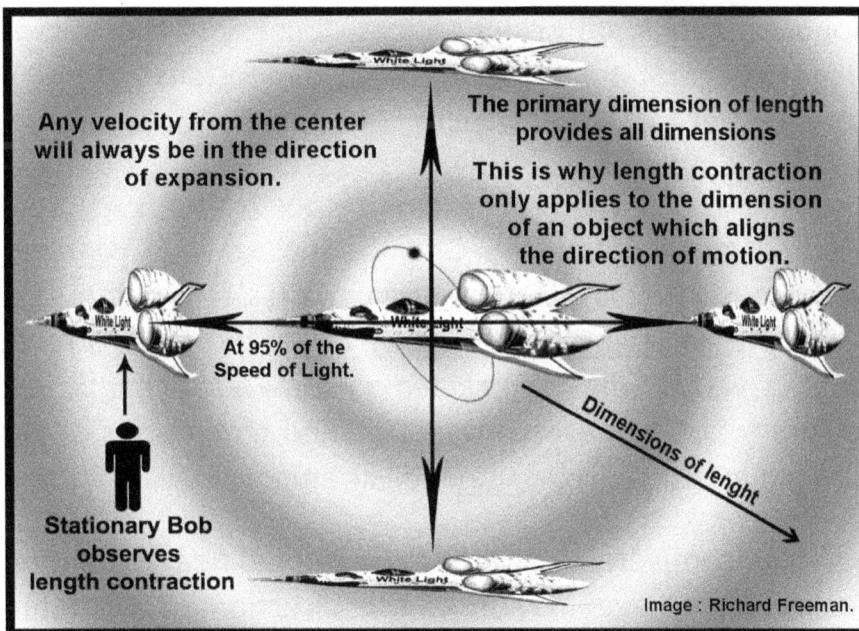

Length contraction <u>only affects the dimension</u> of an object, <u>regardless of it being the length, width or depth</u> of the object, which <u>aligns</u> with the direction of motion. This provides direct evidence that it is truly only one dimension, the dimension of length which provides all dimensions. For example, if Mary's space ship was travelling <u>sideways</u> Special Relativity says **length contraction now contracts the width** of Mary's space ship and Stationary Bob now observes Mary's Space ship as being extraordinary thin and skinny. This is why it requires many atoms providing the dimension of length in many different directions to come together to build a succession of expanding spheres which provide a complete Dimensional Reality of length, width, depth and passing time. The <u>revealing evidence</u> of this is it takes many atoms to come together to fail the double slit experiment. The dimension of length is indeed providing all dimensions.

Because all dimensions are provided by the dimension of length, no matter which dimension aligns with motion it will incur length contraction. Length contraction is a prediction of Einstein's theory of Special Relativity. There is no width or depth contraction within Special Relativity. Length contraction due to motion is <u>additional evidence</u> that dimensions are <u>entangled with their source</u>, that is, Stationary Bob's own dimensions <u>do not normalize</u> Bob's observation of Fast Mary's Space ship. Mary's Space ship very clearly provides its own dimensions.

Being stretched means each tiny Gravitational wave <u>covers more of an object</u> but because the tiny Gravitational waves retain the same Planck length value <u>the stretched distance covered</u> by each stretched, tiny Gravitational wave becomes contracted to the Planck length value which allows Stationary Bob to observe the stretched dimensions of Fast Mary's spaceship as a dimensionally <u>smaller</u> measurement.

Once you have a complete all around Dimensional Reality built from many atoms one can only have motion in the direction of the dimension of length. This is why Special Relativity only applies length contraction to the dimension which aligns with the direction of motion. Motion is obliged to always be the direction of expansion of the tiny Gravitational waves which provide entangled dimensions. All directions from the center always align with the 'one' direction, that of the direction of expansion, which is why length contraction only applies to the dimension of an object, regardless of it being the length, width or depth of the object which faces and aligns with the direction of motion. Actually, this is again confirming that the dimension of length is also providing width and depth.

We have covered Planck length and Planck time but what of Planck mass? With this model, Planck mass equals mass created over a moment of Planck time. Planck time is the passing of a tiny Gravitational wave created by a Quantum vacuum fluctuation which fluctuates to a vacuum of a lower state than the ground-state of dimensional Space which <u>provides a unit of virtual mass.</u> Virtual mass creates regular mass by an amount equal <u>to its interaction with particles.</u> <u>This interaction equals Planck mass</u> which is <u>not</u> upper or lower bound.

Scientists say their calculations from observational evidence show we reside within a <u>flat</u> universe, what do they mean and how can this be within a Natural Universe? On a universal scale each dimension of length extends forever, on a comparatively straight <u>flat plane only around a Planck length thick,</u> like ringlets from rain drops on a pond. However, flat in this instance means two parallel lines extended across the universe remain parallel. Draw two parallel lines on a sheet of paper and imagine they extend across the universe. It is important to note that each line is naturally <u>required</u> to begin at a point which is dimensionally apart from the other line. Now imagine that these two lines are made from photons (light). The lines will remain exactly parallel because each began at the exact same angle <u>from its own dimensional source</u> and there is nothing from its dimensional source to cause a line to curve or deviate. This prevents these parallel lines coming together or to curve like a circle to eventually arrive back at a beginning point. This is called a flat universe which happens because dimensions expand in the form of tiny Gravitational waves which, though curved as a circular wave, travel on a <u>straight path</u> at the speed of light without curving, deviating or bending back on themselves. The waves are <u>driven outwards</u> at light speed by <u>a relentless attraction to a far outer void of completely empty Space</u> which we have called Anti-space. Consequently, the waves will seemingly expand on the same course for eternity. Naturally mass and energy throughout the universe will interfere with this analogy. When Scientists say the universe is flat they are referring to the basic structure of all dimensional Space, not the bumps along the way.

According to Special Relativity, Fast Mary observes herself as being stationary and Bob as having high speed and slow time. This is for the reason that if Mary observes herself as stationary at the center of her entangled Dimensional Reality, Bob relative to her must have speed. However, actual experiments using atomic clocks confirm a moving clock or a clock exposed to stronger Gravity runs slower, suggesting that Fast Mary always has the 'slower time'. If one clock is stationary while another has speed, the clock exposed to more speed runs at a slower rate. When the two clocks are bought 'stationary' together again they will both again run at the same rate but the one which was exposed to higher speed lags behind corresponding to the period at which it was exposed to higher speed.

So how could it be possible for both Fast Mary and Stationary Bob to observe each other as having slower passing moments of time? One would at first wish to believe if Stationary Bob observes Fast Mary as having slow passing moments of time, than Fast Mary must naturally observe Bob as having fast passing moments of time but Special Relativity says that this is not so. One must think about this very carefully for this prediction of Special Relativity is indeed difficult to rationally explain. First of all, keep in mind that Stationary Bob is not really stationary in the sense that the whole Earth is spinning and moving through Space. How can it be that one party has fast time and the other party has slow time but both observe each other as having slow time? I have seen in forums Physicists argue whether or not this is true but the purist to theory will always say it is truly this way.

A stretched or distorted tiny Gravitational wave retains <u>the same</u> single pulse of energy which retains a single Planck unit of length or passing time. For Fast Mary this <u>normalizes</u> her <u>entangled</u> distorted dimensions so as she observes her dimensions as normal and herself as stationary. **Since Bob's** dimensions are <u>only entangled with himself</u>, Stationary Bob's <u>entangled</u> **Dimensional Reality <u>cannot normalize Fast Mary's</u> Dimensional Reality. Consequently, Bob now observes Mary's spaceship <u>as it is</u> within his own entangled Dimensional Reality where Bob observes Mary's spaceship as having less dimensions of length and Mary's time as slow when compared to his own Dimensional Reality which is running at a much faster rate.**

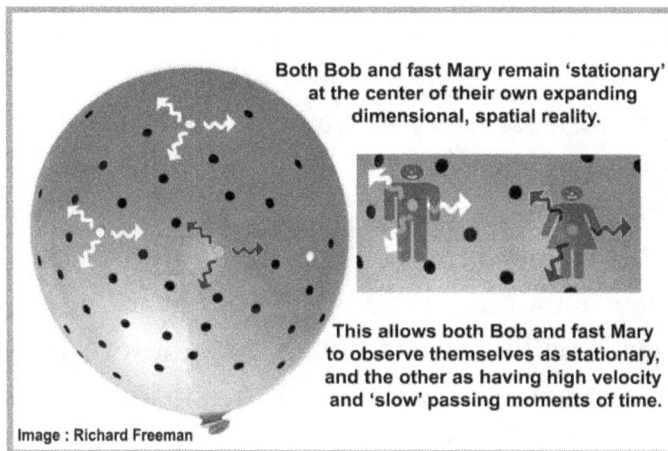

Both Bob and fast Mary remain 'stationary' at the center of their own expanding dimensional, spatial reality.

This allows both Bob and fast Mary to observe themselves as stationary, and the other as having high velocity and 'slow' passing moments of time.

Image : Richard Freeman

Since, Fast Mary's <u>entangled</u> Dimensional Reality normalizes the 'stretched over distance' waves as Planck values she may observe herself as <u>stationary</u> at the center of her <u>own entangled</u> Dimensional Reality. While observing herself stationary at the center of her own entangled Dimensional Reality, Mary now observes the planet which she is travelling to is speeding towards her and observes other objects, including Stationary Bob as having slower time. It acts this way for the <u>same exact reason</u> as free falling to the ground where your own self entangled <u>stretched</u>-by-Gravity dimensions are <u>maintaining a vital</u> single pulse of energy so are contracting the stretched distance covered by a wave to single Planck values which is causing Gravitational length contraction and Gravitational time dilation. Consequently, when free falling, you are now stationary within your own Dimensional Reality and the ground is speeding towards you. The undeniable evidence of being stationary within your own dimensional Space is you will <u>feel</u> no acceleration when free falling.

Mary and Bob may observe things differently because each resides within their own unique, dimensional universe where their own entangled Dimensional Reality is recording dimensions as Planck values. So who really has the slowest rate of time? The slowest rate of time naturally belongs to whoever has the least recorded time when their clocks are bought together at the same speed. That is, if Fast Mary immediately returns to Earth her 'slow' wrist watch may have only advanced less than one minute while the comparatively 'fast' clocks on Earth have advanced 200 years.

There is no benefit for Fast Mary to travel at a speed faster than the speed of light because at the speed of light, like a photon, she can instantly travel to anywhere in the universe and do so without ageing. The problem with light speed travel is in returning to Earth, where hundreds or even billions of Earth years may have passed by while Fast Mary has been gallivanting around the universe within an instant of time. There is no reasoning for Fast Mary to arrive at her destination 'before' she leaves her departure.

Fast Mary cannot exceed the speed of light because the tiny Gravitational waves which are providing her with entangled dimensions to Space are also travelling at the speed of light. At the speed of light, in the direction of motion, her dimensions cannot now escape from her to provide Space with her entangled dimensions. Now she has no entangled, dimensional distance or passing of time. To exceed the speed of light she must travel farther than the dimensional distance of 299,792,458 meters in one second of passing of time, however, at the speed of light Fast Mary now resides where Space effectively has no dimensional meters or seconds of passing time ahead to accomplish this. At the speed of light both dimensional distance and the passing of time are irrelevant since there is no dimensional measurement to exceed in order to travel faster than light.

Multiverse: The notion of multiple universes or Multiverse has surfaced in theorized physics, cosmology and astronomy. Physics can lead to some rather strange predictions without exposing rational or physical reasoning of how such a prediction occurs; we are regularly left to ponder only the equation. 'Sliders' is an American television series based on the notion of different parallel universes. Why do such notions <u>actually appear in theorized physics</u>? My model provides a universe which has multi realms of spatial, classical realities, a different one for every atom or COST unit based on an atom's unique speed and Gravitational position within the universe. The notion of multi universes may be derived from such a characteristic; Stationary Bob, aged 30, can only observe the universe around him as it is within his own Dimensional Universe provided by his own entangled spatial dimensions and Fast Mary, also aged 30, can only observe the universe around her as it is within her own Dimensional Universe provided by her own entangled spatial dimensions. Both Bob and Mary self-observe a normal universe of dimensional Space-time with normal time, however, both actually reside within a vastly different Dimensional Universe with different time and different dimensions which are unique to themselves.

Bob observes Mary as she is within Bob's own observable, unique, Dimensional Reality. Mary observes Bob as he is within Mary's own observable, unique, Dimensional Reality. One may visualize each expanding, classical reality freely moving around within dimension<u>less</u> Space, from where entangled, expanding dimensions freely pass through other entangled, expanding dimensions creating a web of dimensional Space and time. This is similar to the way expanding ringlets from rain drops on a pond freely pass through all other expanding ringlets creating a complete distortion on what was a seemingly transparent smooth surface.

That is, we all observe objects as they are within our very own spatial Dimensional Reality. Objects and events may appear different within different spatial Dimensional Realities because they <u>are really different</u>.

We are all truly individually trapped within our very own universe of classical dimensional Space-time from where we observe each other and every object 'as it exists' within our very own classical spatial reality. That is, all multi or parallel universes passively co-exist for the reason that we all observe all objects as they are within our own universal classical reality. For the reason that all classical realities are based on one's unique speed and Gravitational position within the universe, no two are exactly alike. There may be an almost infinite number of <u>different dimensional</u> universes all of which are continually being renewed at the tiniest of scales and at the amazing speed of light and co-exist as <u>collectively</u> woven together and existing as our home universe of dimensional Space and time.

Because Stationary Bob can only observe from his own, self entangled dimension of time and Space, Stationary Bob observes that it will take Fast Mary a little over 100 years at near the speed of light within his own dimension of time to reach her destination. Fast Mary can only observe from her own, self entangled dimension of time and dimensional Space, Fast Mary observes it will take her just 7 seconds within her own dimension of time and dimensional Space for the reason that she has very few moments of dimensional Space and time between her and her destination.

Stationary Bob ages to be a very old man and dies before Fast Mary arrives at her destination. Fast Mary arrives at her destination, still age 30, a few seconds after her spaceship passed Stationary Bob. This agrees with the predictions of Special Relativity and agrees with the notion of Stationary Bob and Fast Mary each residing within entirely different expressions of dimensional Space and time. For Fast Mary there is not only less self entangled time between herself and the planetary system she is travelling to, there is also less self entangled dimensional Space (dimensions of length) between herself and the planetary system of a far away star.

My Natural Model now has to do what has never been done before; to rationally unravel the irrational, crazy realm of tiny Quantum particles which behave in the most illogical, bazaar and seemingly impossible ways which totally disregard the rules of a Dimensional Reality. Your common sense is attempting to tell you there has to be a rational answer.

13 RELATIVITY AND QUANTUM THEORY.

The theories of Special Relativity, General Relativity and Quantum Theory are brilliantly successful but can they be unified as a theory of everything? Within a Natural Universe Einstein's Relativity describes how objects reside within their own, entangled dimensions of Space and time while Quantum Theory, first developed by Max Planck, best describes how particles reside without entangled dimensions of Space and time. Today's electronics use and depend on the bizarre predictions of Quantum Theory. Although atoms are part of Quantum Theory, atoms are composed of tiny particles. This chapter refers less to atoms which come together in numbers to build a complete Dimensional Reality and more to tiny particles which reside without a Dimensional Reality. Quantum is the realm of very tiny particles.

Within a Natural Universe atoms and all objects made from atoms remain at stationary equilibrium suspended and forever trapped at the center of numerous sets of very tiny units of eternally expanding energy provided by very tiny Gravitational waves which provide these objects with active, entangled dimensions to their Space. Entanglement provides you with your very own unique, entangled Dimensional Reality while all other Dimensional Realities, belonging to all other observers, are **totally ignored.** However, my Natural Model cannot provide tiny Quantum particles with the same permanent, self mechanism to provide entangled dimensions to their Space. Without active, self entangled dimensions to Space, Quantum particles naturally **totally ignore** the dimensions of Space-time and effectively reside within dimension<u>less</u> Space where everyplace is truly in the same dimensionally nowhere place and where all passing of time is reduced to an <u>instant</u>. The tiny particles effectively reside in a <u>fifth dimension</u> where their Space is without <u>self entangled</u> dimensions.

Quantum Theory allows a particle to exist in a foggy situation of possibility, to be anywhere, everywhere or nowhere until returned to classical reality by an act of measurement or observation. It is not a case of not knowing exactly where a particle is but rather a particle is <u>truly at every possible place at the same instant of time until the particle is observed</u>. Sometimes particles are entangled with a partner particle and can instantaneously, faster than the speed of light, <u>imitate the same actions</u> of an entangled partner even if its partner particle is light years away. It is truly as if there is no Space between the entangled particles and is like physically shaking hands with your partner who is on the other side of the Earth. How is this possible? How can a particle be everywhere at the same time? Answer; within an <u>entangled</u> Dimensional Reality a particle <u>cannot</u> and common sense will tell you it's impossible. So what allows the impossible to happen?

No one has ever been able to provide a true rational and <u>correlated</u> answer to this totally irrational, absurd phenomenon. Quantum Theory says a particle may disappear and reappear at another location which has now been observed in experiments inducing some to say; *"you are made from particles so one may at any moment disappear and reappear on far away Mars"*. Why that does not really happen is a mystery to today's science.

[071] Renowned for his effort to prove that Quantum behavior, though bizarre, really takes place, American Physicist Richard Feynman has said; *"We choose to examine a phenomenon which is impossible, absolutely impossible, to explain in any classical way, and which has in it the heart of quantum mechanics. In reality, it contains only mystery"*.

My Natural Model is the <u>only model capable</u> of unraveling, *in a classical way,* this mystery of all mysteries. Quantum particles have <u>no permanent</u> mechanism to provide their Space with the tiny eternally expanding Gravitational waves which provide <u>entangled</u> dimensions to <u>their Space</u>. Quantum particles residing without <u>entangled</u> dimensions to Space **totally ignore** the dimensions of Space and effectively reside within <u>dimensionless Space</u>. Within Space without entangled dimensions everyplace is <u>literally</u> in the same dimensionally undefined place at the same instant of time. A Quantum particle has <u>no entangled</u> dimensions to even provide itself with particle dimensions so resides as a dimensionless fuzzy wave. A Quantum particle resides in Space where every position is truly in the same nowhere position and where <u>no particle can be precisely anywhere.</u> Now a simple act of observing or measuring will momentarily position a particle at a precise dimensional location within your very own entangled Dimensional Reality.

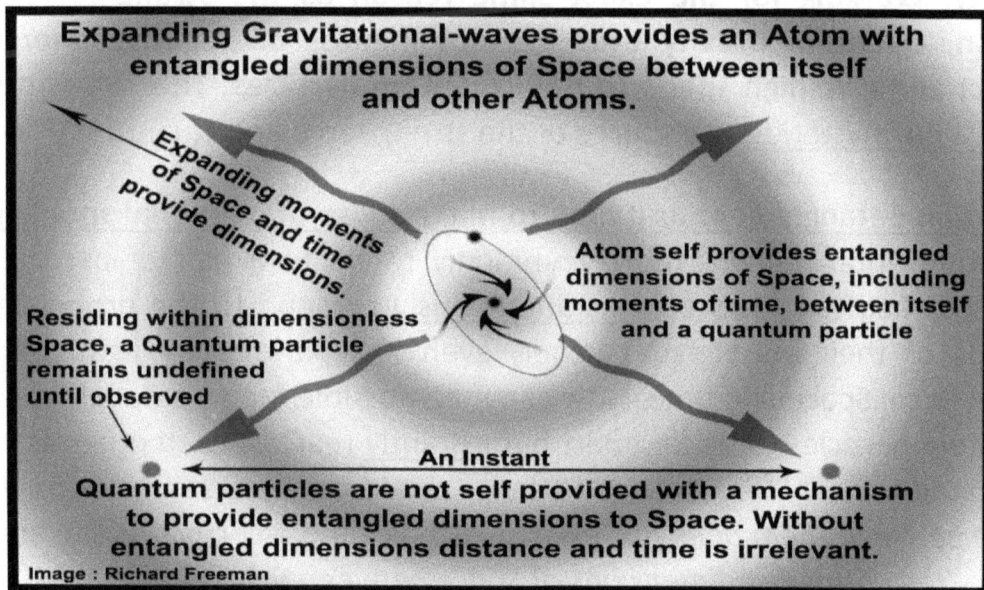

Expanding Gravitational-waves provides an Atom with entangled dimensions of Space between itself and other Atoms.

Expanding moments of Space and time provide dimensions.

Atom self provides entangled dimensions of Space, including moments of time, between itself and a quantum particle

Residing within dimensionless Space, a Quantum particle remains undefined until observed

An Instant

Quantum particles are not self provided with a mechanism to provide entangled dimensions to Space. Without entangled dimensions distance and time is irrelevant.

Image : Richard Freeman

<u>Quantum Gravity is solved with just one paragraph:</u> Naturally this means tiny Quantum particles ignore the dimensions of Space-time and are <u>not influenced in the normal way by Gravity</u> which works by warping the curvature of the dimensions of Space-time. However, <u>without being</u> **restrained** at <u>stationary equilibrium by the energy</u> of entangled dimensions of Space-time, a tiny Quantum particle is free to have a **direct attraction to the strong attraction part of Gravity**. <u>Free of the shackles of being trapped</u> at stationary equilibrium within <u>entangled</u> dimensions allows tiny Quantum particles, including photons of light, to <u>instantly speed away</u> in all directions due to the attraction part of Outer Gravity and for the attraction part of Inner Gravity to <u>redirect their path</u> when near an object of mass. Thus, tiny Quantum particles now act <u>exactly like</u> they are observed to. **This provides Quantum particles with what can now be called <u>Quantum Gravity</u> which works with only the strong attraction part of Gravity and <u>without</u> the dimensions of Space-time.** Since dimensional distance and passing time are now irrelevant the attraction to the far outer void is significant. This is how Gravity operates at the scale of Quantum Theory. <u>Instant speed</u> is a feature of Quantum Gravity and <u>accelerating speed</u> is a feature of Einstein's Gravity. **<u>Working for the exact same reason as the strong force of a proton</u>, Quantum Gravity uses the nothingness of negative energy to attract and give high speed to tiny Quantum particles.** This <u>solves two of today's big mysteries;</u> the source of the strong force and Quantum Gravity.

<u>Quantum Entanglement and how to make sense of irrational spookiness:</u> Quantum entanglement more commonly refers to a pair of particles so closely linked that they can <u>instantaneously</u> respond to the properties or actions of their entangled partner, even if separated by light-years of dimensional Space and time. How can it be possible for an entangled pair of Quantum particles to be far apart but <u>instantly</u> react to each other as if not separated? [108 See video: Einstein's Quantum Riddle] The scientists say it is *"as if the Space between them does not exist"* which is almost correct as the answer will be *'the <u>entangled</u> dimensions of Space between them do not exist'*.

<u>Quantum Entanglement solved with just one sentence:</u> The tiny particles **are not separated by an <u>entangled</u> Dimensional Reality;** it is the **entangled** dimensions of a Dimensional Reality which provides the <u>dimensional</u> separation <u>not the dimensionless</u>, complete nothingness of Space.

A pair of Quantum entangled particles are entangled with each other's opposite entangled orientation of spin mechanism but <u>are not entangled</u> with a mechanism to provide <u>self entangled</u> dimensions to Space. Residing without a mechanism to provide self entangled dimensions to Space allows each particle to <u>ignore</u> the dimensions of Space and <u>instantly</u> react to each other's <u>properties or actions</u> even if perceived to be separated by a vast dimensional distance within an observer's entangled Dimensional Reality.

I have used the word 'evidence' many times throughout this book. I don't know how much evidence adds up to proof but Quantum entangled particles provide extraordinary evidence or <u>real proof</u> if you now prefer, that the universe is only explainable and really works exactly the way of my Natural Universe. Because dimensional Space and passing time cannot be separated <u>instant</u> time requires dimensionless Space, consequently, entangled particles instantly reacting to each other is actually <u>undeniable proof</u> of a realm of dimensionless Space which only the modeling of my Natural Model can very clearly provide. [073] Quantum Theory says all Quantum particles have <u>no properties</u> until they are observed or measured. So this model needs to <u>solve</u> why and how a Quantum particle <u>obtains its properties</u> as it <u>comes into being a particle</u> when observed or measured.

A pair of entangled Quantum particles may each have a spin axis which points up or down, if one of the entangled particles is measured as 'down spin' its entangled partner will always have 'when measured' the opposite 'up spin'. Should the particles be separated and we measure the spin of one particle as being 'down spin' we know the other, even if far away will, <u>when measured</u>, always have the opposite 'up spin'. [075; 090] No worries? One would like to believe that these particles were always oriented that way like a pair of gloves, which is the way Einstein would have liked it to have been.

However, within the world of the mathematics of Quantum Theory, particles <u>do not have defined properties</u> until they are actually observed or measured. An entangled Quantum particle's spin axis points neither down nor up <u>until it is actually being measured</u>. Regardless of 'distance' apart, if one measures the spin of one particle as being 'down spin' the other particle <u>instantly</u>, faster than the speed of light, **reacts** to its entangled partners spin <u>to now be observed or measured</u> with the opposite 'up spin'.

[075; 090; 108] **Albert Einstein had a natural distaste for this prediction from the Quantum world and said he thought of this as** *"spooky actions at a distance".* [072] Scientists have now confirmed such actions really occur with experiments on the ground as well as between Earth and a satellite in Space; an entangled particle can really convey its properties or <u>actions</u> to its entangled partner instantaneously. The problem being is that faster than light speed of 'anything' is prohibited by the current laws of Einstein's Special Relativity and indeed no communication has been sent or received, rather it <u>really is</u> as if these particles are mysteriously not separated by Space, as if both are in the same place, even if observed <u>within one's Dimensional Reality</u> to be very far apart. As many have asked; how could this be possible? It occurs because these particles do not permanently have an <u>entangled</u> Dimensional Reality. **If a particle <u>had to deal with the ramifications of entangled spatial dimensions</u> it would be <u>impossible</u> for a particle to behave this way.** In a Natural Universe tiny Quantum particles act <u>without entangled dimensions</u> so simply <u>totally ignore</u> the dimensions of Space and effectively reside within dimensionless Space where they naturally act <u>exactly like Quantum Theory says</u>. While objects made from <u>many</u> atoms are irreversibly, <u>permanently locked by entanglement</u> to their own unique, dimensional Space-time from where Einstein Relativity excels.

My Natural Model provides the first and likely the only possible realistic answer capable of <u>providing rational reasoning</u> for solving the most bizarre predictions of Quantum Theory. The model makes clear, rational sense of the predictions of Quantum Theory, which are said to make no rational sense to anybody. The solution is clear; instantaneous time cannot occur between objects made from <u>many</u> atoms because they are <u>permanently</u>

entangled with their own mechanism for expressing throughout Space the dimensions of length, width, depth and the passing of moments of time. Instantaneous time occurs between tiny Quantum particles because they are not constrained by residing at the center of numerous sets of very tiny units of expanding energy provided by very tiny Gravitational waves which provides self entangled dimensions of length, width, depth and the passing of moments of time to Space. **Both Special Relativity and General Relativity superbly describe how objects made from many atoms behave within an entangled Dimensional Reality and Quantum Theory superbly describes how Quantum particles behave without an entangled Dimensional Reality.**

To display properties such as up spin or down spin requires the passing of time, consequently, without entangled dimensions it would be impossible for these particles to display their properties until they are provided with, when observed or measured, a mechanism for the passing of time.

Unfortunately, even though Quantum Theory actually predicts the results confirmed by Quantum Entanglement experiments, commonly called a Bell test, science today has no reasoning for 'instantaneous time' or 'dimensionless Space', preferring to have the same opinion as Einstein by relating to the results of these experiments where an entangled pair of particles instantly relate to each other, regardless of distance apart, as irrational 'spookiness'. Within our Natural Universe the entangled particles, although entangled with each other, are not self entangled with a mechanism to permanently generate the tiny Gravitational waves to provide their surrounding Space with the entangled dimensions of length, width and depth or the passing of moments of time. My Natural Model can now truly make rational sense out of irrational 'spookiness'.

A particle's wave silently reflects its particle position which can be revealed by the act of observing or measuring causing a wave to have an interaction with an entangled Dimensional Reality. Observing or measuring acts like observing through a dimensional filter.

Within dimensionless Space all places are in the same place so an entangled particle is in the same place as its entangled partner. Now there are no restrictions on particles <u>instantly copying</u> the properties or actions of their entangled partners. This is where particles <u>are not separated</u> by entangled dimensions. Particles only become separated when their position is exposed, by observing or measuring, <u>from an entangled Dimensional Reality</u>. The true dimensional position of a particle is where it is observed to be within an observer's Dimensional Reality. While fishing I may drive around for hours staring at my echo sounder searching for a school of fish. A fish may be in any place under the sea but is more likely on a part of a reef where you expect it to be. Like a fish in the ocean a particle is where you observe it to be and is more likely found where you expect it to be.

With particles residing within dimensionless Space particles are not dimensionally separated because, without the constraints of dimensions, both distance and passing time, naturally becomes totally irrelevant. Distance is of course real within an entangled Dimensional Reality but is irrelevant without an entangled Dimensional Reality. For example; take a train journey of 170 kilometers which will take 2 hours, when we remove dimensions the very same train journey is now zero kilometers which will take zero time because without dimensions, distance and the passing of time is now totally irrelevant.

So how does this mirror a prediction of Einstein's Special Relativity? At the speed of light an object made from atoms can no longer provide 'entangled' dimensional moments to Space for the reason that it would be travelling at the same speed as its own advancing away (expanding) moments of dimensional Space and time provided by tiny Gravitational waves. Consequently, for objects travelling at the speed of light their Space has no entangled dimensions, which is why Special Relativity predicts that time stops and distances shrink to effectively become zero at the speed of light. So why does time stop and all objects appear to be in the same place at the speed of light? The answer is obviously simple; at the speed of light an object's 'Space' has no entangled dimensions so the object <u>totally ignores</u> the dimensions of Space. Quantum particles have no permanent mechanism to self generate entangled dimensions to surrounding Space so Quantum particles naturally also <u>totally ignore</u> the dimensions of Space and reside within this very same dimensionless state of Space regardless of speed. **Now Quantum Theory and Special Relativity <u>both predict the same thing</u> for the same reason, which is a very neat correlation which <u>only this Physical Reality model can clearly provide understanding for why it is so.</u>**

A pair of Quantum particles cannot self generate the tiny Gravitational waves to provide entangled dimensions of length, width and depth and a passing of time. Consequently, for Quantum particles separation by distance and time is <u>irrelevant</u> which allows each Quantum particle to be instantly aware of its entangled partner's properties. Dimensions are more an entangled property of an atom's COST units and less a property of the nothingness of Space.

This raises a most interesting and beautiful thought; it <u>allows Mother Nature</u> to work with both dimensionless Space and dimensional Space-time. In this way, the realm of our classical reality of dimensions including a passing of time and a dimensionless realm of instant time are bought together to operate in harmony. Operating within a dimensionless realm would have many obvious advantages and remarkably, such harmony has been observed in plants where Quantum particles appear to take advantage of our dimensionless Quantum world to achieve <u>a near perfect efficiency while harvesting light</u>. Within a dimensionless realm of Quantum, light speed is like snail mail. Whether Relativity or Quantum, Mother Nature has it covered. Operating within instant time would obviously provide great speed advantages to the function of computers. The problem I see for a real Quantum computer is that at every point which a particle is required to be observed or measured would create a bottleneck which constrains the speed of a Quantum computer to the spatial Dimensional Reality of our classical world.

This brings to mind an intriguing concept. By blocking the tiny Gravitational waves a pure Quantum drive may have no limitations. A Quantum drive for a spaceship is most probably more <u>pure science friction</u> than ever possible; however, if Scientists were able to design a spaceship with a shield to shield the spaceship and its occupants from their own, <u>entangled</u> dimensions, it opens the possibility of 'instantly' travelling to any place in the universe. One may call this sub-Space travel and even, like Dr. Who, a police phone box could be configured as a 'Time and Relative Dimension in Space' object to do the same job. Dimensionless-Drive provides a far superior form of travel than any slow-old Warp Drive used by Star Trekkers. However, at the very best, this would probably not be completely free of time ramifications. For instance, if one traveled to a planet which was 2000 light years away one would effectively arrive 2000 years into the future of how that planet had appeared from planet Earth.

This should have returned us to the all important measurement of an entangled particle's spin. However, it requires a passing of time to observe, measure or for the particles to even display spin properties, however, the particles have no natural mechanism for providing a classical 'Dimensional Reality' including passing moments of time. Fortunately, we can deduce from the results of experiments and from the predictions of Quantum Theory that the instant one measures or observes the particle it is somehow given a 'Dimensional Reality' including passing moments of time by the simple act of measuring or observing. From our own Dimensional Reality we reside at the center of expansion of entangled dimensional Space-time from where there are always passing moments of time which should make it <u>impossible</u> to observe a realm where there is no passing moments of time. However, there is a way.

For a pair of entangled particles to even <u>display</u> properties such as up spin or down spin obviously requires the passing of time, consequently, without entangled dimensions it <u>may be impossible</u> for entangled particles to display their properties until they <u>are provided</u> with a mechanism for the passing of time. <u>I would have much preferred</u>, at observation, <u>one's own Dimensional Reality provides</u> a Quantum particle with dimensions and the passing of time so as the particle can be observed as it is within one's own Dimensional Reality. **That <u>really is likely</u> all there is to it**, however, we know thanks to Special Relativity that Stationary Bob observes Fast Mary as having slow time and observes her spaceship as being short and stubby. Consequently, observing <u>should not provide an object</u> with one's own entangled dimensions. Observing or measuring simply allows one to observe an object as the object <u>will appear with its own dimensions</u> when observed from one's own entangled Dimensional Reality.

<u>Unfortunately</u>, it seems the act of observing or measuring needs to provide a Quantum particle with a temporary configuration of Space and time, similar to the Quantum vacuum fluctuation of a COST unit, which is able to provide a particle with its own entangled dimensions including a passing of time all of which is needed to display properties, including the properties of spin and mass, and thus allow a Quantum particle's <u>properties</u> to be observed or measured 'as they are' within one's own Dimensional Reality.

One will obviously require a reason as to why 'observing' triggers or activates a Quantum vacuum fluctuation. Conveniently, theories say that Quantum vacuum fluctuations are simply found everywhere, at every minuscule point in Space, regardless of observing, that is, nothing causes a Quantum vacuum fluctuation, they are just always there. However such a notion creates big problems for science such as 'the cosmological constant problem' or 'the vacuum catastrophe' and cannot solve how 'observing' can change a wave into a particle (wave-particle duality) or explain, as in this case, why a Quantum particle has no properties until it is observed or measured. In the Quantum world 'observing' or 'measuring' has to perform many 'physical-like' acts, which a Quantum vacuum fluctuation modeled on the COST can perform. That is, the more mysteries that are solved by the one solution the more credible the solution becomes, which is why one should avoid solving different mysteries with unrelated solutions. The universe should be correlated to act as one, united, single entity.

Consequently, instead of a Quantum vacuum fluctuation being active at every tiny point in Space we require Quantum vacuum fluctuations to be only <u>active</u> at every tiny point in Space which is being observed or measured. There are <u>several very good reasons</u> as to why 'observing' or 'measuring' should activate and provide a Quantum particle <u>with a temporary Quantum vacuum fluctuation</u> which will provide the particle with a <u>temporary</u> configuration of entangled dimensions of Space and time.

(1) For the reason that observing or measuring causes physical-like changes, such changes should occur because of a physical-like phenomenon occurring at the point of observation.

(2) Observing or measuring a Quantum particle should provide a Quantum particle with a configuration of dimensional Space-time which can constrain its wave-like state as a particle-like state. The uncertain, dimensionless wave of potentials of a particle is a ramification of residing within a dimensionless state within non-calibrated, dimensionless Space.

(3) [073] Quantum Theory tells us that an unobserved Quantum particle <u>has no properties until it is observed or measured</u>, which mirrors residing within dimensionless Space without a passing of time. Observing should provide a Quantum particle with a mechanism to provide it with its own reality including the passing of time <u>so as it can display its own properties</u>.

(4) [074] Theories that <u>Quantum vacuum fluctuations</u> are active at every tiny point of Space adds up to a crazy amount of mass-energy, so much so it is

said Space would be way more solid with mass-energy than hard concrete. In mathematical physics this causes what is called 'the cosmological constant problem' or 'the <u>vacuum catastrophe</u>' which should be avoided.

(5) To reconcile how we have permanently provided Dimensional Reality properties to objects made from atoms we require a similar but temporary mechanism, which allows a Quantum particle to display its properties when observed or measured. Properties include spin and mass.

(6) The predictions of Quantum Theory <u>tell us</u> that a Quantum particle has no properties 'until' it is observed or measured, which for the reason that Quantum vacuum fluctuations obviously firmly reside at Quantum scale, is reason why Quantum vacuum fluctuations should also have <u>no properties until observed or measured</u>. That is, the kind of Quantum vacuum fluctuations said to be at every tiny point in Space should not be active until observed or measured. This allows theories to correctly predict that Quantum vacuum fluctuations can be 'observed' at every tiny point of one's dimensional Space-time and <u>solves the 'the vacuum catastrophe'</u> for the reason that Quantum vacuum fluctuations are <u>not actually active</u> at all unobserved or not measured points of Space.

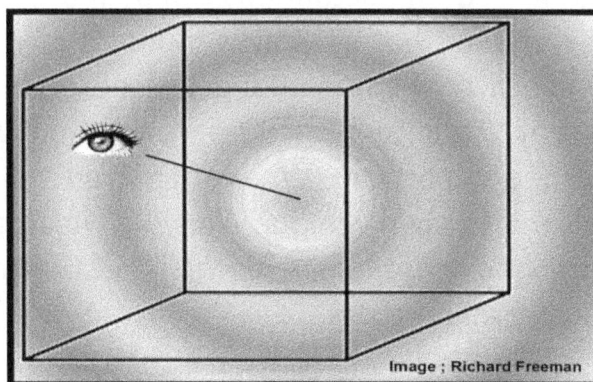

Image ; Richard Freeman

The theory says that at every minuscule point in empty Space there is a tiny Quantum vacuum fluctuation. We will need to rephrase this theory slightly; the theory says that at every minuscule point in empty Space one can 'observe or measure' a tiny Quantum vacuum fluctuation. What all of this means is that at every point in Space where we may possibly observe a Quantum particle we can also expect to observe a tiny Quantum vacuum fluctuation. Although not as loud (powerful) as our permanent, 'amplified' Quantum vacuum fluctuations as with our COST unit (Dark matter) these tiny temporary Quantum vacuum fluctuations which are theorized to be at every tiny point of Space would be expected to act the same way and provide an observed or measured Quantum particle with a mechanism, which provides it with dimensions including a passing of time and the properties of spin and mass. This naturally acts similar to how the COST provides a proton with spin and regular mass.

It is unlikely that the universe provides two separate methods of providing particles with dimensions and a passing of time so as properties can be active and observed. Now we will apply the configuration of a COST unit to the realm of Quantum to solve many of the mysteries of the predictions of Quantum Mechanics. This will add credibility to our configuration of Quantum vacuum fluctuations and credibility to how the COST provides dimensions including the passing of time, mass and Gravity.

Theories say Quantum vacuum fluctuations can be found at <u>every</u> tiny point of Space. However, given that is solves numerous unsolvable cosmic mysteries we will reason how the actual act of observing or measuring a particle triggers a tiny 'Quantum vacuum fluctuation' which is actually uniquely tuned to every particle. When observed or measured every particle will be given its own unique vacuum fluctuation. [094] **Clearly conforming to this *'the physics of nothing'* actually states vacuum fluctuations <u>do exist for all particles</u>.**

I will label these types of Quantum vacuum fluctuations which are created by observing or measuring, as a weakly, interacting COST. 'Weak' for the reason it is <u>not</u> 'amplified' so as to be a regular COST and 'interacting' for the reason that it interacts with an observed or measured Quantum particle and temporarily provides it with an entangled mechanism, which provides an observed Quantum particle with dimensions including a passing of time, allowing a Quantum particle to display its properties 'as they are' within an observers Dimensional Reality. It is important the size of the created Quantum vacuum fluctuation reflects the size or properties of a particle which is truly created from a wave when observing or measuring.

Wave and particle duality: How can simply looking at or attempting to measure a wave cause a physical event which changes a wave like structure into a particle like structure and never even allow self observation of the wave. **What is so special to yourself that allows this to occur?** Wave and particle duality seems an absurdly, irrational event which has been said to be beyond human intelligence. However, a Natural Model provides a rational solution to what allows this to happen. A Natural Model permanently provides us all with our own unique self entangled dimensions which Quantum particles are not permanently provided with. Consequently, wave and particle duality provides amazing evidence or proof if you now prefer, of self entangled dimensions. If this was incorrect Quantum particles would reside within dimensional Space **the same way as we do** and Quantum particles would conform to Einstein's Relativity. Quantum particles clearly operate by a different set of rules. **Wave and particle duality clearly demonstrates the power of entangled dimensions.**

The fuzzy wave state of a particle effectively resides within dimensionless Space where every location is at the same location. Because Space here effectively has no entangled dimensions every part of the wave is effectively located at the same exact nowhere location. When the wave is observed or measured from one's own entangled Dimensional Reality the wave is observed as it is and where it is within your Dimensional Reality.

A dimensionless wave cannot exist within an entangled Dimensional Reality. Observation or measurement drags the wave from its dimensionless state of Space where every part of the wave is located at the

same exact place within an instant of time to where it is observed or measured within an entangled Dimensional Reality from where there are entangled dimensions including a passing of time.

Effectively residing within dimension<u>less</u> Space <u>every part of the wave is located at the 'same' undefined location</u>. When observed or measured from an <u>entangled</u> Dimensional Reality, your own Dimensional Reality <u>correctly measures all facets of the wave as being at the 'same' location</u> which, by **<u>correctly</u> placing every part of the wave at the <u>same</u> dimensional location,** collapses the wave and squeezes it into its <u>tightest possible configuration</u> of a particle. This is why from a Dimensional Reality the wave can <u>only be recorded</u> as a dimensional particle. Observed or measured from a Dimensional Reality a tight particle shape is the closest the whole wave can get to maintain every facet of its wave <u>at the **same** exact location</u>. The very act of attempting to configure <u>every part of a wave to reside at the **same** dimensional location</u> within a Dimensional Reality has collapsed the wave and squeezed it into a round particle which now occupies a clearly defined position within an observers dimensional Space. **This is how a particle actually comes into existence by a simple act of observing or measuring from an <u>entangled</u> Dimensional Reality.**

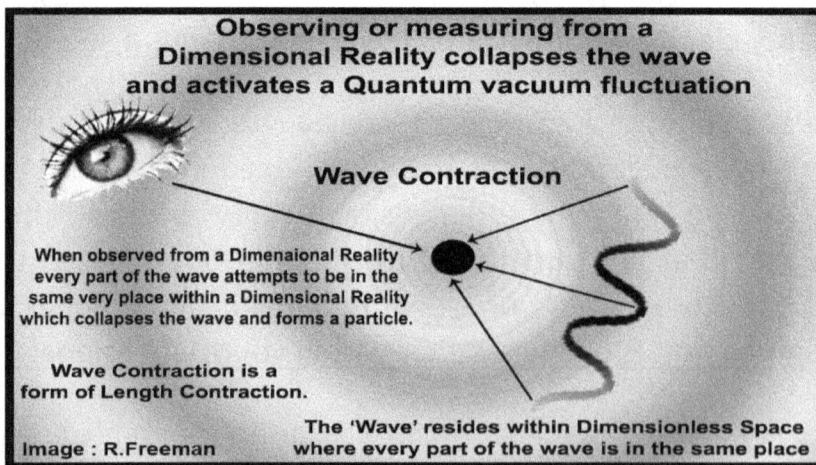

Observing or measuring from a Dimensional Reality collapses the wave and activates a Quantum vacuum fluctuation

Wave Contraction

When observed from a Dimenaional Reality every part of the wave attempts to be in the same very place within a Dimensional Reality which collapses the wave and forms a particle.

Wave Contraction is a form of Length Contraction.

The 'Wave' resides within Dimensionless Space where every part of the wave is in the same place

Image : R.Freeman

<u>Note</u>: The actual progression from a dimensionless wave to a dimensional

particle is primarily caused by wave contraction which naturally occurs for the <u>exact same reason</u> as length contraction within Einstein's Special Relativity but now contracts a wave to a particle. See page 330 how, due to length contraction Stationary Bob observes Fast Mary's spaceship as being short and stubby. Bob observes Fast Mary's spaceship is becoming <u>more particle shape</u>. Because Fast Mary's spaceship has <u>fewer</u> Planck lengths, the transition from Mary's slow-time Dimensional Reality to Bob's <u>regular time</u> Dimensional Reality has caused Stationary Bob to observe Mary's spaceship as being short and stubby. <u>Mirroring Einstein's Special Relativity</u>, Bob, by simply '<u>observing</u>' Mary's spaceship has caused it to become short and stubby within Stationary Bob's Dimensional Reality. One can find similar examples of length contraction online.

A wave state of a particle effectively resides within a fifth dimension where Space is dimensionless and where every facet of the wave is in the <u>same place</u>. Residing in its most natural state the wave has no entangled Planck lengths or passing time. Consequently, when observed or measured from an Dimensional Reality the <u>entangled</u> dimensions of a Dimensional Reality <u>correctly locates</u> every facet of the wave as being in the <u>same place</u> which contracts the wave and squeezes it into its smallest, tightest possible dimensional shape. This is why the actual act of observing or measuring causes 'Wave Contraction' which configures a wave as a particle.

Wave Contraction, exactly like Einstein's Length Contraction, is revealed when observing or measuring an object which is residing within a state of Space which is significantly dimensionally different than one's own. At near light speed Mary's spaceship has slower dimensions which Stationary Bob observes as having fewer dimensions of lengths. My wave contraction occurs because the wave effectively resides in a dimension<u>less</u> state of Space. We cannot observe the wave of a particle for the same reason as Stationary Bob cannot observe Fast Mary's 'long' spaceship. Mary's spaceship resides as both long and short for the same reason a particle resides as both a wave and a particle. Within a Natural Universe one's <u>entangled Dimensional Reality</u> acts as a <u>dimensional filter</u> from where the dimensionless wave can only be recorded as a dimensional particle.

Why does Stationary Bob observe Fast Mary's spaceship as being short and stubby? Answer; Mary's fast spaceship is residing within a state of Space which is <u>significantly dimensionally different</u> than Bob's own state of Space.

Why do we observe or measure a wave as a particle? **The answer requires just one small sentence**: The wave is residing within a state of Space which is <u>significantly dimensionally different</u> than one's own state of Space.

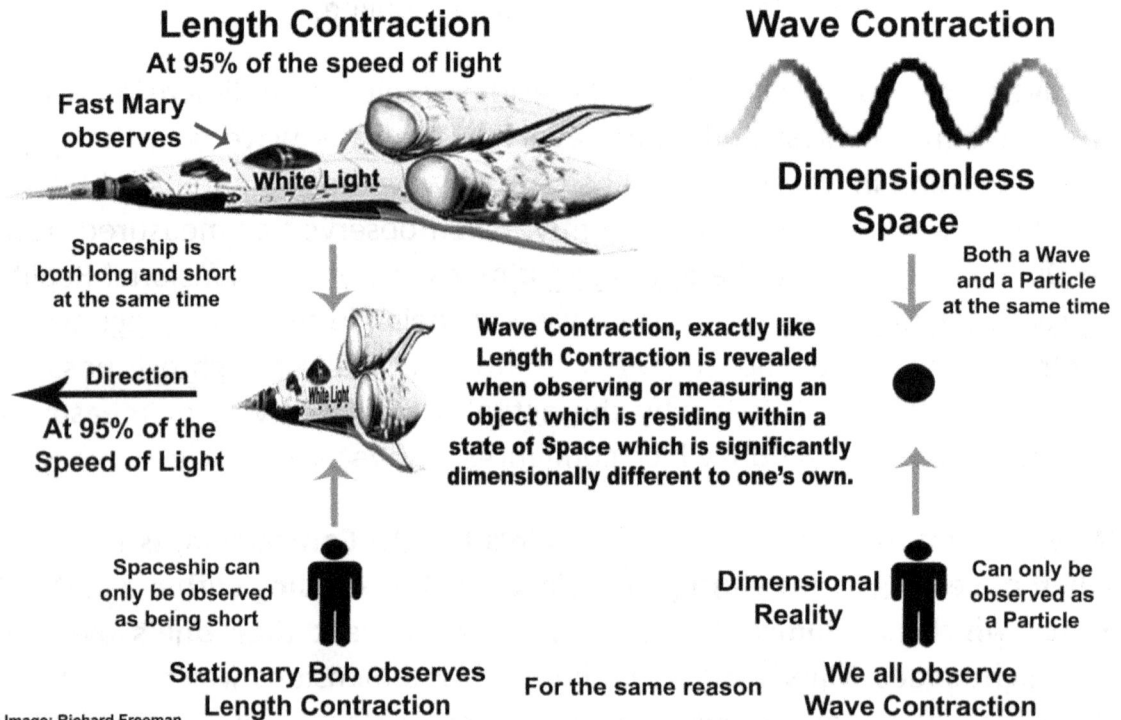

Length Contraction
At 95% of the speed of light

Fast Mary observes

White Light

Spaceship is both long and short at the same time

← Direction
At 95% of the Speed of Light

Spaceship can only be observed as being short

Stationary Bob observes Length Contraction

Image: Richard Freeman

For the same reason

Wave Contraction

Dimensionless Space

Both a Wave and a Particle at the same time

Wave Contraction, exactly like Length Contraction is revealed when observing or measuring an object which is residing within a state of Space which is significantly dimensionally different to one's own.

Dimensional Reality

Can only be observed as a Particle

We all observe Wave Contraction

Unlike length contraction, wave contraction is not speed or direction dependant. This is because a wave is naturally residing in a state of Space which is dimensionless. Because the wave is dimensionless, observation or measurement not only contracts the wave's length but attempts to place every facet of the wave at the exact same location which forms a particle.

[075; 090] I believe Einstein would approve using his Special Relativity to rationally solve the irrational predictions of Quantum Theory. In his later life, Einstein searched for a unified theory which would rationally combine his Relativity with Quantum Theory. Einstein even explored the idea of a fifth dimension and believed Quantum theories could be explained as merely a consequence of a complete unified theory.

My Natural Model clearly says a particle's wave effectively resides within a dimensionless fifth dimension which allows wave and particle duality to be easily and rationally solved. **Rationally solving one of the greatest and irrational mysteries in physics provides clear, powerful evidence that this model is surely correct; we all truly permanently reside within our own entangled Dimensional Reality while Quantum particles are not naturally provided with a permanent Dimensional Reality.** Quantum Theory clearly says a Quantum particle has no properties 'until' observed or measured.

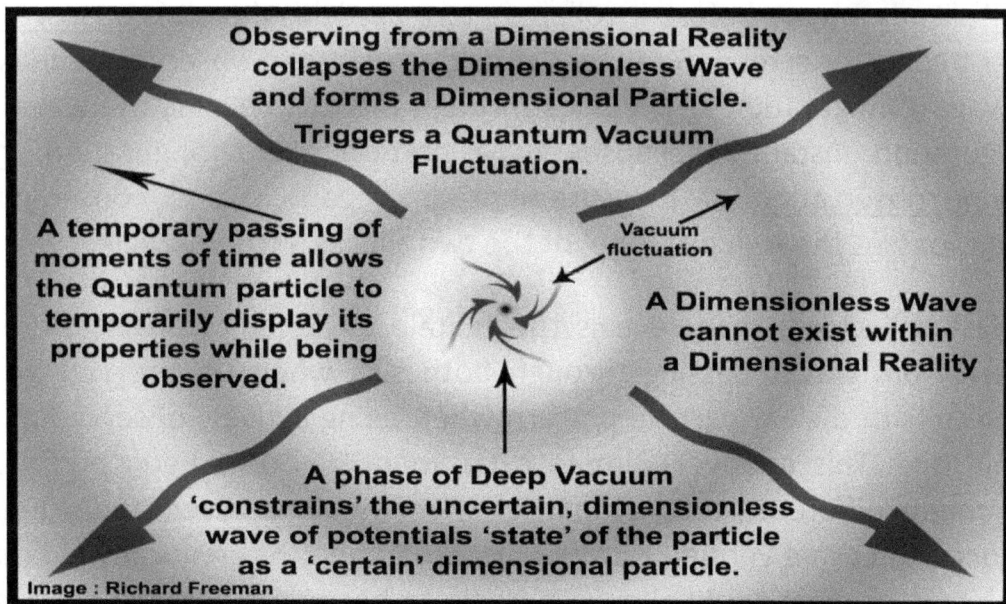

Observing from a Dimensional Reality collapses the Dimensionless Wave and forms a Dimensional Particle.
Triggers a Quantum Vacuum Fluctuation.

A temporary passing of moments of time allows the Quantum particle to temporarily display its properties while being observed.

Vacuum fluctuation

A Dimensionless Wave cannot exist within a Dimensional Reality

A phase of Deep Vacuum 'constrains' the uncertain, dimensionless wave of potentials 'state' of the particle as a 'certain' dimensional particle.

Image : Richard Freeman

The Double Slit Experiment and Wave-particle duality:

The solution to the Double Slit Experiment <u>is naturally exactly the same</u> as solving '*Wave and particle duality*' covered in the previous pages.

The double slit experiment

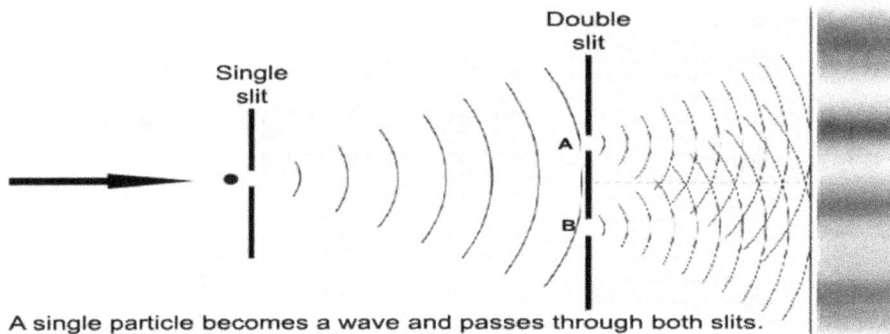

A single particle becomes a wave and passes through both slits.
But if observed at 'A' or 'B' the wave is observed as a particle and
passes through one of the two slits as a particle.

[076] The Double Slit Experiment confirms an unobserved particle acts as a <u>fuzzy wave</u> but when observed instantly becomes <u>a particle</u>. The Double Slit Experiment has revealed to everyone that when a Quantum particle is directed towards two slits an <u>unobserved</u> 'particle' will travel through '<u>both</u>' slits as a wave before it collapses to a particle when it is <u>observed</u> on the last collector screen. Strangely, if one attempts to observe or measure the wave configuration as it passes through the double slits, the wave like configuration instantly collapses to a particle like configuration, <u>only allowing observation</u> or measurement as a particle and consequently actually passes through just '<u>one</u>' of the two slits as a true particle.

The last collector screen reveals the telltale evidence of an interference pattern from waves passing through both slits when unobserved at the double slits or a clear particle pattern when being actively observed at the double slits. This relatively simple experiment has now been performed with many different kinds of particles, many times over and all have produced the same verified results. The experiment has naturally led many to ask how or what could allow a particle to become a wave or a wave to become a particle? How is it possible for a simple and seemingly unrelated act of 'observing' or measuring to instantly collapse a wave form of a particle into a particle 'marble shape' form? How can it be possible to physically change the actual structure of a particle by simply looking at it? It is as if the particle actually knows when it is being observed! This at first appears so bazaar that even the most logical explanation may be difficult to comprehend. We cannot pretend it is not so for the experiments confirm it so but nobody can rationally explain how or what it is that physically causes this irrational, almost magical feat to be rationally performed.

As many within science have already asked, how can a particle actually know it is being observed or measured? It seems to not be of common sense and a notion which one would wish not to believe. I'm looking at a beautiful bunch purple orchids growing outside my window and I would like to think those orchids do not know that I'm observing them any more than a particle could possibly know if it was being observed. These seemingly impossible feats have been proven by science to be correct, which has of course supported much of the enormous success the theory of Quantum Mechanics enjoys.

It is said this is a bizarre weirdness of the physics of Quantum Theory, Scientist cannot explain exactly how or why these seemingly supernatural feats are actually physically performed by particles which reside within the Quantum world. It is said that until a Physicist arrives with a theory which can combine the theories of Relativity and Quantum Mechanics, science has to acknowledge they really do not understand the mystery of what is possibly occurring here. A Natural Model actually allows one to use Relativity to solve this mystery from the realm of Quantum.

The solution to the Double Slit Experiment is naturally exactly the same as solving 'Wave and particle duality'. My Natural Model allows one to finally make rational sense, in layman terms, of the double-slit experiment with the use of just the following sentence. **An unobserved particle resides as a dimensionless wave within dimensionless Space, observing or measuring**

from an entangled Dimensional Reality causes wave contraction which contracts the wave into a dimensional particle. Wave contraction occurs for the exact same reason as Special Relativity's length contraction.

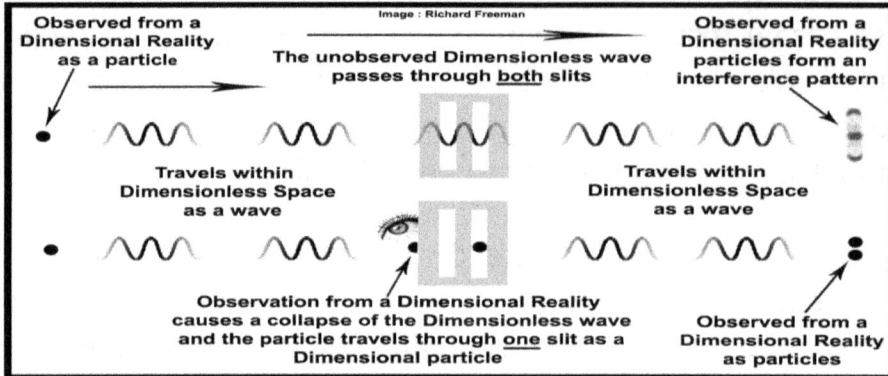

In Space without entangled dimensions every location is in the same place. The wave effectively resides within a fifth dimension which is dimensionless and where every part of the dimensionless wave is located in the exact same place. When viewed from an entangled Dimensional Reality a particle shape is the closest the whole dimensionless wave can get to maintain every facet of its wave in the same exact place. The very act of configuring every part of a dimensionless wave to reside in the same dimensional location within a Dimensional Reality collapses the dimensionless wave and forms a round, dimensional particle. The act of observing or measuring from a Dimensional Reality causes 'Wave Contraction' which configures a wave as a particle. Wave contraction, exactly like Special Relativity's length contraction, is revealed when observing or measuring an object residing in a state of Space which is significantly dimensionally different to one's own.

Exposed to the Double Slit Experiment the unobserved dimensionless wave passes through both slits as a dimensionless wave which builds, from a succession of now observed dimensional particles, an interference pattern at the final point of observation at the detector. The interference pattern is caused from a dimensionless wave residing within dimensionless Space passing through both slits. Now what happens if we attempt to observe or measure the particle at the mouth of the double slits in an attempt to observe how a 'particle' passes through both slits as a wave? The act of observing or measuring from an entangled Dimensional Reality now drags the dimensionless wave out of <u>dimensionless</u> Space into the dimensional state of a particle, within an observer's entangled <u>Dimensional Reality.</u>

An observed particle, because it <u>now has to deal with dimensions</u>, can now only travel through just one slit as a particle. Successions of dimensional particles now build traditional particle impressions at the final point of <u>observation</u> at the detector. The dimension<u>less</u> wave <u>cannot exist</u> within a Dimensional Reality and it is impossible to observe a dimensionless wave from a Dimensional Reality.

Further clarification of a pair of entangled Quantum particles with spin:

I adore the thought of using Albert Einstein's Special Relativity to rationally explain the weirdness of Quantum Theory. Because both theories are very successful they should both be respected and used together. Special Relativity and length contraction; Fast Mary, within her own Dimensional Reality, <u>continues to observe</u> or measure the length of her spaceship as long and normal while, <u>at the same time</u> Stationary Bob, within his own Dimensional Reality, observes or measures Fast Mary's spaceship as being short and stubby. Consequently, Einstein's Special Relativity's length contraction is clearly inferring each particle will <u>continue to reside</u> within dimensionless Space as a wave <u>at the very same time</u> as it is being observed or measured as a particle from within an entangled Dimensional Reality.

Each unobserved 'particle' resides within dimensionless Space as a dimensionless 'wave' which naturally <u>cannot</u> have 'particle' properties. Observing or measuring from an observers own Dimensional Reality causes 'wave contraction' which first collapses the dimensionless wave and <u>squeezes</u> it into a dimensional particle within an observer's Dimensional

Reality. The collapse of the wave has now induced a Quantum vacuum fluctuation in the form of a weakly, interacting COST which provides a passing of time which allows the created particle to display its particle oriented properties. The very 'same' instant a particle is given its property of spin orientation its <u>far away</u> entangled partner is <u>instantly aware</u> of its observed partners spin orientation properties <u>because both 'particles' are also silently residing within dimensionless </u>Space as waves where both are located at the <u>same dimensionless place where they are not separated by dimensions and where time is reduced to an instant</u>. **As experiments have shown, there is no faster than light communication and the particles contained no hidden information. The answer is provided by a Natural Model; since they are not permanently provided with <u>entangled</u> dimensions to Space the particles <u>totally ignore the dimensions of Space.</u>**

Such particles effectively reside within dimensionless Space where time is reduced to an instant and where dimensional distance is **irrelevant**. Now when observed of measured from a Dimensional Reality it becomes as if a Quantum entangled particle was always oriented the way it is determined to be by the act of observing or measuring within a Dimensional Reality and its entangled partner was always oriented the opposite way, even before it was actually determined. **Entangled particles act this way because when a pair of particles are entangled with each other they are reacting directly, <u>in sync</u> with each other, <u>within an instant of time</u>, within dimensionless Space, where without the constraints of dimensions or the passing of time, <u>distance and passing time naturally becomes totally irrelevant.</u>**

In the Quantum world (Quantum Theory), particles are said to not possess specific properties until they are measured and so observed. This is why we have required a method in the form of a weakly, interacting COST which temporarily provides the particles with a mechanism for a passing of time which allows Quantum particles to display properties, including spin and mass. A particle can now be observed, as a particle is, within an observer's

Dimensional Reality. **Within a Natural Universe particles will naturally be given spin and mass from the actual <u>spiraling</u> act of collapsing a wave and <u>physically</u> constraining it as a particle. It will <u>require energy</u> to constrain a fuzzy wave as a particle, energy of which now equals mass.** See page 145.

Even despite the fact that latest Quantum entanglement experiments have provided 'actual evidence' of instant time (between events), Scientists still do not like or understand the notion of instant time or dimensionless Space, preferring to relate to the true and actual results of their experiments as 'spookiness'. **Because Space and time cannot be separated instant time <u>requires</u> dimensionless Space.** One should not write off the notion of instant time (a facet of time) when one says they do not know what time actually is. Because we are all permanently <u>locked by entanglement</u> to a Dimensional Reality, where there is a continuous passing of moments of time, we ourselves have no direct access to instant time.

Particles which do not naturally have a permanent mechanism to generate entangled dimensions to Space reside where Dimensional Reality does not exist. These particles reside as fuzzy waves within an instant facet of time from where the next <u>event</u>, <u>measurement</u> or <u>observation</u> instantaneously occurs. Understanding the fully explainable notion of instant time, a dimensionless place where events can instantly occur, helps to remove much of the 'spookiness' from Quantum Theory (Quantum Mechanics).

Note; when Stationary Bob was observing Fast Mary's spaceship Bob's own Dimensional Reality **was not able to override** Mary's spaceship's own Dimensional properties which is why Bob was able to observe Mary and her spaceship were <u>entwined with different dimensions to his own</u>. What this means, when a particle is being observed it should also be provided, like Fast Mary, with its own entangled mechanism to provide it with a Dimensional Reality which <u>allows it to display its own properties.</u> This will require a mechanism to keep the contracted wave constrained as a particle and a mechanism which provides <u>the observed or measured particle</u> with properties, including a <u>fitted amount</u> of mass and Gravity, while being observed from a Dimensional Reality. A Quantum vacuum fluctuation in the form of a weakly, interacting COST can provide this.

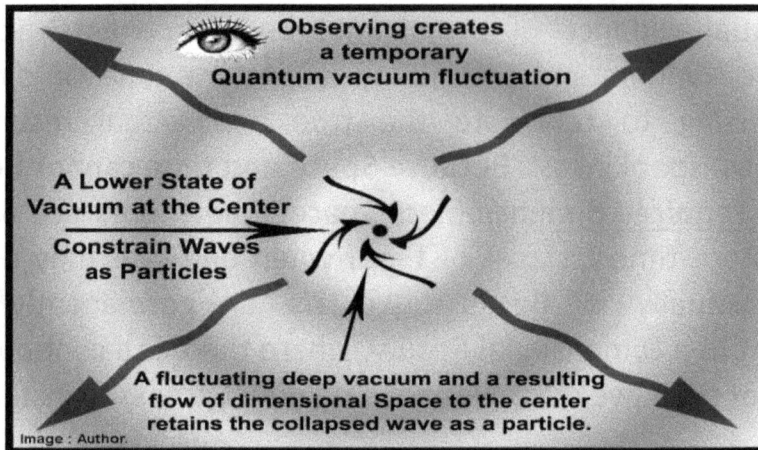

Photons are most influenced by the attraction part of Gravity and <u>do not have</u> a delivery part of Gravity. Although they are restricted by their small mass other small particles are similarly influenced mostly by the attraction part of Gravity. Since this theory [094] clearly says <u>vacuum fluctuations exist for all particles</u> they likely include a mechanism which during observation or measuring <u>gives</u> an observed or measured particle <u>spin and mass.</u> As it will provide a particle <u>with mass</u> this model is not reliant on the Higgs boson.

Because it explains <u>exactly</u> what Quantum particles are observed to do, **Quantum Gravity** is <u>best explained</u> near the beginning of this chapter. However, the position of a subatomic particle is not determinable until it is observed or measured at which time; since it now has an entirely different identity, <u>could now possibly</u> require a form of **Einstein's Gravity** for <u>observed</u> or measured particles. The collapse of the <u>area of the wave</u> to the confines of a particle state <u>induces</u> a lower state of Space around the newly formed particle which ground-state of dimensional Space flows into, thus, activating a <u>fitted</u> vacuum fluctuation in the form of a weakly, interacting COST. The size of the Quantum vacuum fluctuation naturally matches the properties or size of the collapsed wave which is why it is *fitted perfectly* to a particle. The particle is now given spin, mass and Gravity for the same reason a proton is given spin, mass and Gravity. The actual collapse of the wave and flow in of dimensional Space provides spin and the wave <u>bound tightly together</u> into a particle provides the particle with regular mass given by a weak version of <u>the strong force</u>. Like the COST of a proton, the fitted vacuum fluctuation maintains a wave bound tightly together as a particle and provides tiny Gravitational waves which now provide the bound together particle with a <u>fitted</u> Gravity delivery system which <u>may possibly</u> have some <u>very small</u> influence on only observed or measured particles.

Quantum Theory and disappearing particles: Weird, strange, mind bending and bizarre are just a few of the words commonly used to describe the peculiar world of Quantum Theory which describes Matter at microscopic scales. The acceptance by science of Quantum Theory is of course due to its enormous success in explaining the properties of particles at microscopic scales. However this acceptance has come at the cost of science accepting some bizarre occurrences may be happening within the universe.

For instance, the assumption of Quantum Theory that a particle can be in many places at the very same instant or may disappear and reappear someplace different has wacky ramifications. You and I are made from particles so a true ramification of the understanding of Quantum Theory is that you or I may at any time disappear and instantly reappear on far away Mars or even the whole Earth may disappear and instantly reappear on the other side of the Galaxy. Scientists accept this weird notion of Quantum Theory for the reason that science today has no reasoning as to why the atoms which you and I are made from should allow one to be <u>permanently locked by entanglement</u> to the traditional, dimensional realm of Relativity.

Fortunately, within a Natural Universe, we are able to offer a sure way to dilute some of this supposed factual madness because the nucleons of all atoms are trapped by their COST component at an anchored center point of entangled moments of expanding dimensions of Space and time, stabilizing objects made of Matter (atoms) within the traditional, dimensional world of Albert Einstein's Relativity. Within a Natural Universe you will be glad to know that we can be assured that all atoms which we are made from are collectively secured at the center of an expansion of a Dimensional Reality by Relativity, meaning you and I, regardless of what the current Quantum physics say, are absolutely unable to disappear and instantly reappear on far away Mars. Tiny Gravitational waves expanding from all spinning protons and neutrons come together to allow atoms to collectively build an entangled, Dimensional Reality which permanently locks us all to our own unique Dimensional Reality from which there is no escape.

Although it is said, as a ramification of Quantum Theory, a particle may disappear and reappear, Quantum Theory does not tell us where a disappearing particle goes to when it disappears, why it disappears or how it manages to disappear and reappear. It seems that one should not ask these questions for the reason that nobody or no theory can explain how this disappearing and reappearing act occurs; some say that this disappearing and reappearing act is merely a weirdness confined to the Quantum world of physics and to even ask for a rational answer is in itself irrational. Consequently, many Physicists will say that it is silly to ask such questions and will proclaim that it is that way simply for the reason that the physics say so, however, Albert Einstein was never happy with the "silly" ramifications of Quantum Theory.

When one understands that Quantum particles do not permanently possess a mechanism for providing entangled dimensions of length, width and depth and a rhythm of a continuous passing of moments of time the answer becomes obvious. An unobserved Quantum particle effectively resides within a fifth dimension of dimensionless Space, as a dimensionless wave, with no mechanism to provide the passing of time. <u>If an unobserved particle had to deal with dimensions it could never behave so bizarrely</u>. A <u>dimensionless</u> wave cannot exist within an entangled <u>Dimensional</u> Reality and a <u>dimensional</u> particle cannot exist within dimensionless Space.

A particle may disappear between events. Where does the particle go when it disappears? The particle does not disappear from existence; it exists within an instant of time within a dimensionless state from where it simply <u>disappears</u> from one's own entangled Dimensional Reality where there is permanent passing moments of time. **An unobserved particle disappears because an instant will not fit into a passing moment of time and it is impossible for a dimensionless state to display itself within a Dimensional Reality.** The particle now resides in a configuration of a dimensionless wave of potentials and without a coordinated spatial system to uniquely determine a position. Existing without a coordinated, spatial system to uniquely determine an exact position, the particle in wave form is free of the 'constraints' of a coordinated, spatial system. A particle resides as a dimensionless wave until its <u>next event</u>, <u>measurement</u> or <u>observation</u>, which again requires the passing of time, which allows it to be observed or measured again as a regular particle within a coordinated spatial system where it may be recorded within an entangled Dimensional Reality at a certain, positional coordinate.

A coordinate system is a system which uses one or more numbers, or coordinates, as a way to uniquely determine a position. Hence a coordinate system cannot exist within non-calibrated, dimensionless Space, which is why particles residing within dimensionless Space reside as uncertain, dimensionless wave of potentials. That is, a wave of potentials may take a potential path and appear as a particle at any observable position.

Residing without a continuous, 'balanced', regulated mechanism for the passing of time is essentially where events can instantly occur. Events can instantly occur because there is no balanced, continuous, passing rhythm of time between events. With Quantum particles it is as if they are not separated, because without being self entangled within the constraints of a Dimensional Reality, dimensional distance, like passing time, naturally becomes irrelevant.

Dimensionless Space is much the same as frequently expressed as <u>sub-Space</u> in episodes of Star Trek which utilizes technology <u>theorized by real scientific hypothesis</u>. Star Trekkers use <u>sub-Space</u> as a means to provide near instantaneous contact with places that are light years distant. Spaceships disappear from dimensional Space-time into windows of subspace and mysteriously reappear at any convenient place. Faster than light restriction does not apply if one knows how to utilize sub-Space.

Remember how (page 319, 320) Fast Mary and Stationary Bob, <u>during the very same event,</u> each observed light to travel very different dimensional distances at the <u>same</u> constant speed. Mary and Bob can see each other during the event but each observes the light travel a different distance. **This can occur because Mary and Bob each reside within their own <u>unique, entangled</u> Dimensional Reality where the same photons of light from the <u>same event are observed</u> to travel dimensionally different paths and a different distance at the very same speed.** An unobserved Quantum particle silently resides in a fifth <u>dimensionless</u> dimension from where it can be extracted at an observed position from within a Dimensional Reality. That same particle may be observed at a <u>different</u> position within a different Dimensional Reality. This can only occur if we all reside within our very own unique, entangled Dimensional Reality and the unobserved Quantum particle silently resides in a fifth <u>dimensionless</u> dimension where it can really be in any position which is required to match an observers entangled Dimensional Reality. Welcome to the wonderful fifth dimension. <u>So where does the particle go when it disappears? The particle silently resides within a fifth dimension where everything is dimensionless.</u>

Departure and Arrival; Residing within the fifth dimension and within an instant of time allows a Quantum particle to disappear (departure) from one's own entangled Dimensional Reality and travel anywhere within an instant (of time and Space) and reappear (arrival) again at any 'new' position, without ever being between the departure and arrival positions at a moment which matches one's own <u>unique</u>, entangled, Dimensional Reality of a coordinated system with passing of moments of time.

While residing within an instant and from where dimensions cannot be expressed, the particle simply 'skips' and so Quantum leaps between departure and arrival, while paying no heed to dimensions such as length (distance and so dimensional Space) or the 'passing of moments of time'. Its positions between departure and arrival were recorded within an instant of time which <u>cannot exist</u> within a passing dimensional moment of time. A particle existing within an instant and Quantum leaping <u>one's own</u> dimensions of length, width, depth or a passing of time is only 'rationally' and 'physically' possible within the concepts of my Natural Universe.

[077] Scientists have actually observed particles Quantum leaping our traditional, Dimensional Reality of the dimensions of length, width, depth and a passing of time. Expressed as a *"bizarre type of quantum movement"* by *Lead researcher Hui Zhao from the University of Kansas:* **'The electrons in some way defied regular physics and moved directly from the top layer of material to the bottom layer without ever being in the middle layer'.**

Each layer would be expected to have a dimension (thickness) which the electrons would be expected to have travelled through. However, when residing within a dimensionless realm the electrons know nothing of the dimension of the thickness of the middle layer and simply traveled directly from their departure layer to their arrival layer. Moving without the dimensional restrictions of a coordinated spatial system allows the electrons to seemingly skip, the dimension of the middle layer by <u>instantly</u> moving directly from the top layer to the bottom layer without passing through the coordinates of the middle layer <u>at a moment of time</u> which matches the passing of time of a Dimensional Reality.

'The cosmological constant problem' also known as the vacuum catastrophe: The first question to ask; is there any evidence that empty Space has an underlying amount of energy? Theories, mathematical calculations and more importantly experimental <u>measurements</u> at the Large Hadron Collider imply that 'empty' Space may contain an enormous amount of 'hidden' energy. To find this 'hidden' energy we must explore the hidden world of Quantum Theory and a phenomenon known as 'Quantum vacuum fluctuations' which provide a temporary change in the amount of energy at a point in Space, as explained in Werner Heisenberg's uncertainty principle. Quantum vacuum fluctuations allow the creation of Matter particle and anti-particle pairs of virtual particles which quickly annihilate to produce a tiny pulse of energy.

[074] Quantum vacuum fluctuations have energy and vacuum fluctuations are said to be 'found' within every minuscule 'bit' of Space. Apparently, when one adds all of this 'hidden' energy within say a cubic meter of empty Space one can calculate a mass-energy total as being way more than a cubic meter of solid Matter, like the stuff house bricks are made from, and possibly as much as that of our Sun. Now this is all said to be true and fascinating, the problem is though, all of this energy should cause the universe to expand Galaxies away from each other at a vastly greater rate than truly observed. Now, by taking the true observed expansion rate in consideration and the 'true' observation of the amount of energy within a cubic meter of empty Space, the true amount of energy within empty Space is said to be as little as that of just one atom.

The <u>actuality</u> that energy within empty Space is really only an incredibly small fraction of the known energy of vacuum fluctuations said to be within empty Space is a conundrum for science. This is called 'the cosmological constant problem' or 'the vacuum catastrophe' which has been described as *"the worst theoretical prediction in the history of physics"*.

This <u>is a reason as to why</u>, within a Natural Model, Quantum vacuum fluctuations should not be 'actively' everywhere but only where they are 'actively' observed to be. With such reasoning Quantum vacuum fluctuations are found at every tiny point of Space just as theories predict, however they remain <u>hidden, like the wave state of a particle,</u> and have no properties until observed or measured from a Dimensional Reality. Theories predicting Quantum vacuum fluctuations are found at every tiny point of Space has allowed this modeling to create the COST in the massive numbers required to align with the COST masquerading as Dark matter.

Within a Natural Model, a COST unit (Dark matter) is a permanent configuration of Space-time which consists of an 'amplification' of Quantum vacuum fluctuations by the confinement or <u>squeezing of Space</u>. This phenomenon has been observed in experiments.

<u>The same rules should apply:</u> With this model, the predictions provided by Quantum Theory and experiment evidence that a Quantum particle has no properties until it is observed or measured is strong evidence that 'observing' activates a temporary Quantum vacuum fluctuation, which provides a Quantum particle with a coordinated, spatial system including a passing of time, which allows a Quantum particle to display properties including mass. This can also explain why <u>should-be-massless</u> neutrinos are 'observed' to have mass. That is, 'observed or measured' from a Dimensional Reality neutrinos unavoidably acquire a small amount of mass from the phase of virtual mass of a tiny, temporary vacuum fluctuation which maintains a collapsed neutrino-wave as a neutrino-particle with mass. Like compressing a small spring, mass is a reflection of the energy required to constrained the neutrino-wave as a neutrino-particle.

It is said that vacuum fluctuations can be found at every tiny point of Space which suggests vacuum fluctuations are everywhere. However, in order to avoid 'the vacuum catastrophe', it is <u>required</u> that vacuum fluctuations are not necessarily active 'everywhere' one is not observing.

An unobserved Quantum particle is said to have no properties until it is observed or measured. Quantum vacuum fluctuations are very much a part of the world of the Quantum realm and so the <u>same rules should apply</u>. Thus, when one applies this prediction from Quantum Theory the Quantum vacuum fluctuations <u>silently reside</u> in a fifth silent, <u>dimensionless</u> dimension and like particles are 'only' activated and given specific properties when they are actively observed or measured from a <u>Dimensional</u> Reality. Similar to wave and particle duality, Quantum vacuum fluctuations exist at every tiny point of Space but are <u>only part of your Dimensional Reality</u> when actively observed or measured.

Consequently, if one observes or measures a Quantum fluctuation and then uses the 'Planck length' to determine how many of these tiny parcels of energy are in a cubic meter of empty Space, one can calculate an absurdly big number, which fowls up the 'cosmological constant' and causes the problem known as 'the vacuum catastrophe'.

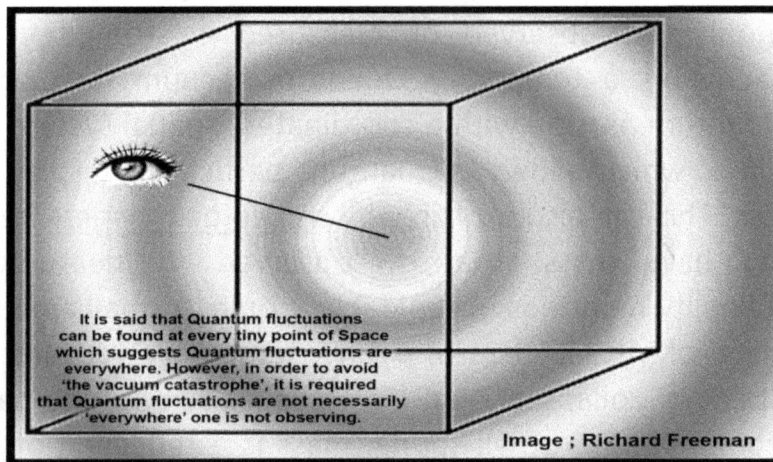

It is said that Quantum fluctuations can be found at every tiny point of Space which suggests Quantum fluctuations are everywhere. However, in order to avoid 'the vacuum catastrophe', it is required that Quantum fluctuations are not necessarily 'everywhere' one is not observing.

Image ; Richard Freeman

The actual number though is only the 'one' within ones entangled, Dimensional Reality is the one which is **being observed or measured.** How does observing or measuring activate a Quantum vacuum fluctuation? [015] The best clue is possibly provided by Prof. Dr. Alfred Leitenstorfer and his team at the University of Konstanz. The Scientists say they have directly detected Quantum vacuum fluctuations. Their experiments also clearly detected something strange when surrounding Space was **squeezed**; the fluctuations became significantly louder and so amplified. Consequently, squeezing Space around the fluctuations very significantly increased their intensity. Now consider my wave contraction (page 360 to 362). The wave existed within dimensionless Space but when observed or measured from a Dimensional Reality one's own Dimensional Reality squeezed the wave into a tight particle. Now if one applies the same rules to a dormant Quantum vacuum fluctuation existing within dimensionless Space, when observed or measured the Quantum vacuum fluctuation is squeezed by entangled dimensions into becoming active and the more it is squeezed the more intense it becomes. **This at least provides good, evidence based reasoning why these types of Quantum vacuum fluctuations may have no properties until they are activated by observing or measuring and merely lay hidden from a Dimensional Reality when not being observed or measured.**

What does all this mean? All that mass-energy, as much as that of our Sun, from many billions of vacuum fluctuations mathematically found within a cubic meter of empty Space may not be actively there and their energy silently resides, like a fuzzy wave, within a dimensionless realm only to be exposed, like a particle, by an act of observing or measuring from an active Dimensional Reality. There is now not a problem with the cosmological constant and not a vacuum catastrophe. Untold energies may silently reside within a dimensionless realm hidden from one's Dimensional Reality.

Solving Hierarchy Problems.

In theoretical physics, a hierarchy problem occurs when modeling provides a very much different value from its effective or actual measured value. One of these unsolved problems asked many times over is ***"why is Gravity so weak in relation to other fundamental forces"?*** Scientists know the strong force creates at least 99% of all mass and they know Gravity and mass share a common, bound together relationship. The more mass an object has the more Gravity an object has. However, for some <u>unknown reason</u> the strong force is about 100 trillion trillion trillion times stronger than Gravity. Gravity being so ridiculously weak has confounded Physicists as it appears to bear no relationship to other forces which makes no sense in theoretical physics. For example a small fridge magnet is able to create an electromagnetic force able to pick up a paper clip against the Gravitational pull exerted by the whole of planet Earth. In an attempt to solve this Gravity has been theorized to weakly bleed through from an unknown dimension. [115] Others say we don't feel the full effect of Gravity because part of it spreads to extra dimensions. Only the Natural Model can provide a correlated and sensible answer to this long time mystery. The <u>attraction</u> part of Gravity is sensibly derived from the source of the <u>strong force</u> which is the source of regular mass which sensibly <u>reconciles Gravity with the source of mass</u>. The attraction part of Gravity is obviously <u>very strong</u> but the variable delivery part of Gravity, provided by tiny Gravitational waves, **has a <u>restraining part</u> which operates in the <u>opposite direction of a falling object</u> causing Gravity to be <u>very weakly delivered</u>. This solves why Gravity is <u>far weaker</u> than other forces.**

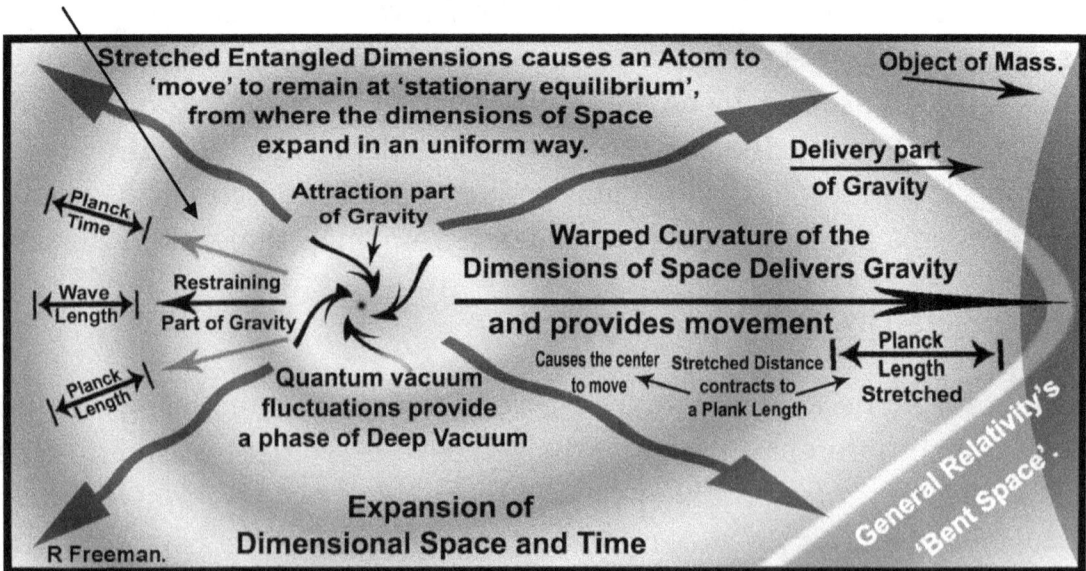

The restraining aspect of the delivery part of Gravity is <u>most important</u>, since **without** this restraining element, Gravity may be 100 trillion trillion trillion times stronger and directly mirror the strong force. The dire consequences of this would likely be if you dropped a hammer the resulting impact crater may destroy you. Even worst, this super strong Gravity would very likely collapse the Earth to a Black Hole the size of a coin.

Some have argued that science's understanding of Gravity is mistaken. [026,] [070] Unfortunately for Scientists the Large Hadron Collider, the world's largest and most powerful <u>particle</u> accelerator, has failed to find any trace of many of the proposed theories put forward to solve problems and mysteries of the cosmic kind. The failure of the Large Hadron Collider to discover many of the desperately sought after particles required for many cherished theories has led some Physicists to suggest we may reside within an unnatural universe. Even though the tiny deep vacuum phase of the COST <u>can solve these mysteries relating to Gravity</u>, it is impossible to use a particle collider to crash a particle into a phase of deep vacuum nothingness, a particle would simply pass through a state where there is nothing there. [015; 016 a false vacuum?] However, measurements conducted during experiments have provided evidence this phase of deep vacuum exists which is now said to be one of the biggest unsolved mysteries in physics.

Within a Natural Universe the Gravity we feel on Earth is a two part mechanism so is difficult to generalize the absolute strength of Gravity. The two parts of Gravity include a strong <u>attraction part</u> and a variable <u>delivery part.</u> The strong attraction part is both the actual source of all Gravity and the source of the strong force which <u>sensibly reconciles</u> all Gravity with the regular mass which the strong force creates. The variable delivery part <u>sensibly mirrors</u> Einstein's General Relativity and relates to tiny expanding Gravitational waves which provide the dimensions of Space-time. Gravity is only applied by the amount the strong attraction part of Gravity attracts and so stretches and warps the curvature of the tiny Gravitational waves.

<u>Why I have quantized Einstein's Gravity, Space and Time:</u> A Natural Model uses Planck length at the delivery part of Gravity where Gravity is applied in variable amounts by stretching the curvature of very tiny Gravitational waves. **The tiny Gravitational waves quantizes Gravity and the dimensions of Space and time.** The wavelength of the Gravitational waves provides and preserves both tiny Planck lengths and Planck time to Space. A stretched wave contracts to a Planck length which causes Gravitational length contraction which causes objects to fall. A stretched wave takes longer to pass which is why Gravity slows time but because Planck time is preserved one's own rate of time appears unchanging.   *020 Link shows how: '***Planck time and Planck length are linked to the ultimate limit of quantization and directly to Gravity'.** Planck length is the distance light travels in Planck time. Smooth Gravity is the <u>combined</u> resource of many trillions of sets of tiny Gravitational waves which are providing the delivery part of Gravity.

For light to be a constant light must always travel Planck Length in Planck time. Gravity slows time so ideally one should show how Gravity slows time at the smallest of scales which is, of course, Planck time. As Gravity slows time and light always travels Planck length in Planck time we should now use Planck Length with Gravity since if we did not the speed of light may not be a constant. Space has dimensions which Gravity stretches and warps the curvature of. Gravity shortens the dimensional distant between objects and brings objects together. For Gravity to do this it is best to show how Gravity does this at the smallest possible dimensional distance which again is ideally Planck length. I have called this Gravitational length contraction where it is again best to use 'length' as being a Planck length. Planck length is where smooth Gravity becomes quantized. There is <u>nothing measurable</u> below Planck scale. Where Planck scales ends is where Quantum Gravity takes over. **Because very tiny Quantum particles ignore the dimensions of Space-time, Quantum Gravity works with only the strong attraction part of Gravity and without being quantized by the dimensions of Space-time.** As these particles are so very tiny and don't come together like atoms to build large objects this is now not a problem but rather an asset.

The source of Einstein's Gravity is a tiny phase of virtual mass which provides negative energy and mimics a very tiny fluctuating, in a state of collapsing and reappearing Black Hole. The attraction part of Gravity within every proton and neutron of every atom is always operating at 100%, consequently, if one measures Gravity at the <u>attraction</u> part, Gravity is massively stronger than Gravity is 'applied' here on planet Earth. Because the binding <u>negative energy</u> phase of virtual mass responsible for the strong force is exactly the same negative energy phase of virtual mass responsible for the attraction part of Gravity, one may wish to express the attraction part of Gravity as being as strong as the <u>strong nuclear force</u>. The strong force simply uses the same source which is the attraction part of Gravity to create regular mass with a <u>near interaction</u> with the tiny particles which make up the protons and neutrons at the nucleus of all atoms. Thus, Gravity is passively sourced as a <u>byproduct</u> of a mass generation mechanism. The pressure inside of a proton peaks ten times higher than inside a neutron star. To provide such tremendous pressure and agitate a proton's particles at near light speeds the tiny negative energy phase of virtual mass, which provides the attraction part of Gravity, <u>needs to be a state</u> which mimics a minuscule fluctuating, in a state of appearing and collapsing, tiny Black Hole. The variable delivery part of Gravity is the part which passively does all of the heavy lifting for Gravity and is the part which conforms to Einstein's General Relativity.

To conform to current theories one may wish to measure Gravity at the attraction part and where it is strongest at very short range. It is said this hierarchy problem is complicated by the involvement of Einstein's General Relativity; I have shown why General Relativity's 'bent Space' is based <u>solely</u> on the variable delivery part of Gravity. Thus, this hierarchy problem probably results from the way science measures Gravity. Here on Earth, if we measure Gravity by the <u>variable delivery part of Gravity</u>, which is the way General Relativity applies Gravity, one may well conclude that Gravity is mysteriously many times weaker than other forces. However, the actual attraction part of Gravity, which is where science should measure Gravity to solve this problem, is actually many times stronger. This problem is now solved by measuring Gravity where it should be measured for this problem; this is at the source of the <u>attraction part of Gravity</u> which sensibly creates the strong force which creates a proton's mass-energy and is <u>exactly what provides</u> the unbreakable bond between mass and Gravity. (The <u>attraction part of Gravity sensibly links</u> Einstein's Gravity with Quantum Gravity).

<u>Unlike</u> Quantum Gravity, Einstein's Gravity acts solely on the dimensions of Space-time and provides no direct particle to particle attraction like a true force. An atom being influenced by Gravity remains stationary at the center of the expansion of its own entangled dimensions of Space-time from where it <u>feels no force</u>. An atom remains at 'stationary equilibrium' unless it's stationary position is blocked by other atoms. Blocked by other atoms an atom is now <u>forced away</u> from its moved away stationary position which is the only kind of force felt due to Gravity. This force is felt as acceleration.

General Relativity and Quantum Theory should be compatible at scales of the smallest and at scales of the largest but in present form they are not. Each of these theories has ramifications which causes it to fail in areas where the other theory excels. Currently, one of the major problems in theoretical physics is merging the theory of General Relativity, which relates to Gravity, with Quantum Theory, which describes <u>fundamental forces</u> acting on the atomic scale. All attempts by science, including those most talented, have failed to seamlessly and rationally combine Gravity with the physics of Quantum Theory which deals with the very small.

[078] <u>Scientists say</u> Einstein's <u>General Relativity becomes nonsensical</u> when it is scaled down or retained to Quantum scale. Similarly, Quantum Theory is said to suffer from disastrous results when scaled up to the vast dimensions of our amazing universe where energy of Quantum field theory becomes so great that it creates a <u>monstrous Black Hole</u> which causes the universe <u>to fold in on itself</u>. This is not a small issue; Mother Nature obviously allows

Gravity to ascend smoothly and seamlessly from the smallest of scales to the largest. Although both of these theories enjoy success within their own realms, each theory cannot fully complement the other. So what could be really occurring? A Natural Universe has both Einstein's Gravity and Quantum Gravity. Einstein's Gravity works by warping the curvature of the dimensions of Space and time. Because a Natural Model's Quantum Gravity <u>needs</u> to work without the dimensions of Space and time, Quantum Gravity works with only the strong attraction part of Gravity and <u>cannot be quantized</u> in the same way as Einstein's Gravity.

Einstein's Gravity: Within a Natural Universe Einstein's Gravity has a <u>delivery part</u> which delivers Gravity with tiny Gravitational waves which provide entangled dimensions to Space. Gravity is delivered when the curvature of the waves become stretch and warped towards the attraction part of Gravity. The distance across a stretched wave contracts to maintain Planck length value which causes length contraction. Because they are caused by the fluctuations of Quantum vacuum fluctuations the tiny Gravitational waves begin at the scale of Quantum and expand continuously at the speed of light to universal scale. This naturally allows Gravity to be delivered to universal scales. It is <u>nonsensical</u> to constrain the tiny Gravitational waves to Quantum scales and to do so would cause General Relativity to become, <u>like the Scientists say</u>, *"nonsensical"*.

<u>The attraction part</u> of Inner Gravity is derived from the source of the <u>strong force</u> and can be combined to create a Black Hole, however, if one was to attempt to scale this part of Gravity to <u>universal scale</u> one would create, <u>like Scientists say</u>, *"a monstrous Black Hole"* made from negative energy which would become part of a far outer dimensionless void of nothingness.

As one can see the Natural Model always provides a sensible solution to many unsolved mysteries. With many of these mysteries I believe it is not myself solving the mystery, rather <u>the model itself</u> clearly provides me with a <u>very obvious</u> and realistic answer which is why I believe in the model.

14 COSMIC MICROWAVE BACKGROUND.

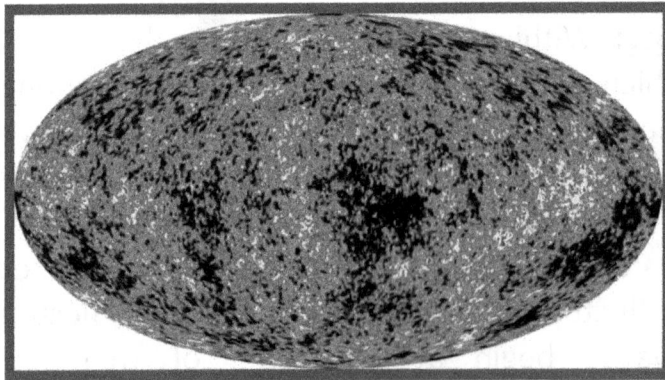

Image: Credit: NASA/WMAT science Team.

For some odd reason Scientists have <u>never attempted</u> to explain the Cosmic Microwave Background radiation by way of a model where the universe sensibly creates 'its own Matter and correctly deals with its created Anti-matter. Strangely, Scientists wish to believe from no given or <u>theorized source</u> all energy and all Matter simply, near instantly, materialized into being. The Big Bang effectively began everywhere and everywhere expanded, consequently, everywhere should look the same. I have read it many times; the Big Bang should have been born perfectly smooth.

All current theories and physics fail at the seemingly infinitely tiniest point which had the resemblance of a singularity at the very beginning of the Big Bang. Thus, nothing sensible can be reasoned of how or why the Big Bang began. A Natural Model has no Big Bang at its beginning but progresses to an early time which would have resembled an early time after a Big Bang.

The Cosmic Microwave Background radiation is a black-body radiation that fills the universe and can be detected in every direction. The CMB image truly has a captivating story in time to tell. The radiation of the CMB represents photons which wavelength has been stretched from less than a micron to 1.063 mm. The stretching to this wavelength is caused from the photons travelling within an <u>expanding</u> universe for billions of years. The CMB image is said to effectively provide a snapshot of the oldest light from when the universe is thought to be just 380,000 years old. The CMB is the cause of the off channel noise (snow) on our old cathode ray tube TV sets.

Scientists say the black-body spectrum of the Cosmic Microwave Background radiation is direct evidence of their Big Bang. [043] The CMB radiation is said to have been first created when all Matter was first tightly constrained as hot plasma. Scientists say there is no doubt there was a Big Bang for they know of no other model which could possibly provide this extremely hot plasma and radiation and in a way which produced its <u>very slight temperature variations</u>. **The photons of the CMB radiation were set free when protons attracted an electron and formed hydrogen atoms.**

Why are we compelled to explain the existence of the Cosmic Microwave Background radiation with a model which did not begin with a Big Bang? [008; 009] Today the stringent laws of physics and as <u>every associated experiment</u> has confirmed when creating Matter from energy one must create an <u>exactly</u> mirror amount of Anti-matter. Consequently, the way a Big Bang creates all Matter its mirror Anti-matter would have indisputably, completely annihilated all Matter. Every particle Physicist knows this is true. Consequently, since the Cosmic Microwave Background radiation exists, its existence should be possible to explain within a universe which correctly deals with its obligated mirror amount of Anti-matter.

Within my Natural Universe Matter is created in a most similar, hot, dense, uniform state at the poles of numerous Super Massive Black Holes. It is important to create Matter at these precise locations for several reasons; it is <u>absolutely essential</u> to provide a method to separate Anti-matter from Matter and an <u>accessible storage bin</u> for the compulsory mirror amount of Anti-matter which <u>will be created</u>. Without this storage bin, in the form of Super Massive Black Holes, all Matter will be completely annihilated <u>as it would have been if there was a Big Bang</u>. A Big Bang cannot <u>first</u> create the <u>vital</u> Gravitational scaffolds of a Dark Disk Galaxy with a central Super Massive Black Hole. Within a Natural Universe, newly created Matter is released in dense 'blobs' <u>inside</u> of a Galaxy's Dark matter Gravitational scaffold which provided the precise conditions ideal for star formation.

[043] For a Natural Universe the near perfect Black-body spectrum of the CMB radiation actually provides astonishing, direct evidence of how Matter was at first <u>tightly constrained</u> within dense plasma at an astoundingly hot 10 trillion Kelvin. The near perfect Black-body spectrum was attained within <u>multiple closed</u> dense plasma cores at the poles of numerous Super Massive Black Holes where particle and anti-particle annihilation was creating numerous photons and electrons. Inside of the extremely dense, <u>closed</u> plasma cores when photons were emitted they were immediately reabsorb by other particles providing <u>thermal equilibrium</u>. **Note; Matter is actually <u>first</u> created in a region where there is extremely strong Gravity.**

Quasar jets from the poles of Super Massive Black Holes are observed as being both very <u>short</u> and extraordinary <u>long.</u> [006] NASA's Swift Satellite was launched in 2004. The Swift team says they have discovered Quasar jets are *"**made of protons and electrons**"*. [116] Astronomers have now discovered a Quasar, J2054-0005, powered by a SMBH and existing in the <u>early universe</u> which they say is *"**spewing out molecular gas, the raw material needed to form new stars**"*. <u>Conforming to these two discoveries</u> protons released from Quasar jets **very quickly cooled** to about 3000 Kelvin and attracted an electron to form hydrogen atoms and two hydrogen atoms now make molecular hydrogen (gas) which a Quasar now spills out. **All of which will provide <u>documented evidence</u> to why the CBM radiation image revealed regions which are slightly cooler and regions which are slightly warmer.**

Unlike a Big Bang a Natural Model is now able to explicitly explain, **<u>without having to develop a complex theory,</u>** why Gravity was stronger and weaker within different regions from where the CMB radiation was released which created slightly cooler and slightly warmer regions of radiation. A Big Bang is first filled with all (hot) Matter for all time which requires 380,000 years to <u>cool from within</u> to form the first atoms. A Natural Universe ejects hot protons <u>directly into a cold environment</u> where atoms **very quickly formed**.

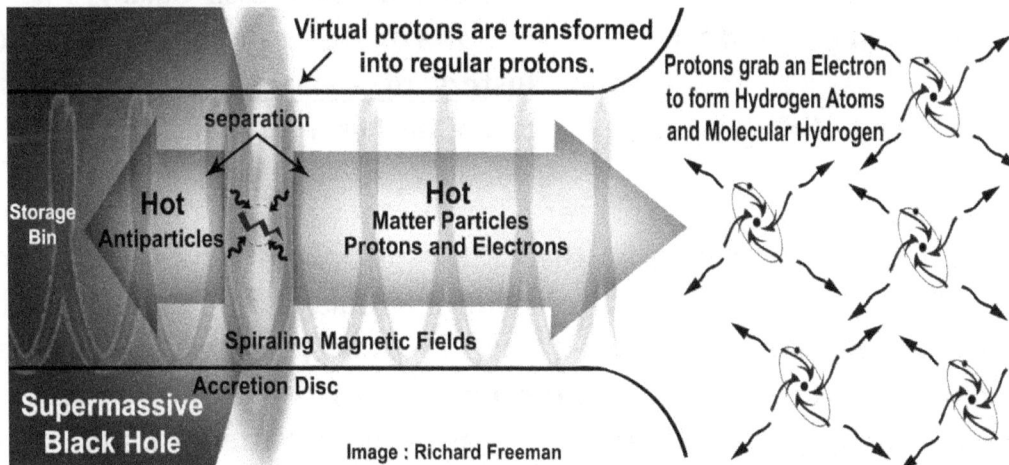

Virtual protons are transformed into regular protons.

Protons grab an Electron to form Hydrogen Atoms and Molecular Hydrogen

separation

Hot Antiparticles

Hot Matter Particles Protons and Electrons

Storage Bin

Spiraling Magnetic Fields

Accretion Disc

Supermassive Black Hole

Image : Richard Freeman

In a Natural Universe, protons and electrons were first explosively set free of dense plasma cores <u>or</u> carried as plasma by very <u>short</u> powerful, spiraling magnetic fields <u>at polar regions</u> before being released <u>close to the strong Gravity</u> of Super Massive Black Holes (SMBH) and their Dark Disk Galaxies. The protons **very quickly cooled** to about 3000 Kelvin and attracted an electron to form hydrogen atoms. Now the plasma became a gas <u>setting free the photons</u> of the CMB radiation. Called the **Sachs-Wolfe Effect,** the photons now <u>lost some of their energy</u> due to escaping from the <u>stronger</u> Gravitational force <u>near</u> to a <u>SMBH</u> and its Dark Disk Galaxy. **Due to this loss of energy the photons became <u>slightly cooler</u> causing forming <u>clusters</u>**

of Galaxies to create slightly cooler regions of background radiation.

As the powerful spiraling magnetic fields became <u>increasingly</u> very much longer protons and electrons were <u>carried</u> as plasma and released farther above and below the Gravity from the plane of a Dark Disk Galaxy and its central Super Massive Black Hole. The released protons **quickly cooled** and attracted an electron to form hydrogen atoms at which time <u>the escaping photons</u> of the CMB radiation were <u>already nearly free</u> of the <u>strongest Gravity</u> of a Super Massive Black Hole and its Dark Disk Galaxy. The set free photons now <u>lost very little energy</u> escaping from <u>weaker</u> Gravity so remained <u>slightly warmer</u> than the photons released nearer to the <u>stronger Gravity</u> of a Super Massive Black Hole and its Dark Disk Galaxy. **Photons released far above and far below the plane of the Gravitational scaffold of the Dark Disk Galaxies caused <u>clusters of Galaxies</u> to be commonly surrounded by <u>slightly warmer</u> regions of background radiation.**

When Galaxies were at first nearer to each other and creating heavy loads of new Matter the photons of the CMB radiation became slightly cooler when released near the Gravity of a forming Galaxy but remained slightly warmer when released farther from the Gravity of a forming Galaxy. **Travelling at the speed of light the photons of the CMB radiation now naturally filled Space with ever so slight temperature variations which provided the CMB image its speckled not uniform appearance.**

So why have Scientists deduced that the CMB radiation is evidence of a Big Bang? The key to understanding the all important slight temperature variations of the CMB radiation is Gravity. The photons of the CMB radiation were set free when protons cooled and attracted an electron and formed hydrogen atoms. When the photons of the CMB radiation escaped from where there was <u>strong</u> Gravity the photons lost energy and became slightly cooler than photons escaping from regions of <u>weaker</u> Gravity.

The temperature of the CMB radiation is today 2.725 Kelvin. The cooler regions are said to be only 0.0002 Kelvin cooler than the warmer regions but it <u>allowed Scientists to deduce</u> there were cooler regions which once had more Gravity. If Matter came from a Big Bang the regions with more Gravity were now <u>theorized</u> to be regions which were denser with Matter and would have been regions which form stars and Galaxies. The slightly warmer regions of CMB radiation are where the photons escaped from less Gravity where regular Matter was now <u>theorized</u> to be less dense and where Matter was less likely to create stars and Galaxies.

Scientists say this is proof that, after a Big Bang, when cooling allowed hydrogen atoms to form, their Big Bang had already distributed itself into dense and slightly less dense regions of hydrogen.

In a Natural universe much of the Matter released in regions with less Gravity also eventually fell to a Galactic plain and created stars, while the void like regions between Galaxies grew larger as Galaxies moved apart due to Outer Gravity. Scientists know the Space between Galaxies remains relatively void of Matter. See page 211; CMB radiation [052].

Ideally, I needed to acquire actual, factual, observed evidence that Matter really existed in such a state of extremely hot plasma at the poles of Super Massive Black Holes. This is where primeval cores of Quasars lurked at near the beginning of the universe. I also required evidence that Matter was first tightly constrained where it was held at <u>thermal equilibrium.</u> The Massive Gravity of Super Massive Black Holes and the powerful spiraling magnetic fields found at their poles would provide the ideal tools to tightly constrain Matter. [038] It is theorized that the temperature of Matter just 0.0001 of a second after the Big Bang was an astonishing <u>10 trillion Kelvin</u> and in a dense plasma state which provided a near perfect black-body spectrum.

Such extreme temperatures <u>disagree</u> with the current 'theories' for the cores of Quasars where the Natural Model creates its Matter. [005] Amazingly, researchers have now made a truly mind-boggling observation; they discovered temperatures <u>inside the jets</u> of Quasar 3C 273 were of <u>10 trillion Kelvin which is an astounding 100 times hotter than current theories for Quasars allow</u>. However, the unexpected observation exactly matches the temperature of Matter at a similar time after the proposed Big Bang and is, of course, exactly where my Natural Model creates its Matter. [046] I have also uncovered direct evidence Matter is explosively released from these regions which indicates that Matter had been first tightly restrained. It is unlikely a coincidence that this <u>direct observation</u>, even though it <u>defies</u> current theories, truly mirrors the state of Matter at the proposed Big Bang.

10,000,000,000,000 K.[005] The measured so <u>factual</u> temperature of a Quasar. 10,000,000,000,000 K.[038] The theorized temperature at 0.0001 of a second after the Big Bang when anti-protons were <u>annihilating</u> with protons.

Within a <u>Natural Universe</u>, hot Matter is first retained by powerful spiraling magnetic fields. The magnetic fields break, and in a <u>documented and observed</u> way, Matter and radiation is explosively released in dense pulses from polar regions of Super Massive Black Holes. Matter is now correctly <u>slowed</u> by the Gravitational influence of its parent SMBH and its Dark Disk Galaxy allowing Matter to cool to create many halo stars before falling to a Galactic plane where it forms the first stars of a developing Galactic Bulge.

Theorists of the <u>Big Bang</u> had to first devise a way for all Matter to escape from a mysterious, super hot, singularity like point. Theorists proposed an inflation theory. Because <u>no known-to-man form of energy</u> could cause Matter to escape from a singularity like event theorists required a powerful unknown-to-man form of energy to appear everywhere from a never to be revealed source. The astoundingly powerful energy of inflation caused all Matter from a <u>Big Bang</u> to speed apart faster than the speed of light which plainly violates today's laws of physics but allowed Matter to escape from <u>a singularity like</u> event. Scientists know this tremendous escape speed would have caused Matter to quickly become too far apart for the <u>puny Gravity of individual atoms</u> to ever bring this Matter back together. All Black Holes are also said to contain <u>singularities</u> from which <u>nothing can escape</u>, since, like the Big Bang the escape speed needs to exceed the speed of light.

[043] Theorists now required clever physics to provide a way to give the early universe imperfections or else the Big Bang Universe would be unable to clump <u>speeding apart Matter</u> into denser regions where Gravity will allow stars and Galaxies to form. Galaxies have to form where Matter was dense and not speeding apart and becoming less dense **like it would have been from a Big Bang.** Theorists turned to tiny vacuum fluctuations to provide tiny imperfections which <u>grow and expanded</u> during inflation and are said to eventually provide the imperfections which would allow Gravity to take over and Galaxies to form. When Scientists obtained data from satellites such as COBE, WMAP and Planck the obtained data created the CMB image which show cooler regions where there had been more Gravity and warmer regions which once had less Gravity. Galaxies were naturally thought to form in the regions where there was more Gravity which were believed to be dense with regular Matter.

This confirmed the early universe actually had the configuration which it <u>needed to have</u> for Galaxies to form. Scientists say the CMB image is a revelation for it shows Matter from their proposed Big Bang very quickly distributed itself in an <u>irregular</u> manner and into denser regions where Galaxies will form. Scientists now say their Big Bang was not born smooth and the CMB confirms their hypothesized theories are true and no one should doubt there was a Big Bang. The truth is the universe was always this way, <u>regardless of how it began,</u> since if it was smooth and uniform there would be no Galaxies. There are no such hassles for a Natural Universe which simply makes its Matter exactly where a Galaxy will form.

[079] The European Space Agency's Planck Space telescope has now revealed the highest precision image of the CMB which uncovered bothering mysteries; the fluctuations and temperatures in the CMB at large angular scales <u>do not match predictions</u>. Scientists say Gravity cannot produce the

observed fluctuations and suggest that <u>a division of new physics will be required</u> to provide reasoning for how the fluctuations were produced.

A Natural Universe requires no new division of physics to be invented for it is easy to clearly understand that the CMB image with its speckled, not quite uniform appearance is the <u>only possible way</u> this image could have been. The CMB radiation image along with the observation that Quasar activity peaked in the early universe indicates Matter production very likely <u>peaked at a similar time to the proposed Big Bang</u>, however, Matter production most probably began well before the proposed Big Bang and may still be occurring today. The CMB radiation simply reflects the way Matter was created within an early universe. **The CMB image closely matches the way a Natural Model releases Matter and forms hydrogen atoms in regions of <u>both</u> stronger and weaker Gravity.**

One of the very first deep Space images from the James Webb Space Telescope has astounded Scientists by revealing mature Galaxies existing less than 500 million years after the Big Bang. Scientist have now said even if their proposed vacuum fluctuations grow the required imperfections at the maximum allowable rate, it's very difficult to get enough Galaxies that evolved early enough to be consistent with JWST's recent observations.

[091] With the recently launched James Webb Space Telescope many are saying the Big Bang theory now has real trouble. The problem is the current theories clearly required longer time from the proposed Big Bang to form many of the newly observed Galaxies. Scientists say the Galaxies are *"so massive they should not be possible under current cosmological theory"*. The JWST has revealed a Galaxy of similar size of our Milky Way with a redshift of 10 placing it less than 500 million years after the Big Bang. **Given its maturity the Galaxy may well have originated from a time <u>before</u> the proposed Big Bang.** *"Even if you took everything that was available to form stars and snapped your fingers instantaneously, you still wouldn't be able to get (a Galaxy) that big that early (after a Big Bang),"* says Michael Boylan-

Kolchin, a cosmologist at the University of Texas at Austin.

It really is just basic common sense that Matter had to be most dense in regions where Galaxies will form, that can't be debated. A Natural Universe does not require hypothetical theories with long drawn out processes to provide theoretical imperfections which allow Galaxies to form. When the CMB image is applied to our Natural Universe there is less of a revelation and much more of an expectation. Because Matter within a Natural Model is actually created where a Galaxy will form, Matter could have 'never been' distributed completely smooth and uniform throughout our universe. This evidence is clear and observable. Because of its appearance, the CMB image also provides evidence that all Matter was created at the hearts of regular Galaxies and carried as plasma within spiraling magnetic fields. Released at the end of both short and long spiraling magnetic fields the plasma cooled which allowed Hydrogen atoms to form causing the plasma to become a gas and setting free the photons of the CMB radiation. When I look at the CMB image it is easy for me to see my Natural Model in it.

The universe is said to be 13.8 billion years old. As far back as can be directly observed, including the startling James Webb Space Telescope images of Galaxies from near the beginning of the universe, practically all Matter is contained within numerous Galaxies like tiny islands within incredibly vast oceans of near empty Space. This actual observation strongly infers nearly all regular Matter has always resided within Galaxies and surely agrees with Galaxies creating their own Matter. [059; 091] The Natural Universe has no Big Bang allowing **the very first Galaxies to form before the proposed Big Bang which easily conforms to the recent JWST discoveries which has confounded Astronomers by finding mature Galaxies too close to a Big Bang to form by current theories which is *"completely upending existing theories about the origins of Galaxies"*.** [104.]

[092] Back in 2017 astronomers discovered a Quasar 13.1 billion light-years away and existing 690 million years after the Big Bang. Quasar ULAS

J1342+0928 is powered by a Super Massive Black Hole (SMBH) 800 million times the mass of our sun. It was argued the large size of the distant SMBH is too difficult to explain so soon after the Big Bang. Now that the JWST has discovered Galaxies as far back as just 300 million years after the proposed Big Bang it has become exceedingly difficult for the current Big Bang theory but there are obviously no problems for a Natural Universe.

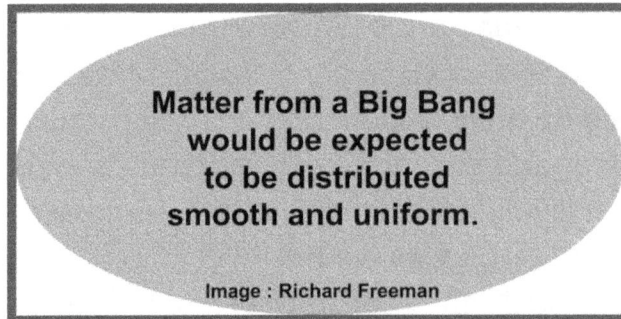

Matter from a Big Bang would be expected to be distributed smooth and uniform.

Image : Richard Freeman

With a Natural Model the distribution of Matter does not require clever physics to allow it to develop from the above image to the speckled not uniform appearance below.

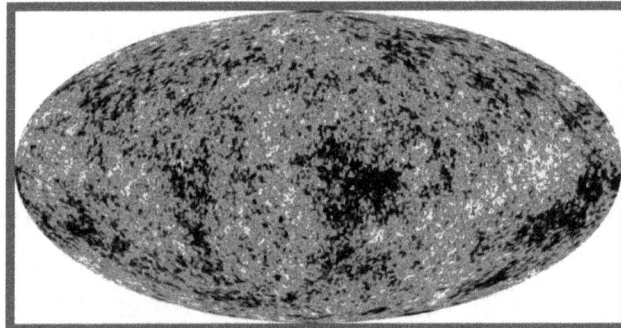

Image: Credit: NASA/WMAT science Team.

The photons of the CMB radiation were released, for fully explained and <u>observable</u> reasons, from regions of stronger and weaker Gravity which is why the radiation filled Space with ever so slight temperature variations giving the CMB radiation image it's not completely uniform appearance.

15 CONCLUSION.

080 See link to: ***Five mysteries the Standard Model of Particle Physics can't explain:*** Short answers from the Natural Model;

1. ***Why do neutrinos have mass?*** Neutrinos unavoidably gain mass from observing or measuring or interacting with a Dimensional Reality. Related to wave and particle duality.

2. ***What is Dark matter?*** Dark matter is a virtual proton.

3. **Why is there so much Matter in the universe?** Within spiraling magnetic fields three anti-quarks were ripped from virtual protons and locked away within Super Massive Black Holes. The beginning of the strong force allowed virtual protons to cling on to the corresponding three Matter-quarks allowing virtual protons to become regular protons, thus, creating a visible, asymmetry universe dominated by Matter.

4. ***Why is the expansion of the universe accelerating?*** The acceleration is due to Outer Gravity; Outer Gravity is applied in unison with Inner Gravity as one of the same. Galaxies are simply falling outwards.

5. ***Is there a particle associated with the force of Gravity?*** No, Gravity is not conveyed by a force carrying particle. Gravity is a two part mechanism supplied by a tiny fluctuating phase of virtual mass. [015] Based on a Quantum vacuum fluctuation a virtual proton fluctuates to a tiny, <u>negative energy</u> density phase of empty Space called virtual mass.

A Universe provided by Mother Nature.

[059] The page heading reads; **SCIENTISTS PUZZLED BECAUSE JAMES WEBB IS SEEING STUFF THAT SHOULDN'T BE THERE** *"the models just don't predict this...."* and *"will likely rewrite, long-held understandings about the origins of our universe".* The Natural Universe clearly predicts what the James Webb Space telescope is seeing by explaining <u>exactly why</u> Galaxies <u>very quickly self evolved</u> their regular disk shapes.

[123] <u>The Smoking Gun</u>: Page heading reads; '**James Webb Space Telescope sees lonely supermassive black hole-powered Quasars in the early universe'.** Anna-Christina Eilers, assistant professor of physics at Massachusetts Institute of Technology, said in a statement. *"It's difficult to explain how these Quasars could have grown so big if they **appear to have nothing to feed from.**"* Eilers concluded. *"If there's not enough material around for some Quasars to be able to grow continuously, that means there must be some other way that they can grow that we have yet to figure out."*

<u>Notably</u> this discovery is of <u>five of the very earliest known Quasars.</u> Since the discovery exactly matches the modeling of a Natural Universe it provides the <u>conclusive evidence</u> I was hopeful the James Webb Space Telescope would expose. With current ideas, in order to exist, these early Quasars <u>need to be</u> surrounded by massive amounts of fuel in the form of regular Matter from the said Big Bang. Within a Natural Universe there is no mystery, since, early Quasars and their Super Massive Black Holes are <u>at first</u> created and fuelled by <u>invisible</u> virtual protons (Dark matter) which is <u>exactly why</u> there appears to be, as Eilers says, *"nothing to feed them".*

No cosmic fog from a Big Bang = no Big Bang! Because the Big Bang creates <u>all of its Matter for all time,</u> neutral hydrogen is said to have filled the early universe with a cosmic fog. Called the Cosmic Dark Ages, even the largest telescope <u>would not detect</u> the earliest Galaxies as <u>they are hidden</u> within a surrounding thick haze of neutral hydrogen created from the Big Bang. <u>A Natural Universe</u> did not make all of its first atoms from a single event like the Big Bang. Within a Natural Universe Galaxies <u>self create</u> their own hydrogen atoms to make their own stars. The first Galaxies quickly made stars from their <u>earliest ejection</u> of <u>dense pulses</u> of Matter from the poles of their Super Massive Black Hole which then ejected more Matter to make more of a growing Galaxy's stars. Consequently, the earliest Galaxies are not hidden within the surrounding fog of neutral hydrogen from a Big Bang.

Thus, the Space around early Galaxies remains clear and transparent which **is <u>very likely</u> what has unexpectedly allowed** the James Webb Space Telescope to see early Galaxies which <u>should have been hidden from view</u> in the Big Bang's cosmic fog. For example; [124a]James Webb Space Telescope has detected unexpected light from a very distant Galaxy called JADES-GS-z13-1. Kevin Hainline, from the University of Arizona Says; *"We <u>really</u> <u>shouldn't have</u> found a Galaxy like this, given our understanding of the way the universe has evolved. We could think of the early universe as shrouded with a thick fog that would make it exceedingly difficult to find even <u>powerful lighthouses</u> peeking through, yet here we see the beam of light from this Galaxy piercing the veil".* [124b]Astronomers have now used the Murchison Widefield Array (MWA) <u>radio</u> telescope to peer deep into the cosmic past in search of neutral hydrogen's signature wavelength; however, they failed to find a trace of the Big Bang's neutral hydrogen's cosmic fog.

I have done my utmost to model a universe based on <u>direct evidence and actual observations</u> arguably far more than possible with the ramifications of a rather vague Big Bang beginning. Unlike the Big Bang, the Natural Model **obeys the laws of physics** and <u>at the very creation of Matter</u> correctly accounts for its <u>mandatory</u> creation of Anti-matter.

Unlike the Big Bang universe, which requires baffling, dark mythical-like energies to somehow appear from a never to be revealed source, my

Natural Universe does not require several different unexplainable and all powerful mystery forces or energies at play in the universe. A Natural Model explains how Gravity, <u>amplified from</u> squeezing numerous Quantum vacuum fluctuations, slowly arose from seemingly empty Space and began building the Natural Universe. Thus, Gravity began the universe and today continues to power the entire universe. The Natural Model shows why the <u>source of both mass and Gravity</u> cancels out the mass and energy of Matter so as the sum-total mass-energy of the universe remains zero.

Rather than having many seemingly unrelated and problem plagued theories, the Natural Universe advances with clear, <u>unprecedented correlation</u> from the creation of Dark matter to the creation of Super Massive Black Holes and from the creation of atoms to the formation of all Galaxies. In doing so the model fluently and comprehensively explains how Super Massive Black Holes formed, how Matter was created, the source of all Gravity, why Galaxies are accelerating, the source of all mass, the source of dimensional Space-time, the source of time, the source of time dilation and the source of length contraction. The model also clearly and <u>rationally</u> unravels the mysteries of Quantum by explaining exactly why a Quantum particle <u>really is in all places</u> at the same time, Quantum particle entanglement, wave and particle duality, Quantum Gravity and much more.

Arguably, the evidence clearly favors a Natural Universe. To explain the very largest, I have focused on the very smallest. The model has exposed the hidden blueprint for the beginnings of a beautifully <u>fully correlated</u> and Natural Universe, where all acts with clear and connected reasoning from the smallest of scales to the largest as one single entity.

Understanding how the dimensions of Space-time arise from our very own atoms as an <u>entangled entity</u> has allowed all aspects of both Einstein's Special Relativity and General Relativity to be rationally and physically explained. The nonsensical predictions of Quantum Theory become sensible with the understanding that tiny Quantum particles naturally reside <u>without entangled</u> dimensions to their Space and so <u>totally ignore</u> the dimensions of Space and the passing of time which is exactly why Quantum particles behave as predicted by Quantum Theory. Thus, many of today's deepest mysteries have finally been unmasked within a practical, unified model. The Natural Universe is now complete and portrays a rational universe where many of today's deepest mysteries have been unmasked in a way which finally provides a realistic unified model.

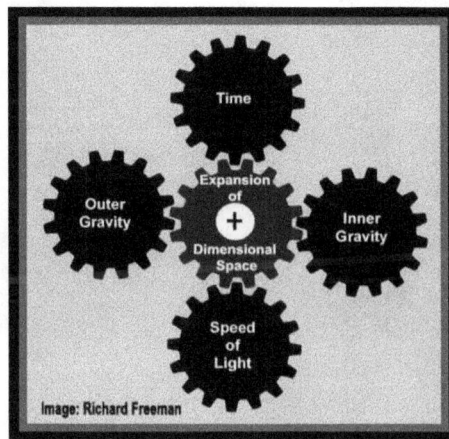

Image: Richard Freeman

The Natural Model has evolved over endless hours of deliberation and decades of deep research and is the only model which seamlessly shows how the theories of Special Relativity, General Relativity and Quantum Theory operate snugly alongside each other. I thoroughly believe the model now physically and comprehensively explains how the whole universe <u>really</u> works and provides an astonishing amount of direct evidence that it is true.

We are incredibly privileged to live our lives within a tiny window of passing time, on an astoundingly beautiful and <u>exceptional</u> planet within an amazing solar system, which resides as part of the gorgeous Milky Way Galaxy, within a simply dazzling home universe. Please take pleasure in enjoying the wonders which our simply stunning, home universe provides.

Richard Freeman.

Image captured from my place of work.

♥ 'and I think to myself, what a wonderful world' ♥

Like the far and boundless reaches of the universe, the power of thought and the ability to dream with our astonishing human brain, can be both overwhelmingly impressive and agonizing humbling.

Almost all of my working days and nights have been alone at sea. Indeed, the ocean has provided everything for me and my wife Beverley to raise a family at a beautiful place near the ocean. I marvel at all of its astonishing creatures. My pet dolphin 'Dolly' brings me back a fish when it falls off my hook, vigorously chases the sharks away and rolls over for a scratch under her chin. Just her presents truly turns a difficult day into a day of joy. Urged on by the sounds of the ocean, my mind is always motivated and free to take me to untold places and speculate about the many mysteries of the night sky and what more amazing mysteries may lie beyond. I would wonder what could really be going on far out there, how did it all begin and will it ever end? Now enjoying my retirement years I still find it difficult to even comprehend the unimaginable distances within our amazing universe.

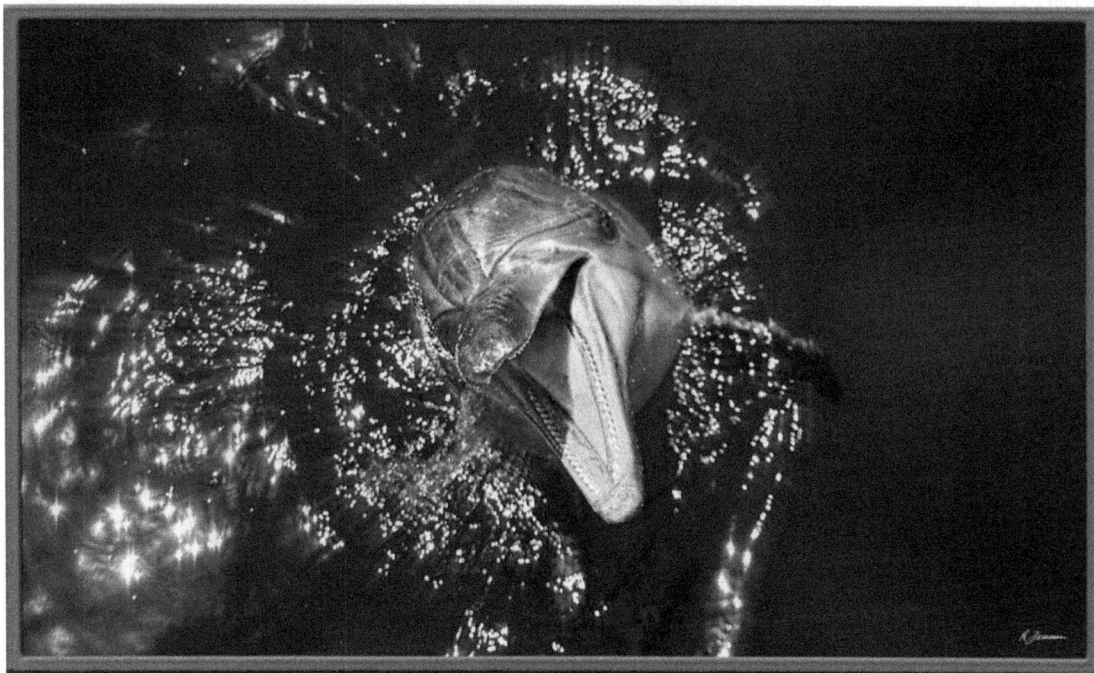

"Dolly" my amazing at sea companion for well over 35 years.

Reference Addresses.

001 *https://www.bbc.com/news/science-environment-56491033 Machine finds tantalising hints of new physics.* 95% of contents of universe is unknown. *https://science.nasa.gov/astrophysics/focus-areas/what-is-Dark energy* **Search**: Dark energy, Dark Matter.

002 *https://news.mit.edu/2019/Physicists-calculate-proton-pressure-distribution-0222* **Search**: *Physicists* calculate proton's pressure distribution for first time. **This core pushes out from the proton's center, while the surrounding region pushes inward.** *https://www.popularmechanics.com/science/energy/a20730787/pressure-proton-neutron-star/* **Search:** Pressures Inside a Proton Are More Extreme Than Inside a Neutron Star.

003 *https://profmattstrassler.com/articles-and-posts/particle-physics-basics/the-structure-of-matter/protons-and-neutrons/* Protons and Neutrons: **Search:** The Massive Pandemonium in Matter. *https://profmattstrassler.com/articles-and-posts/largehadroncolliderfaq/whats-a-proton-anyway/* **Search**: Of Particular Significance

004 *https://www.cbc.ca/radio/quirks/may-22-solving-our-sand-crisis-nuclear-quasicrystals-voyager-hears-an-interstellar-hum-and-more-1.6035006/what-evidence-for-Dark matter-has-been-found-in-the-milky-way-1.6035008* **Search:** What evidence for dark matter has been found in the Milky Way?

005 *https://www.sciencedaily.com/releases/2020/07/200701160132.htm* **Search:** What is a quasar. *https://earthsky.org/space/new-record-most-distant-quasar-black-hole-j0313-1806/* : **Search:** A new record for the most distant quasar *https://www.dailymail.co.uk/sciencetech/article-3515338/ The-quasar-jets-break-laws-physics-Streams-100-times-hotter-thought-possible-10-TRILLION-degrees.html https://www.space.com/32467-black-hole-jets-hotter-than-expected.html* **Search:** Black Hole Jets Hotter Than Expected.

006 *https://www.csmonitor.com/2006/1019/p14s01-cogn.html* **Search:** More clues in the cosmic jet mystery. *https://www.sciencenews.org/article/jet-set-astronomers-identify-makeup-quasar-streams* **Search:** Jet Set: Astronomers identify the makeup of quasar streams. (Identified the particles in the jets as electrons and protons).

007 *https://www.sciencedaily.com/releases/2019/12/191219074632.htm* **Search:** ESO observations reveal black holes' breakfast at the cosmic dawn.

008 *https://www.abc.net.au/news/science/2016-06-23/antimatter-explainer/7487354* **Search:** The antimatter mystery: Annihilation and a universe that shouldn't exist.

009 https://www.hexapolis.com/2015/08/15/new-study-confirms-that-matter-and-antimatter-are-perfect-mirror-images-of-one-another/ **Search:** New study confirms that matter and antimatter are perfect mirror images of one another.

010 https://www.sciencefocus.com/space/antimatter-where-did-it-go/ **Search:** What is antimatter, and why is it missing from the Universe today?

011 *https://www.youtube.com/watch?v=Ztc6QPNUqls* Video: (Most of) **Search:** 'Your Mass is NOT From the Higgs Boson'. Empty Space gives us most of our mass.

012 *https://www.duhoctrungquoc.vn/wiki/en/Graviton* **Search:** Wiki Graviton. This article is about the hypothetical particle.

013 *https://viewspace.org/interactives/unveiling_invisible_universe/dark_matter/bullet_cluster* **Search:** Dark matter Bullet Cluster.

014 *https://physicsworld.com/a/the-casimir-effect-a-force-from-nothing/* **Search:** The Casimir effect: a force from nothing. *https://www.youtube.com/watch?v=_DXHrp6-LZI* **Search:** Video: The Heisenberg Uncertainty Principle Part 2: Energy/Time and Quantum Fluctuation *https://www.scientificamerican.com/article/are-virtual-particles-rea/* **Search:** Are virtual particles really constantly popping in and out of existence? Or are they merely a mathematical bookkeeping device for quantum mechanics?

015 *https://www.sciencealert.com/Physicists-say-they-ve-managed-to-manipulate-pure-nothingness* **Search:** Physicists Say They've Manipulated 'Pure Nothingness' And Observed The Fallout. **An astonishing phenomenon.** *https://www.sciencedaily.com/releases/2017/01/170118132244.htm* **Search:** Traffic jam in empty Space Physicists study the quantum vacuum. " While the fluctuation amplitudes positively deviate from the vacuum noise at temporally increasing speed of light, a slowing down results in an astonishing phenomenon: the level of measured noise is lower than in the vacuum state -- that is, the ground state of empty Space".

016 *https://www.livescience.com/34052-unsolved-mysteries-physics.html* **Search: Scroll to Do we live in a false vacuum?**

017 *https://spaceplace.nasa.gov/Gravitational waves/en/* **Search:** What Is a Gravitational Wave?
https://www.forbes.com/sites/startswithabang/2018/12/15/ask-ethan-are-Gravitational waves-themselves-affected-by-gravity/?sh=1940f2482f3f **Search:** Ask Ethan: Are Gravitational Waves Themselves Affected By Gravity?

017a *https://www.forbes.com/sites/startswithabang/2019/03/02/ask-ethan-why-dont-gravitational-waves-get-weaker-like-the-gravitational-force-does/?sh=1a1fe50c2f58A* **Search:** Ask Ethan: Why Don't Gravitational Waves Get Weaker Like The Gravitational Force Does?

018 *https://www.livescience.com/19268-quantum-double-slit-experiment-largest-molecules.html* **Search:** Largest Molecules Yet Behave Like Waves in Quantum Double-Slit Experiment

019 *https://www.livescience.com/32216-what-is-relativity.html* **Search:** What Is Relativity?

020 *https://hal.archives-ouvertes.fr/hal-03669505/document* **Search:** The Planck time and Planck length are linked to the ultimate limit of quantization and directly to gravity. *https://www.phys.unsw.edu.au/einsteinlight/jw/module6_Planck.htm#:~:text* **Search:** The Planck scale: relativity meets quantum mechanics meets gravity.

021 *https://astronomy.swin.edu.au/cosmos/p/Planck+Time* **Search:** Planck Time. Light travels Planck Length in Planck Time. *http://www.phys.unsw.edu.au/einsteinlight/jw/module6_Planck.htm* **Search:** The Planck scale: relativity meets quantum mechanics meets gravity. Tiny Virtual Black Hole

022 *https://www.forbes.com/sites/startswithabang/2019/02/22/the-wimp-miracle-is-dead-as-Dark matter-experiments-come-up-empty-again/?sh=7440c2726dbc* **Search:** The 'WIMP Miracle' Hope For Dark Matter Is Dead.

023 *https://lweb.cfa.harvard.edu/seuforum/bh_whatare.htm* **Search:** A black hole is a one-way exit from our universe. The object's mass and gravity remain behind.

024 *https://news.harvard.edu/gazette/story/2015/02/mysterious-link-between-galaxy-and-black-hole/* **Search:** Mysterious link between galaxy and black hole.

025 *https://aasnova.org/2020/09/30/the-link-between-black-holes-and-their-galaxies/* **Search:** The Link Between Black Holes and Their Galaxies.

026 *https://www.quantamagazine.org/what-no-new-particles-means-for-physics-20160809/* **Search:** What No New Particles Means for Physics. *https://www.forbes.com/sites/startswithabang/2019/02/12/why-Supersymmetry-may-be-the-greatest-failed-prediction-in-particle-physics-history/#6015f74569e6* **Search:** Why Supersymmetry May Be The Greatest Failed Prediction In Particle Physics History.

027 *https://www.quantamagazine.org/complications-in-physics-lend-support-to-multiverse-hypothesis-20130524/* **Search:** Is Nature Unnatural? We may reside in an unnatural universe where natural does not make sense. Supersymmetry particles allow the discovered light Higgs boson to exist, however, the LHC failed to find any Supersymmetry particles.

028 *https://profmattstrassler.com/articles-and-posts/largehadroncolliderfaq/whats-a-proton-anyway/* **Search:** What's a Proton, Anyway? Not just three quarks. Protons have zillions of quarks anti-quarks and gluons.

029 *https://www.forbes.com/sites/startswithabang/2016/08/03/where-does-the-mass-of-a-proton-come-from/#63327b7a2e1d* **Search:** Where Does The Mass Of A Proton Come From? The way quarks bind into protons is fundamentally different from all the other forces. One of the longest-standing mysteries of physics where the mass of matter comes from.

030 *https://www.flickr.com/photos/hydrogenrandd/6256628954* **Search:** flickr ELECTRON MICROSCOPE HYDROGEN ATOM *https://www.jpl.nasa.gov/edu/news/2019/4/19/how-Scientists-captured-the-first-image-of-a-black-hole/* **Search:** Image of Black Hole.

031 *https://en.wikipedia.org/wiki/Proton_spin_crisis* **Search:** Proton Spin Crisis.

032 *https://insidetheperimeter.ca/the-building-blocks-of-the-universe-phiala-shanahan-public-lecture-webcast/* **Search:** The Building Blocks of the Universe — Phiala Shanahan public lecture webcast. *https://www.youtube.com/watch?v=YkymlrUL0Sw* *Video:* **Search:** Phiala Shanahan Public Lecture: The Building Blocks of the Universe

033 *https://www.scientificamerican.com/article/the-mysteries-of-the-world-s-tiniest-bits-of-matter/*   The Mysteries of the World's Tiniest Bits of Matter. **Search:** We wonder how exactly gluons do the work of binding quarks in the first place. *https://www.bnl.gov/physics/NTG/linkable_files/pdf/SciAm-Glue-Final.pdf* **Search:** The glue that binds us. Details of how gluons function remain surprisingly mysterious.

034 *https://nautil.us/the-physics-still-hiding-in-the-higgs-boson-10955/* **Search:** The Physics Still Hiding in the Higgs Boson. Physicists understand little about the omnipresent Higgs field.

035 *https://www.quantamagazine.org/complications-in-physics-lend-support-to-multiverse-hypothesis-20130524/* **Search:** Is Nature Unnatural? However, in order for the Higgs boson to make sense with the mass (or equivalent energy) it was determined to have, the LHC needed to find a swarm of other particles, too. None turned up.

036 *https://profmattstrassler.com/2012/07/05/a-new-era-dawns/* **Search:** A New Era Dawns. The found Higgs: there is an additional intrinsic contribution whose source is not known or easily discovered… so don't expect an answer anytime soon. As for why its mass is so big — or small — we do not know.

037 *https://angelsanddemons.web.cern.ch/antimatter/trapping-antimatter.html* **Search:** Trapping antimatter.

038 *https://www.astro.ucla.edu/~wright/BBhistory.* **Search:** Brief History of the Universe. Temperature 0.0001 seconds after the Big Bang.

039 *https://www.astron.nl/most-detailed-ever-images-of-galaxies-revealed-using-lofar/* **Search:** Most detailed-ever images of galaxies revealed using LOFAR.   *https://www.sciencealert.com/breathtaking-new-images-are-the-most-detailed-yet-of-other-galaxies* **Search:** Breathtaking New Images Reveal Several Distant Galaxies in Unprecedented Detail. *https://www.bbc.com/news/science-environment-57998940* **Search:** Astronomers see galaxies in ultra-high definition.

040 *https://www.newscientist.com/article/dn16095-its-confirmed-matter-is-merely-vacuum-fluctuations/* **Search:** It's confirmed: Matter is merely vacuum fluctuations.

041 *https://www.youtube.com/watch?v=YkymlrULOSw* Video: At time 4.37 **Search:** Phiala Shanahan Public Lecture: The Building Blocks of the Universe. Protons and neutrons are made from quarks and gluons, that's it. There is no deeper structure to be found.

042 *https://www.youtube.com/watch?v=J3xLuZNKhlY* Video: **Search:** Empty Space is NOT empty.

043 *https://www.youtube.com/watch?v=9ee_2n4FuqQ* video: **Search:** The Cosmic Microwave Background explained. *https://www.forbes.com/sites/startswithabang/2018/03/15/why-isnt-our-universe-perfectly-smooth/?sh=2ea676657bc6* Why Isn't Our Universe Perfectly Smooth?

044 *https://spaceflightnow.com/news/n0607/25quasar/* **Search:** New view of quasar emerges.

045 *https://www.sciencealert.com/black-hole-magnetic-field-weaker-than-expected-v404-cygni* **Search:** New Study on Black Hole Magnetic Fields Has Thrown a Huge Surprise at Astronomers.

046 *https://www.universetoday.com/383/a-new-view-of-quasars/* **Search:** A New View of Quasars. Magnetic fields actually extend and penetrate right through the surface of the 'collapsed central object'. Spinning magnetic field lines spool up, winding tighter and tighter until they explosively unite, releasing huge amounts of energy that power the jets.

047 *https://www.eso.org/public/news/eso1921/* **Search:** ESO Observations Reveal Black Holes' Breakfast at the Cosmic Dawn *https://www.eso.org/public/images/eso1921b/* **Search:** Artistic impression of a distant quasar surrounded by a gas halo. Astronomers have spotted enormous hydrogen halos around 12 quasars which are located over 12.5 billion light-years from Earth.

048 *https://astronomy.com/magazine/news/2020/03/do-all-galaxies-have-Dark matter* **Search:** Do all galaxies have dark matter? Galaxies NGC 1052-DF2 and NGC 1052-DF4 lack any significant amount of Dark matter.

049 *https://astronomy.swin.edu.au/cosmos/i/Intergalactic+Medium* **Search:** Intergalactic Medium. X-ray observations revealed large quantities of metals mixed in with the hydrogen and helium. These metals could only have been made by stars within the Galaxies. Medium is enriched through the action of galactic winds.

050 *https://fermi.gsfc.nasa.gov/science/constellations/pages/bubbles.html* **Search:** Fermi Bubbles. A completely unexpected discovery like the Fermi Bubbles is a special treat.   *https://www.universetoday.com/146359/astronomers-find-the-source-of-the-huge-bubbles-of-gas-flowing-out-of-the-milky-way-still-no-idea-what-caused-them/* **Search:** Astronomers Find the Source of the

Huge Bubbles of Gas Flowing Out of the Milky Way, Still No Idea What Caused Them.

051 *http://news.bbc.co.uk/2/hi/science/nature/2181455.stm*   **Search:** Einstein's theory 'may be wrong' *https://www.theage.com.au/national/einsteins-relativity-theory-hits-a-speed-bump-20020808-gduh2d.html*   **Search:** Einstein's relativity theory hits a speed bump.   It's entirely possible that the speed of light would have got greater and greater as you go back (through time) towards the Big Bang. Paul Davies, a Physicist at Macquarie University, Sydney.

052 *https://cosmosmagazine.com/science/physics/was-the-speed-of-light-faster-at-the-beginning-of-the-universe/*   **Search:** Was the speed of light faster at the beginning?

053 *https://astronomy.swin.edu.au/cosmos/s/supermassive+black+hole*   **Search:** Supermassive Black Hole.   How Super Massive Black Holes were made is a mystery.

054 *https://astronomy.com/magazine/ask-astro/2013/07/supermassive-black-holes*   **Search:** Is there a correlation between the mass of a supermassive black hole and the amount of stars within its galaxy? If so, what is it? *https://aasnova.org/2020/09/30/the-link-between-black-holes-and-their-galaxies/*   **Search:** The Link Between Black Holes and Their Galaxies.   But is this a statistical fluke, or is there a physical reason for the connection?

055 *http://www.astro.sunysb.edu/fwalter/AST101/stellarpops.html#:~:text*   **Search:** Stellar Populations.   Globular clusters, elliptical galaxies, and the bulges and halos of spiral galaxies are Population II.

056 *https://astronomy.swin.edu.au/cosmos/h/halo*   **Search:** Halo.   A stellar halo is an spherical population of old stars.

057 *https://www.nationalgeographic.com/science/article/supernovae* Supernovae. **Search:** Shock wave that compresses clouds of gas to aid new star formation. *https://science.nasa.gov/pencil-nebula-supernova-shock-wave*   **Search:** The Pencil Nebula Supernova Shock Wave.   Shock wave sweeps up surrounding material.

058 *https://www.sciencedaily.com/releases/2017/08/170803103134.htm*   **Search:** Our solar system's 'shocking' origin story. Evidence implies the formation of our own star the Sun and its solar system was triggered by a shock-wave from an exploding supernova.

059 *https://futurism.com/the-byte/scientists-puzzled-james-webb-stuff*   **Search:** SCIENTISTS PUZZLED BECAUSE JAMES WEBB IS SEEING STUFF THAT SHOULDN'T BE THERE.   "THE MODELS JUST DON'T PREDICT THIS".

060 *https://lup.lub.lu.se/luur/download?func=downloadFile&recordOId=8912003&fileOId=8912097*   **Search:** Nucleosynthesis in accretion disks around black holes. **Accretion disks produce at most 10–4 times the amount of the same elements that stars produce.**

061 *https://science.nasa.gov/astrophysics/focus-areas/what-is-Dark energy*   **Search:** Dark energy, Dark Matter. Scientists were certain the expansion of the universe had to be slowing. This link provides a good read from NASA.

062 *https://www.newscientist.com/definition/general-relativity/*   **Search:** General relativity.   John Wheeler "Space-time tells matter how to move; matter tells space-time how to curve".

063 *https://newscenter.lbl.gov/2011/07/17/kamland-geoneutrinos/*   **Search:** What Keeps the Earth Cooking?   Radioactive decay alone is not enough to account for Earth's heat energy.

064 *https://windows2universe.org/headline_universe/speed_gravity.html&edu=high*   **Search:** It Looks Like Einstein Was Right! Speed of Light and Gravity are Equal!

065 *https://pilotswhoaskwhy.com/2021/03/14/gnss-vs-time-dilation-what-the/*   **Search:** What is Time Dilation and How Does it Affect GPS.

066 *https://aeon.co/ideas/the-most-wonderful-words-in-science-we-have-no-idea-yet*   **Search:** The Most Wonderful words in science: 'We have no idea... yet!

067 *https://phys.org/news/2019-07-vacuum-fluctuations-perspective.html*   **Search:** Measuring light and vacuum fluctuations from a time flow perspective.

068 *https://www.abc.net.au/science/bigquestions/s460740.htm*   **Search:** In Conversation with Paul Davies and Phillip Adams. Time is in our very souls, isn't it?

069 *https://www.bnl.gov/newsroom/news.php?a=22870* **Search:** The Glue that Binds Us All. By speeding the particles up, Scientists can slow down the gluon fluctuations.

070 *https://www.quantamagazine.org/crisis-in-particle-physics-forces-a-rethink-of-what-is-natural-20220301/* **Search:** A Deepening Crisis Forces Physicists to Rethink Structure of Nature's Laws.

071 *https://www.informationphilosopher.com/quantum/mystery/* **Search:** The One Mystery of Quantum Mechanics.

072 *https://www.science.org/content/article/china-s-quantum-satellite-achieves-spooky-action-record-distance* **Search:** China's quantum satellite achieves 'spooky action' at record distance.

073 *https://www.sciencealert.com/reality-doesn-t-exist-until-we-measure-it-quantum-experiment-confirms* **Search: Reality Doesn't Exist Until We Measure It, Quantum Experiment Confirm.**

074 *https://profmattstrassler.com/articles-and-posts/particle-physics-basics/quantum-fluctuations-and-their-energy/* **Search:** Quantum Fluctuations and Their Energy.

075 *https://www.aps.org/publications/apsnews/200512/history.cfm* **Search:** Einstein's quest for a unified theory.

076 *https://www.youtube.com/watch?v=Q1YqgPAtzho* **Search:** DR. QUANTUM - DOUBLE SLIT EXPERIMENT. (Video) *https://www.popularmechanics.com/science/a22280/double-slit-experiment-even-weirder/* **Search:** The Logic-Defying Double-Slit Experiment Is Even Weirder Than You Thought.

077 *https://www.sciencedaily.com/releases/2017/03/170315125604.htm* **Search:** Quantum movement of electrons in atomic layers shows potential of materials for electronics and photonics.

078 *https://medium.com/@akhilendrasingh_15688/general-relativity-and-quantum-mechanics-21ea9450325c* **Search:** General Relativity and Quantum Mechanics.

079 *https://www.youtube.com/watch?v=9ee_2n4FuqQ* **Search:** The Cosmic Microwave Background explained (video) *https://www.space.com/33892-cosmic-microwave-background.html* **Search:** What is the cosmic microwave background?

080 *https://www.symmetrymagazine.org/article/five-mysteries-the-standard-model-cant-explain* **Search:** Five mysteries the Standard Model can't explain.

081 *https://www.sciencedaily.com/releases/2020/07/200701160132.htm* **Search:** Beacon from the early universe.

082 https://home.cern/science/physics/dark-matter **Search:** Dark matter. CERN : Universe has 6 times more Dark matter than visible Matter.

083 *https://www.nasa.gov/content/3c353-giant-plumes-of-radiation* **Search:** NASA image of Galaxy 3C353.

084 *https://www.livescience.com/what-is-singularity* **Search:** What is a singularity?

084a *https://science.howstuffworks.com/science-vs-myth/everyday-myths/relativity8.htm* **Search:** How Special Relativity Works.

085 *https://public.nrao.edu/news/black-hole-jets-rapid-wobble/* **Search:** Black Hole's Tug on Space Pulls Fast-Moving Jets in Rapid Wobble.

086 *https://home.cern/science/physics/Supersymmetry* **Search:** Supersymmetry.

087 *https://owlcation.com/stem/Are-Gluons-Massless-Is-There-a-Big-Problem-with-Little-Physics-and-Other-Gluon-Questions* **Search:** Are Gluons Massless? Is There a Big Problem With Little Physics?

088 *https://www.quantamagazine.org/higgs-boson-mass-explained-in-new-theory-20150527/* **Search:** A New Theory to Explain the Higgs Mass.

089 *https://www.sciencedaily.com/releases/2020/07/200701160132.htm* **Search:** Beacon from the early universe.

090 *https://phys.org/news/2014-06-einstein-quantum-mechanics-hed-today.html* **Search:** Einstein vs quantum mechanics, and why he'd be a convert today.   https://blogs.scientificamerican.com/observations/einstein-and-the-quantum/. Search: Einstein and the Quantum.

(cont

091 *https://www.scientificamerican.com/article/jwsts-first-glimpses-of-early-galaxies-could-break-cosmology/* **Search:** JWST's First Glimpses of Early Galaxies Could Break Cosmology.   https://futurism.com/the-byte/scientists-puzzled-james-webb-stuff **Search:** SCIENTISTS PUZZLED BECAUSE JAMES WEBB IS SEEING STUFF THAT SHOULDN'T BE THERE "THE MODELS JUST DON'T PREDICT THIS".

092 *https://astronomical.fandom.com/wiki/ULAS_J1342%2B0928* **Search:** ULAS J1342+0928.

093 *https://svs.gsfc.nasa.gov/11534* **Search:** Galaxy Formation.

094 *https://faculty.washington.edu/seattle/physics541/%202010-reading/virtual-3.pdf* **Search:** THE PHYSICS OF NOTHING. Empty matter and the full physical vacuum Peter I P Kalmus Department of Physics, Queen Mary & Westfield College, London E1 4NS

095 *https://astronomy.com/magazine/ask-astro/2021/02/ask-astro-why-do-galaxies-have-spiral-arms* **Search:** Ask Astro: Why do galaxies have spiral arms? Do smaller objects have them too?

096 *https://svs.gsfc.nasa.gov/30680* **Search:** Active Galaxy Hercules A: Visible & Radio Comparison.

097 *https://www.youtube.com/watch?v=Douno3m0s8Y* **Search:** This Is What Centaurus A Galaxy Looks Like From Earth (In Radio) *https://www.nustar.caltech.edu/page/relativistic_jets#:~:text* . **Search:** Relativistic Jets. First address is a Video.

098 *https://en.wikipedia.org/wiki/Centaurus_A* **Search:** Wikipedia Centaurus A. Centaurus A was identified as a source of cosmic rays of highest energies.

099 *https://www.cfa.harvard.edu/news/new-insights-how-spiral-galaxies-get-their-arms* **Search:** New Insights on How Spiral Galaxies Get Their Arms.

100 *https://en.wikipedia.org/wiki/Zero-energy_universe* **Search:** Zero energy universe.

101 *https://www.wired.com/2016/08/sorry-folks-lhc-didnt-find-new-particle/* **Search:** Sorry, Folks. The LHC Didn't Find a New Particle After All.

102 *https://www.nature.com/articles/nchem.2186* **Search:** First there was hydrogen.

103 https://www.space.com/20930-Dark matter.html **Search**: What is dark matter? *https://www.youtube.com/watch?v=e7FNMWKuYcl* (video) **Search**: What is Dark Matter? How The Universe Works.

104 *https://edition.cnn.com/2023/02/22/world/webb-telescope-massive-early-galaxies-scn/index.* **Search:** Webb telescope makes a surprising galactic discovery in the distant universe.

105 https://www.sciencenews.org/article/jwst-distant-galaxies-surprises-space **Search:** JWST's hunt for distant galaxies keeps turning up surprises.

106 *https://www.space.com/large-radio-galaxy-alcyoneus-discovery* **Search:** Astronomers discover massive radio galaxy 100 times larger than the Milky Way.

107 *https://www.livescience.com/space/astronomy/james-webb-telescope-spots-thousands-of-milky-way-lookalikes-that-shouldnt-exist-swarming-across-the-early-universe* **Search:** James Webb telescope spots thousands of Milky Way lookalikes that 'shouldn't exist' swarming across the early universe.

108 https://www.youtube.com/watch?v=068rdc75mHM&t=2s Video: **Search:** Einstein's Quantum Riddle. (worth watching).

109 https://www.youtube.com/watch?v=0yiDGFLQvls&t=3s **Search:** CSSM and CoEPP: The Standard Model and Beyond.

110 https://www.met.edu/blog/The_Mysterious_Properties_of_the_Graviton#:~:text **Search:** THE MYSTERIOUS PROPERTIES OF THE GRAVITON.

111a https://bigthink.com/hard-science/negative-energy-wormholes-warp-drives/#:~:text **Search:** What is negative energy? And can it give us wormholes and warp drives? https://en.wikipedia.org/wiki/Warp_drive **Search:** Wikipedia Warp drive.

111b https://www.youtube.com/watch?v=wC38DvKPJtk   Video: **Search:** Warp Drive Progress: NASA Moves Us Toward Faster-Than-Light Travel

112 https://www.livescience.com/space/astronomy/james-webb-telescope-discovers-2-of-the-oldest-galaxies-in-the-universe **Search:** James Webb telescope discovers 2 of the oldest galaxies in the universe.

113 https://www.wired.com/story/the-jwst-has-spotted-giant-black-holes-all-over-the-early-universe/   **Search:** The JWST Has Spotted Giant Black Holes All Over the Early Universe.

114 https://phys.org/news/2008-09-dark-disk-galaxy.html   **Search:** A dark matter disk in our Galaxy.

115 https://home.cern/science/physics/extra-dimensions-Gravitons-and-tiny-black-holes   **Search:** Extra dimensions, Gravitons, and tiny black holes.

116 https://www.space.com/molecular-gas-fountain-ancient-universe-black-hole   **Search:** Scientists find black hole powering a molecular gas fountain in the ancient universe.

117 https://www.quantamagazine.org/what-goes-on-in-a-proton-quark-math-still-conflicts-with-experiments-20200506/   **Search:** What Goes On in a Proton? Quark Math Still Conflicts With Experiments.

118 https://www.livescience.com/space/space-exploration/baby-quasars-spotted-by-james-webb-telescope-could-transform-our-understanding-of-monster-black-holes   **Search:** Baby quasars' spotted by James Webb telescope could transform our understanding of monster black holes.

119 https://www.sciencedaily.com/releases/2017/01/170118132244.htm   **Search:** Traffic jam in empty space. Physicists study the quantum vacuum.

120 https://www.youtube.com/watch?v=PCuyCJocJWg   (video) **Search: Why This Stuff Costs $2700 Trillion Per Gram -**Antimatter at CERN.

121 https://www.colorado.edu/today/2023/02/22/webb-telescope-spots-super-old-massive-galaxies-shouldnt-exist#:~:text **Search:** Webb telescope spots super old, massive galaxies that shouldn't exist – University of Colorado Boulder.

122 https://www.cfa.harvard.edu/news/unexpectedly-massive-black-holes-dominate-small-galaxies-distant-universe   **Search:** Unexpectedly Massive Black Holes Dominate Small Galaxies in the Distant Universe.

123 https://www.space.com/james-webb-space-telescope-ancient-black-hole-quasar   Search: James Webb Space Telescope sees lonely supermassive black hole-powered quasars in the early universe**.**
124a https://science.nasa.gov/missions/webb/nasas-webb-sees-galaxy-mysteriously-clearing-fog-of-early-universe/   Search: NASA's Webb Sees Galaxy Mysteriously Clearing Fog of Early Universe. **124b** https://www.space.com/neutral-hydrogen-dark-ages-of-universe.html Search: When Did the 'Dark Ages of the Universe' End? This Rare Molecule Holds the Answer.
125 https://www.astronomy.com/science/weird-object-milky-way-antimatter-fountain/   **Search:** Weird Object: Milky Way Antimatter Fountain.

The film 'Richard Freeman's Natural Universe' is now available at streaming services including Amazon Prime Video: